Functional Foods as a New Therapeutic Strategy 2.0

Functional Foods as a New Therapeutic Strategy 2.0

Guest Editors

Ivan Cruz-Chamorro
Guillermo Santos Sánchez

MDPI

Basel • Beijing • Wuhan • Barcelona • Belgrade • Novi Sad • Cluj • Manchester

Guest Editors
Ivan Cruz-Chamorro
Facultad de Enfermería
Universidad de Castilla-La Mancha
Albacete
Spain

Guillermo Santos Sánchez
Instituto de Investigación en Ciencias de la Alimentación
CIAL (CSIC-UAM)
Madrid
Spain

Editorial Office
MDPI AG
Grosspeteranlage 5
4052 Basel, Switzerland

This is a reprint of the Special Issue, published open access by the journal *Nutraceuticals* (ISSN 1661-3821), freely accessible at: https://www.mdpi.com/journal/nutraceuticals/special_issues/0EFX59NVEN.

For citation purposes, cite each article independently as indicated on the article page online and as indicated below:

Lastname, A.A.; Lastname, B.B. Article Title. *Journal Name* **Year**, *Volume Number*, Page Range.

ISBN 978-3-7258-3643-7 (Hbk)
ISBN 978-3-7258-3644-4 (PDF)
https://doi.org/10.3390/books978-3-7258-3644-4

Cover image courtesy of Ivan Cruz-Chamorro

Contents

About the Editors

Ivan Cruz-Chamorro

Ivan Cruz-Chamorro is professor at the University of Castilla-La Mancha, Spain, specializing in biochemistry and molecular biology. He began his research career in 2009, with the completion of his graduate studies at the Department of Biomolecular Sciences at the University of Urbino, Italy. Since 2012, he has been a member of the Molecular Neuroimmunoendocrinology Laboratory of the Institute of Biomedicine of Seville. In 2018, he earned his PhD from the University of Seville with a dissertation titled "Evaluation of the Health Effects of Plant Bioactive Peptides: Immunomodulatory, Antioxidant, and Metabolic Effects", achieving the highest grade. His research focuses on studying the beneficial effects of naturally derived bioactive peptides, particularly their immunomodulatory, antioxidant, and metabolic properties. He has published numerous articles in renowned scientific journals, covering topics such as the anti-inflammatory effects of lupin protein hydrolysates and the role of melatonin in multiple sclerosis, among others. In addition to being an Editorial Board Member of the *Nutraceuticals* journal, he is an invited editor of different Special Issues dedicated to enhancing knowledge about natural compounds and their impact on human health. Ivan has supervised a doctoral thesis, is a professor of biochemistry, and has participated in several funded research projects, making significant contributions to advancing knowledge in his field of expertise. Ivan collaborates with different international groups, and he is currently exploring the effects of different natural matrices on diseases such as metabolic-associated fatty liver disease, multiple sclerosis, and cancer models.

Guillermo Santos Sánchez

Guillermo Santos Sánchez, a graduate in Human Nutrition and Dietetics, is a postdoctoral researcher in the food proteins group of the Instituto de Investigación en Ciencias de la Alimentación (CIAL-CSIC) research center. In 2018, Guillermo received a Training Program for Academic Staff (FPU) grant from the Ministry of Education, Culture, and Sports (Spain) to pursue doctoral studies at the University of Seville in the Molecular NeuroImmunoEndocrinology group of the Institute of Biomedicine of Seville (IBiS). Guillermo's PhD project studied the biological effects (immunomodulators, antioxidants, lipid-lowering agents, etc.) of protein hydrolysates from plant sources (such as lupin, hemp, and wheat) on atherosclerosis and fatty liver disease associated with metabolic dysfunction (MAFLD). In 2020, he spent six months at the Department of Pharmaceutical Sciences at the University of Milan (Italy), with the support of an Erasmus+ mobility grant, analyzing the molecular mechanisms underlying the hypocholesterolemic effects of lupin peptides. In December 2022, Guillermo earned an international PhD degree from the University of Seville with the highest mark. To date, he has published 27 scientific articles (25 in Q1 and 12 in D1), being the first author in 19 of them, along with two book chapters. Guillermo has also participated in 6 international and 10 national conferences, delivering both oral presentations and posters, and he is a member of the Spanish Society of Biochemistry and Molecule Biology (SEBBM). In addition, he has been awarded the best article award from the Faculty of Medicine of the University of Seville (2022) and the 10th edition of the Seville City Council's "Young Award for Scientific Culture 2022".

Preface

Functional foods provide health benefits beyond basic nutrition and have emerged as a promising strategy for preventing and managing non-communicable diseases (NCDs). These include obesity, cardiovascular diseases, cancer, and type II diabetes, which are among the leading causes of mortality worldwide, accounting for over 41 million deaths annually. Functional foods contain bioactive compounds capable of modulating physiological functions and reducing disease risks. In recent years, researchers have explored a wide range of plant-derived compounds, marine bioactives, and commercially available nutraceuticals that may serve as therapeutic agents for metabolic, inflammatory, and degenerative disorders.

This reprint of the Special Issue delves into the latest advancements in functional food science, presenting insights from recent studies on bioactive ingredients, their mechanisms of action, and their potential applications in human health. The chapters explore various food sources—including quinoa, avocado, edible wild plants, marine algae, and plant-derived compounds such as ursolic acid—and their roles in managing diabetes, controlling obesity, protecting cardiovascular health, and reducing neuroinflammation. Additionally, this reprint of the Special Issue explores the therapeutic potential of commercially available supplements, such as Dekosilhue®, melatonin, and caffeine-free fat loss formulations, evaluating the scientific backing for their use as nutraceuticals.

This reprint of the Special Issue aims to serve as a comprehensive reference for nutritionists, healthcare professionals, food scientists, and industry stakeholders interested in the intersection of food technology and health sciences.

Interdisciplinary collaboration between nutrition science, medicine, and food technology will be instrumental in translating laboratory findings into real-world dietary solutions. We hope that this reprint will inspire further exploration and innovation in this field, ultimately contributing to the global effort to combat NCDs through evidence-based nutrition.

Ivan Cruz-Chamorro and Guillermo Santos Sánchez
Guest Editors

nutraceuticals

Editorial

Functional Foods as a New Therapeutic Strategy 2.0

Guillermo Santos-Sánchez [1,*] and Ivan Cruz-Chamorro [2,*]

1 Instituto de Investigación en Ciencias de la Alimentación, CIAL (CSIC-UAM), 28049 Madrid, Spain
2 Facultad de Enfermería, Universidad de Castilla-La Mancha, 02071 Albacete, Spain
* Correspondence: g.santos@csic.es (G.S.-S.); ivan.cruz@uclm.es (I.C.-C.)

Non-communicable diseases (NCDs), including obesity, cardiovascular conditions, cancer, and type II diabetes, are the primary causes of mortality worldwide, accounting for over 41 million deaths annually. NCDs are often associated with a dietary pattern that is considered a risk factor due to its relatively high content of fat, refined sugar, salt, and cholesterol [1]. For this reason, dietary approaches are cost-effective and safe interventions that can provide short- and long-term benefits for the management of these NCDs. In this sense, the use of 'functional foods' could be considered a useful strategy to prevent or improve these medical outcomes, and it has gained significant research attention in the fields of food health and technological innovations [2,3].

This Special Issue, entitled "Functional Foods as a New Therapeutic Strategy 2.0", provides an overview of some potential ingredients and foods that could be used in future therapeutic functional foods and nutraceuticals. Among the foods/ingredients studied in the five original articles and six reviews that make up this Special Issue are (1) vegetables such as quinoa (contribution 1), avocado (contribution 2), and some edible plants (*Momordica balsamina, Trifolium pratense, Achillea millefolium*, etc.) (contributions 3–5); (2) plant-derived compounds like ursolic acid (contribution 6); (3) commercially available supplements such as Dekosilhue® (contribution 7), melatonin (contribution 8), and Phoenix caffeine-free (Legion®) (contribution 9); and (4) marine products such as marine organisms, edible seaweeds (contribution 10), and marine macroalgae (contribution 11).

Taco et al. highlight the potential of quinoa leaves, which consistently show high levels of polyphenols and α-amylase inhibition after 80 days of cultivation, regardless of variety, time, or production site. These leaves exhibit strong free radical scavenging activity and α-amylase inhibition, suggesting potential benefits for managing diabetes and oxidative stress-related diseases. Compared to quinoa seeds, quinoa leaves offer valuable health-promoting properties. The authors noted that further research could unlock the potential of quinoa leaves as a dietary vegetable, functional food, or source of bioactive compounds for the food and pharmaceutical industries.

The avocado (*Persea americana*) seed contains 64% of the phenolic compounds found in the entire fruit, indicating its potential for use in managing obesity, a significant risk factor for metabolic disorders. Research conducted by Mokhele et al. explored the impact of avocado seed powder on obesity caused by a high-fat diet in rats. After 6 weeks, rats consuming a high-fat diet with avocado seed powder gained less weight than those in the control group. Furthermore, triglyceride levels were lower in the treatment group. Due to its phytochemicals and trace elements, avocado seed powder shows promise as a natural solution for reducing obesity. Nonetheless, more extensive studies are necessary to understand its long-term effects, mechanisms, and safety for human use before it can be recommended as a regular supplement.

Received: 29 January 2025
Accepted: 7 February 2025
Published: 11 February 2025

Citation: Santos-Sánchez, G.; Cruz-Chamorro, I. Functional Foods as a New Therapeutic Strategy 2.0. *Nutraceuticals* **2025**, *5*, 7. https://doi.org/10.3390/nutraceuticals5010007

On the other hand, three narrative reviews in this Special Issue deal with the bioactive chemical components of edible and wild plants and their potential use in folk medicine. Red clover (*Trifolium pratense*), native to Southeast Europe and Anatolia, has traditionally been used to treat upper respiratory conditions. Zukić et al. highlight the bioactive compounds in red clover, with a focus on its commercially available isoflavone extracts, which have proven effective in managing menopausal symptoms. These symptoms include hot flashes, mood changes, sleep disturbances, and long-term risks such as osteoporosis, cardiovascular disease, and cognitive decline. The review analyzed eight randomized controlled trials involving 8769 menopausal women. The results showed that isoflavones which are structurally similar to 17β-estradiol activate estrogen receptors, alleviating common menopausal symptoms. Compounds like biochanin A and formononetin reduce hot flashes, improve lipid profiles, maintain bone density, and exhibit anticancer and cognitive benefits. Polyphenols such as daidzein and genistein also address hot flashes, lipid metabolism, and atherosclerosis. The authors concluded that red clover isoflavones hold promise for managing menopause-related health concerns.

The review from Thiaw et al. explores the therapeutic and nutritional potential of *Momordica balsamina*, commonly known as balsam apple. Widely used in traditional medicine, its various parts—leaves, fruits, roots, and stems—contain bioactive compounds such as polyphenols, flavonoids, terpenes, and carotenoids. These molecules demonstrate diverse biological activities, including antioxidant, anti-inflammatory, anti-diabetic, anticancer, and antimalarial properties. The leaves of this plant are particularly rich in micronutrients, proteins, and amino acids, making them a valuable nutritional resource. The review highlights the chemical composition of this plant, as well as its biological functions and nutritional benefits, while emphasizing the need for conservation strategies, further exploration of its cosmetic applications, and innovative methods for synthesizing its bioactive compounds.

Finally, and continuing in the field of wild edible plants, Fantasma et al. carried out a narrative review that summarizes the phytochemical, nutritional, and pharmacological properties of wild plants commonly found in the central Italian Apennines and the Mediterranean region, such as *Achillea millefolium*, *Borago officinalis*, and *Foeniculum vulgare*. These plants are commonly used to treat ailments like digestive issues, colds, and circulatory problems, as well as more specific conditions such as hypertension and hyperglycemia. Their effects, which are discussed in the review, are attributed to phytochemicals like phenols, polyphenols, flavonoids, and carotenoids, which are gaining attention for their diverse biological effects and health benefits.

In addition to plants, fruits, and seeds, there are compounds that are extracted from plant sources for their health-promoting properties. This is the case with ursolic acid, a triterpenoid found in plants like cranberries, which has strong anticancer potential. Kornel et al. conducted a review of recent evidence on the effect of this acid on colorectal cancer. Research over the past decade suggests that it inhibits the proliferation of colon cancer cells and induces apoptosis. Furthermore, limited animal studies also indicate reduced tumor volumes and angiogenesis.

On the other hand, there are also products on the market that have been shown to have therapeutic effects. Dekosilhue® (DKS), for example, is a phytocomplex composed of a mix of thirteen plant extracts that is used as a dietary supplement to promote carbohydrate and lipid metabolism and has recently been reported to possess anti-neuroinflammatory activity. The study conducted by Borgonetti et al. evaluated DKS for its potential to reduce microglial senescence, produced by chronic low-grade systemic inflammation, a key factor in obesity.

Senescence was induced in a microglial cell line (BV2 cells) using intermittent lipopolysaccharide stimulation, and treatment with DKS (100 g/mL) decreased βgalactosidase activity, restored senescence-associated heterochromatin foci, and improved cell viability. DKS also reduced transcription factor (NF-κB) expression by 20%, a key regulator of the senescence-associated secretory phenotype. Additionally, DKS increased the viability of neuronal cells (SH-SY5Y cells) exposed to a senescent BV2-conditioned medium. These findings suggest that DKS exhibits senotherapeutic properties, with the potential to serve as an adjunctive intervention for obesity and related neurological disorders.

The effectiveness of caffeine-free fat loss supplements in enhancing thermogenesis remains unclear. The study conducted by Lafontant et al. compared the effects of a caffeinated (CAF) and non-caffeinated (NCAF) fat loss supplement on resting energy expenditure (REE), hunger, and hemodynamic variables in 25 healthy adults. CAF significantly increased participants' REE compared to the placebo at all time points, while NCAF reduced hunger more effectively than CAF and the placebo at 120 min post ingestion. CAF raised systolic and diastolic blood pressure, but neither supplement affected participants' heart rate. NCAF showed its potential to work as a fat loss aid through appetite suppression.

The study from Santos-Sánchez et al. explored the effects of melatonin (MLT) on lipid metabolism and cardiovascular risk in ApoE-deficient mice fed a Western diet. Mice were treated intragastrically with 2 or 9 mg/kg of MLT for 12 weeks. Although their body weight remained unchanged, the higher MLT dose significantly reduce low-density lipoprotein (LDL) cholesterol (LDL-C) levels and improved cardiovascular risk indexes. Additionally, MLT lowered hepatic total cholesterol and LDL-C levels, decreased leukocyte and lymphocyte populations, and enhanced antioxidant status. These findings suggest that MLT could be considered a functional ingredient that prevents or treats the development of cardiovascular diseases derived from a high cholesterol intake.

Seaweed is emerging as a sustainable and nutrient-rich food source, offering high levels of the vitamins, minerals, and proteins essential for human health. Seaweeds could serve as an alternative protein source, addressing global food security concerns and dietary needs, particularly within plant-based diets [4]. The review by Cotas et al. explores the growing interest in seaweed as a food source and nutraceutical. Seaweeds are recognized for their sustainability and potential to combat climate change, but their tendency to contain both beneficial and harmful components necessitate regulations to ensure consumer safety. Variations in composition among seaweed species influence their suitability as food or nutraceuticals, emphasizing the need for careful species selection and monitoring. The review underscores the importance of developing safe, innovative, and sustainable seaweed-based food products while addressing regulatory and safety challenges to improve human health.

Taking a more specific approach, the review by Yamaguchi highlights the potential of the marine alga *Sargassum horneri* to be a functional food factor with therapeutic and preventive effects against various metabolic disorders, including osteoporosis, diabetes, inflammation, and cancer. *S. horneri* water extracts promote bone health by stimulating osteoblastic bone formation and inhibiting osteoclastic bone resorption, preventing bone loss from aging and diabetic conditions. They also reduce serum glucose and lipid levels, suppress adipogenesis, and alleviate inflammation. Additionally, *S. horneri* extract inhibits the growth of bone-metastatic breast cancer cells (MDA-MB-231) and suppresses NF-κB signaling, which is linked to inflammation, in osteoblastic and macrophage cells. This multifunctional bioactive component shows promise for developing health supplements to prevent and manage metabolic disorders.

The research presented in this Special Issue underscores the growing potential of functional foods and nutraceuticals to become viable strategies for preventing and managing NCDs. These diverse studies highlight the expanding scope of dietary interventions in improving metabolic, cardiovascular, and inflammatory conditions. While these findings reinforce the promise of bioactive compounds in addressing key health challenges, further clinical studies are necessary to translate these findings into safe and effective dietary solutions. As functional food science continues to evolve, interdisciplinary collaboration between nutrition, medicine, and food technology will be crucial in developing innovative approaches to enhancing global health.

Author Contributions: G.S.-S. and I.C.-C.: Writing—original draft. All authors have read and agreed to the published version of the manuscript.

Conflicts of Interest: The authors declare no conflicts of interest.

List of Contributions

1. Taco, V.; Palmieri, C.; Borja, D.; Villacrés, E.; Duez, P.; Nachtergael, A. Qualitative Analysis by High-Performance Thin-Layer Chromatography–Bioautography of Ecuadorian Chenopodium quinoa Willd. Leaves: Influence of Variety, Phenological Stage, and Place of Cultivation on Free Radical Scavenging and α-Amylase Activity. *Nutraceuticals* **2024**, *5*, 1.
2. Mokhele, S.; Aboyade, O.; Katerere, D.R. Obesity Prevention Effects of Avocado (*Persea americana*) Seed Powder in High-Fat Diet-Induced Obesity in Rats. *Nutraceuticals* **2024**, *4*, 417–429.
3. Thiaw, M.; Samb, I.; Genva, M.; Gaye, M.L.; Fauconnier, M.-L. *Momordica balsamina* L.: A Plant with Multiple Therapeutic and Nutritional Potential—A Review. *Nutraceuticals* **2023**, *3*, 556–573.
4. Zukić, M.; Taljić, I.; Banjari, I. Effectiveness of Commercial Red Clover (*Trifolium pratense* L.) Products for the Treatment of Symptoms in Menopausal Women—A Narrative Review. *Nutraceuticals* **2024**, *4*, 430–449.
5. Fantasma, F.; Samukha, V.; Saviano, G.; Chini, M.G.; Iorizzi, M.; Caprari, C. Nutraceutical Aspects of Selected Wild Edible Plants of the Italian Central Apennines. *Nutraceuticals* **2024**, *4*, 190–231.
6. Kornel, A.; Tsiani, E. Effects of Ursolic Acid on Colorectal Cancer: A Review of Recent Evidence. *Nutraceuticals* **2024**, *4*, 373–394.
7. Borgonetti, V.; Sasia, C.; Morozzi, M.; Cenci, L.; Galeotti, N. Protection against Microglia Senescence by the Dietary Supplement Dekosilhue® in BV2 Cells: A New Perspective for Obesity and Related Complications. *Nutraceuticals* **2023**, *3*, 250–261.
8. Santos-Sánchez, G.; Álvarez-López, A.I.; Ponce-España, E.; Álvarez-Ríos, A.I.; Lardone, P.J.; Carrillo-Vico, A.; Cruz-Chamorro, I. Melatonin Modulates Lipid Metabolism and Reduces Cardiovascular Risk in Apolipoprotein E-Deficient Mice Fed a Western Diet. *Nutraceuticals* **2024**, *4*, 260–272.
9. Lafontant, K.; Broeckel, J.; Phillips, K.; Alsayed, Y.; Ayers-Creech, W.A.; Ohigashi, Y.; Solis, J.; Bale, C.; Parks, A.; Dankel, S. The Acute Effects of a Commercially Available Caffeinated and Caffeine-Free Thermogenic Dietary Supplement on Resting Energy Expenditure, Hunger, and Hemodynamic Responses. *Nutraceuticals* **2024**, *4*, 82–93.
10. Cotas, J.; Tavares, J.O.; Silva, R.; Pereira, L. Seaweed as a Safe Nutraceutical Food: How to Increase Human Welfare? *Nutraceuticals* **2024**, *4*, 323–362.
11. Yamaguchi, M. The Marine Alga *Sargassum horneri* Is a Functional Food with High Bioactivity. *Nutraceuticals* **2024**, *4*, 181–189.

References

1. Taheri Soodejani, M. Non-communicable diseases in the world over the past century: A secondary data analysis. *Front. Public Health* **2024**, *12*, 1436236. [CrossRef] [PubMed]

2. Santos, H.O.; Tinsley, G.M. Is breakfast consumption detrimental, unnecessary, or an opportunity for health promotion? A review of cardiometabolic outcomes and functional food choices. *Diabetes/Metab. Res. Rev.* **2024**, *40*, e3684. [CrossRef] [PubMed]
3. Cruz-Chamorro, I.; Álvarez-López, A.I.; Santos-Sánchez, G.; Álvarez-Sánchez, N.; Pedroche, J.; Millán-Linares, M.d.C.; Lardone, P.J.; Carrillo-Vico, A. A Lupin (*Lupinus angustifolius*) Protein Hydrolysate Decreases the Severity of Experimental Autoimmune Encephalomyelitis: A Preliminary Study. *Int. J. Mol. Sci.* **2024**, *26*, 32. [CrossRef] [PubMed]
4. Pathan, A.; Mahajan, M.M.; Jain, P.G.; Zambad, S.P.; Bhandari, G.S.; Darak, A.D.; Ahire, E.D.; Rajora, A.K.; Surana, K.R. Nutraceutical Properties of Seaweed Vitamins. In *Vitamins as Nutraceuticals: Recent Advances and Applications*; Wiley Online Library: Hoboken, NJ, USA, 2023; pp. 167–184.

nutraceuticals MDPI

Article

Qualitative Analysis by High-Performance Thin-Layer Chromatography–Bioautography of Ecuadorian *Chenopodium quinoa* Willd. Leaves: Influence of Variety, Phenological Stage, and Place of Cultivation on Free Radical Scavenging and α-Amylase Activity

Verónica Taco [1,2], Claudio Palmieri [2], Dayana Borja [1], Elena Villacrés [3], Pierre Duez [2] and Amandine Nachtergael [2,*]

[1] Faculty of Chemical Sciences, Central University of Ecuador (UCE), Av. Universitaria, Quito 170129, Ecuador; vjtaco@uce.edu.ec (V.T.)

[2] Unit of Therapeutic Chemistry and Pharmacognosy, Faculty of Medicine, Pharmacy and Biomedical Sciences, University of Mons (UMONS), Avenue du Champ de Mars, 25, 7000 Mons, Belgium; pierre.duez@umons.ac.be (P.D.)

[3] Nutrition and Food Quality Department, National Agricultural Research Institute, Santa Catalina, Quito 170518, Ecuador; elena.villacres@iniap.gob.ec

* Correspondence: amandine.nachtergael@umons.ac.be; Tel.: +32-65372212

Abstract: The present study aimed to qualitatively assess the influence of *Chenopodium quinoa* Willd. varieties (INIAP-Tunkahuan, INIAP-Pata de Venado varieties and Chimborazo genotype), phenological stages (40, 60, and 80 days), and places of cultivation (Pichincha and Chimborazo Ecuadorian provinces) on the leaf and seed phenolic composition and biological properties. Their nutraceutical potential was assessed through qualitative analyses of (i) their polyphenols by high-performance thin-layer chromatography (HPTLC); and (ii) their free radical scavenging (quenching of 2,2-diphenyl-1-picrylhydrazyl free radical, DPPH$^\bullet$) and α-amylase inhibitory properties (iodine visualization of starch hydrolysis) by HPTLC–bioautography. Compared to seeds, the quinoa leaf methanolic extracts present a high content of polyphenols with free radical scavenging activity, and compounds with an α-amylase inhibitory property; both biological activities indicate a remarkable potential of quinoa leaves, which may be relevant for the treatment of diabetes but also for the chemoprevention and/or treatment of pathologies related to oxidative stress. In quinoa leaves harvested after 80 days of cultivation, regardless of the place of production and the variety, a high content of bioactive compounds was observed. Future research is undoubtedly needed to further promote quinoa leaves as a dietary vegetable or to develop them into a nutritional supplement. This would empower quinoa smallholders in Andean regions to promote the sustainable development of this culture in its places of origin.

Keywords: amaranthaceae; DPPH$^\bullet$; polyphenols; antioxidant detection; α-amylase enzyme inhibitor screening

Academic Editors: Ivan Cruz-Chamorro and Guillermo Santos Sánchez

Received: 8 November 2024
Revised: 16 December 2024
Accepted: 24 December 2024
Published: 27 December 2024

Citation: Taco, V.; Palmieri, C.; Borja, D.; Villacrés, E.; Duez, P.; Nachtergael, A. Qualitative Analysis by High-Performance Thin-Layer Chromatography–Bioautography of Ecuadorian *Chenopodium quinoa* Willd. Leaves: Influence of Variety, Phenological Stage, and Place of Cultivation on Free Radical Scavenging and α-Amylase Activity. *Nutraceuticals* **2025**, *5*, 1. https://doi.org/10.3390/nutraceuticals5010001

1. Introduction

Quinoa (*Chenopodium quinoa* Willd., Amaranthaceae) is an important crop recognized as an ally for global food security in the 21st century and stands out among the crops for its resistance to abiotic stresses such as drought, hot temperature, high altitude, and saline soils. It has a broad genetic diversity, which allows it to adapt to various tough environments, such as highlands and frost [1–3]. Quinoa seeds are rapidly gaining global popularity as a functional food and nutraceutical because they contain a variety of bioactive compounds (flavonoids, saponins, peptides, phytoecdysteroids, phytosterols, lectins, among others) with antioxidant, cytotoxic, antidiabetic, and anti-inflammatory broad spectra properties, demonstrated in vitro [4–7].

Today, biologically active compounds isolated from plants are still important sources for modern drug and nutraceutical formulations. Koseoglu Yilmaz et al. previously reported the phenolics, flavonoids, and biological activities of acetone and methanol extracts of a scarcely studied species, *C. album subsp. album var. micropyhllum* (note: the botanical status of this plant is not recognized in www.worldfloraonline.org (accessed on 12 December 2024). Acetone and methanol extracts showed similar DPPH free radical scavenging activities, whereas the cupric ion-reducing capacity of the acetone extract was the highest. The methanol extract was found to contain significant amounts of hesperidin and rutin [8].

Peru, Bolivia, and Ecuador are the leading quinoa-producing countries in the Andean region, as they have the largest quinoa cultivation areas and are key centers of domestication and biodiversity [9,10]. While quinoa is now grown in many countries worldwide, preserving agrobiodiversity remains a crucial task. This preservation not only safeguards the plant's genetic diversity but also protects the cultural heritage of the indigenous farmers in the Andean region [11]. Today, quinoa faces growing global demands. Sudden spikes in consumer interest led to rising prices, prompting land-use changes at both farm and national levels, resulting in rapid production increases. This phenomenon, known as a "boom", is often followed by a "bust", where prices and production sharply decline [12]. These effects are beginning to surface in Ecuador. Quinoa boom and bust cycles could have unforeseen consequences for consumers, including food insecurity in the regions of production and origin, simplified local food systems, and a potential reduction in the genetic diversity that has been preserved for centuries. Addressing this challenge requires a more inclusive approach that fosters active decision-making and evaluates future scenarios. One way to enhance the sustainable development of quinoa is by diversifying its use through research that explores the nutraceutical potential of other plant parts, such as the leaves, and by empowering smallholders to engage in sustainable production practices [11,12]. In this context, only limited information is available on the nutraceutical potential of quinoa leaves, even though their polyphenolic content is reported to be even higher compared to seeds [5,13]. The young quinoa leaves (harvested in the phenological vegetative stage, around 90 days of cultivation) are used as vegetables in some places in Asia and North and South America [13,14]. The advantages of promoting the consumption of quinoa leaves would be to (i) reduce the growth time before harvesting, i.e., a maximum of 3 months for leaves compared to 6 to 9 months for seeds, which, for producers, would lessen the risks of climatic threats; (ii) diversify the economy of quinoa producers; and (iii) promote health, as quinoa leaves may help in chemoprevention of cancer and other diseases related to oxidative stress; indeed, a previous in vitro study reported a probable high bioavailability of quinoa leaf polyphenols with cytostatic and anti-invasive effects on prostate cancer cells [13].

Polyphenolic compounds are a diverse and widely occurring group of plant phytochemicals; they are potent antioxidants and reactive oxygen species (ROS) quenchers [15,16] through both direct and indirect mechanisms, notably hydrogen donation, metal chelation, antioxidant enzyme upregulation, nuclear factor erythroid 2-related factor 2 (Nrf2) activation, sirtuin activation and/or pro-oxidant enzyme inhibition [17]. Such properties presumably imply preventive and/or therapeutic effects on a wide range of diseases, from cancer to neurodegenerative disorders [18]. Among polyphenols, flavonoids have been extensively investigated and are already successfully applied to the treatment of osteoarthritis and certain vascular diseases; their interest has led to the development of new product delivery systems likely to increase their efficacy [17,19]. Also, a high number of in vitro and in vivo studies indicate that a diet rich in flavonoids may contribute to reducing the incidence of diabetes; flavonoids notably reduce glycemia, inhibit α-amylase and α-glucosidase, and protect the pancreas [20]. The inhibition of starch-hydrolyzing enzymes delays the elevation of postprandial blood sugar levels. Most of the commercially available synthetic inhibitors, such as acarbose, miglitol, and voglibose, strongly inhibit α-amylase and α-glucosidase, leading to colonic fermentation of undigested saccharides, abdominal distention, meteorism bloating, flatulence, and possibly diarrhea [21,22]. Therefore, there is interest in screening natural sources for novel enzyme inhibitors that may present fewer side effects [5]. Natural sources investigated for flavonoids' antidiabetic potential, specifically quercetin and kaempferol, include apples, berries, red onions, cherries, grapefruits, teas, cruciferous vegetables, *Hypericum ascyron* L., and *Myrica rubra* (Lour.) Siebold & Zucc [20,23].

When studying natural sources of active compounds, such as quinoa leaves, it is necessary to know the influence of production factors on their quality and efficacy, i.e., on their polyphenolic profiles and their biological activities. Indeed, a versatile analytical system is required to study such influences [24]. In this sense, high-performance thin-layer chromatography (HPTLC) and bioautographic assays on HPTLC plates play a relevant role in rapidly fingerprinting and screening plant extracts for various bioactivities, e.g., antibacterial, antifungal, antioxidant, or enzyme inhibition [25,26]. HPTLC–bioautography, applied to these complex matrices, enables the fast detection and localization of the compounds responsible for tested activities. Furthermore, this technique is effective, relatively inexpensive, and can be performed in small research laboratories without access to sophisticated equipment [27]. A distinct advantage of HPTLC assays resides in the chromatographic separation that precedes the application of detection reagents, which reduces or avoids interferences, notably from colored compounds such as leaf chlorophylls.

Ecuador exhibits significant morphological variability in quinoa crops, making it important to determine the chemical profile of compounds with biological activities.

Our previous study [28] allowed us to optimize the parameters necessary for obtaining reproducible results in the assay of α-amylase inhibitory activity, focusing on extracts derived from a single quinoa variety. Based on these data, the present paper aims to study the variabilities inherent to *Chenopodium quinoa* cultivars, culture stations, and phenological stages. Additionally, this study proposes a qualitative method for the rapid detection of phenolic compounds and associated bioactivities in quinoa leaves, while minimizing interference from colored molecules such as chlorophyll, which can affect results obtained using spectrophotometric methods.

2. Material and Methods

2.1. Plant Material

In this study, INIAP-Tunkahuan (IT) and INIAP-Pata de Venado (PV) quinoa varieties and the Chimborazo (C) quinoa genotype from Ecuador were investigated. The INIAP varieties, selected for their low saponin content, were grown at the Instituto Nacional de Investigaciones Agropecuarias (INIAP), Santa Catalina Experimental Station, Pichincha Province (altitude 3050 m.a.s.l., 0°22′01″ S 78°33′17″ W; during quinoa leaves growth and harvest, average temperature 12 °C, relative humidity 78%, rainfall 20 mm) and in Calpi, Chimborazo Province (altitude 3060 m.a.s.l., 1°38′48″ S 78°43′47″ W; during quinoa leaves growth and harvest, average temperature 14 °C, relative humidity 74%, rainfall 37 mm). The Chimborazo genotype has a high saponin content and is cultivated in Calpi; it is endemic to Ecuador and exclusively cultivated in Chimborazo Province, whose name is due to its proximity to the Chimborazo volcano. The cultivation cycles of INIAP-Tunkahuan and INIAP-Pata de Venado varieties and Chimborazo genotype are 150–170, 130–150, and 240–270 days, respectively. The phenological growth of the different quinoa varieties and genotype is presented in Figure 1, according to previous descriptions of stages [27,28]; Table 1 describes the investigated samples with their phenological stages at harvest times. After harvest, quinoa leaves were washed with distilled water, and the excess water was removed with a paper towel. They were then lyophilized, ground, and stored at −20 °C. Seed quinoa samples were stored at −20 °C and ground before analysis. All material was powdered in a laboratory mill (PX-MFC 90 D, Kinematica AG, Malters, Switzerland) and sieved to obtain particle sizes ≤ 0.5 mm.

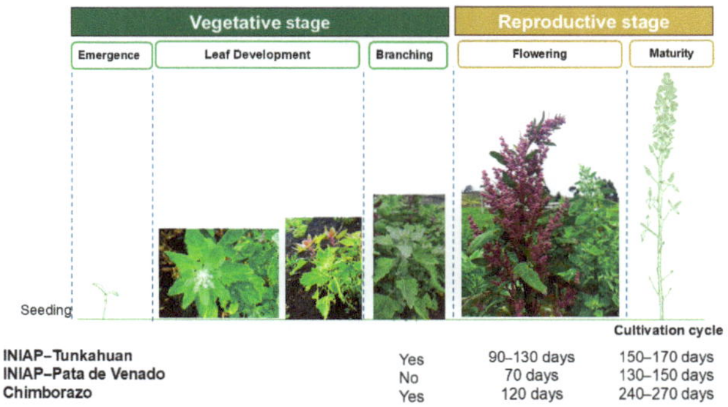

Figure 1. Phenology growth of quinoa plant (adapted from Sosa-Zuniga et al. [29] and Yzarra et al. [30]).

Table 1. Description of quinoa methanolic extracts, sweet and bitter varieties, investigated in this work.

Quinoa	Description	Place of Cultivation	Variety or Genotype	Characters	Phenological Stage	Track N° in Chromatogram
Sweet	Quinoa leaves (QL)	Pichincha (P)	INIAP-Tunkahuan (IT)	Green leaves	40 days [(a)]	1
					60 days [(b)]	2
					80 days [(c)]	3
			INIAP-Pata de Venado (PV)	Green leaves	40 days [(a)]	4
					60 days [(b)]	5
					80 days [(c)]	6
		Chimborazo (C)	INIAP-Tunkahuan (IT)	Green leaves	60 days [(b)]	7
					80 days [(c)]	8
			INIAP-Pata de Venado (PV)	Green leaves	60 days [(b)]	9
					80 days [(c)]	10
	Quinoa seed (QS)	Pichincha (P)	INIAP-Tunkahuan (IT)	White seeds	6 months	a, b, c
			INIAP-Pata de Venado (PV)	White seeds	5 months	d, e, f
Bitter	Quinoa leaves (QL)	Chimborazo (C)	Chimborazo (C)	Mix of yellow, green, and red leaves	60 days [(b)]	11
				Mix of yellow, green, and red leaves	80 days [(c)]	12
				Yellow leaves (y)	80 days [(c)]	13
				Green leaves (g)	80 days [(c)]	14
				Red leaves (r)	80 days [(c)]	15

[(a)] Vegetative stage: six true leaves are visible on the plant. [(b)] Vegetative stage: amounts of leaves significantly increase. [(c)] Vegetative stage: large amounts of leaves.

2.2. Chemicals and Reagents

2-Aminoethyl diphenylborinate (97%) (NP reagent), α-amylase from *Bacillus licheniformis* (Cat. No. A4582-5mL), soluble starch (ACS reagent), iodine (≥99.8%), 2,2-di (4-tert-octylphenyl)-1-picrylhydrazyl (DPPH•), rutin hydrate (≥94%), quercetin hydrate (≥95%), kaempferol (≥90%), and chlorogenic acid (≥95%) were purchased from Sigma-Aldrich (Darmstadt, Germany); hyperoside (≥98%) and isoquercitroside (≥99%) were obtained from Extrasynthese (France); polyethyleneglycol 400 (PEG 400), methanol (99%), absolute ethanol (≥99.8%), methyl ethyl ketone (GPR reactapur), formic acid (98%), and ethyl acetate (ACS reagent) were purchased from VWR Chemicals. Water was obtained using milli-Q grade water Plus Millipore (18.0 MΩ·cm).

2.3. Extraction

To prepare the extracts, 0.150 g of quinoa leaf powder or 0.300 g of quinoa seed powder were mixed with 3 mL of methanol–water (80:20, *w*/*w*), vortexed for 2 min, heated at 40 °C for 1 h, sonicated at room temperature (RT) for 30 min, and centrifuged (4000× *g*, 20 min, 25 °C; UniCen MR, Herolab, Germany) [31]. The supernatants were stored in an amber glass 5 mL vial with a screw cap and analyzed the following day.

2.4. Preparation of Solutions

Standard solutions were separately prepared in methanol at the following concentrations: 0.8 mg/mL of quercetin, 1.6 mg/mL of rutin, 0.5 mg/mL of hyperoside, 1.0 mg/mL of isoquercitroside, 1.0 mg/mL of kaempferol, and 1.2 mg/mL of chlorogenic acid. For polyphenols detection, a 10 g/L solution of diphenylboric acid 2-amino ethyl ester, and a 50 g/L solution of PEG 400 were prepared in methanol. For radical scavengers' detection, a 0.05% (*w*/*v*) DPPH• solution was prepared in methanol and protected from light. The α-amylase solution (~5 U/mL) was prepared by mixing 50 μL of α-amylase stock solution with 20 mL of water and diluting it to the required concentrations with ethanol–water (10:90, *v*/*v*). A homogeneous solution of starch was prepared by dissolving 1 g of starch in 40 mL of water (70 °C, stirring at 350 rpm for 30 min), adding water up to 100 mL, and cooling to room temperature slowly by stirring. That solution was viscous and difficult to apply using the automatic Camag® Derivatizer; therefore, 10 mL of ethanol was added, and the solution was stirred for 2 h at RT to reduce its viscosity [32].

2.5. High-Performance Thin-Layer Chromatography

HPTLC was performed according to the general chapter 2.8.25 of the *European Pharmacopoeia* 10 [32]. Fingerprinting of polyphenolic compounds was performed according to the procedure reported by Liu et al. [31]; antioxidant and α-amylase inhibitory zones were visualized using the protocols reported by Agatonovic-Kustrin et al. [33] with some optimizations [32]. Chromatographic layers were HPTLC silica gel 60 F_{254} plates, size 20 × 10 cm (Merck, Darmstadt, Germany). Samples were applied by spray, using an Automatic TLC Sampler (ATS 4, Camag, Basel, Switzerland); 4, 6, and 8 μL of the sample were sprayed onto the plate for the polyphenols detection protocol, the free radical scavenging assay (DPPH•), and the α-amylase inhibition assay, respectively; 15 bands of 8 mm were applied per plate, 8 mm from the plate's lower edge; the plates were equilibrated under a 33% relative humidity and developed over a pathway of 70 mm from the lower edge in an Automated Multiple Development chamber (AMD2, Camag) with formic acid–water–methyl ethyl ketone–ethyl acetate (10:10:30:50, *v*/*v*/*v*/*v*). The plates were dried automatically for 5 min after development and heated at 105 °C for 60 min using a TLC Plate Heater (Camag). All reagents were sprayed with the Derivatizer Camag®, selecting the appropriate nozzle. Upon derivatization, the plates were documented as digital images under short-wave UV light (254 nm), long-wave UV light (365 nm), and white light using the TLC Visualizer 2. The Camag® systems were driven by visionCATS software, version 2.5.

Post-Chromatographic Derivatization

Polyphenolic compounds: 2 mL of aminoethyl diphenylborinate solution was applied on the warm plate (green nozzle, level 3) followed by 2 mL of polyethyleneglycol 400 solution (blue nozzle, level 2).

Free radical scavenging assay (DPPH•): 2 mL of DPPH solution was applied on a warm plate (green nozzle, level 3). Images were recorded 90 s, 30 min, and 100 min after derivatization.

α-Amylase inhibition assay: 3 mL of α-amylase solution was applied (yellow nozzle, level 4) on a cooled plate that was then incubated at 37 °C for 30 min in a humid chamber. Then, 2 mL of starch solution was applied (yellow nozzle, level 6), and the plate was incubated at 37 °C in a humid chamber for 10 min and treated with iodine vapors for 2 min (1 g of solid iodine in a lidded 27.0 × 26.5 × 7.0 cm TLC development chamber).

3. Results and Discussion

3.1. Polyphenols HPTLC Fingerprinting

Phenological stages, cultivation sites, and varieties were evaluated by HPTLC methods for their influence on polyphenolic profiles, ROS scavenging, and α-amylase inhibition of quinoa leaves' polar extracts. The HPTLC fingerprints of quinoa leaves and quinoa seeds were compared, considering that quinoa seeds have been extensively studied in recent years for their nutritional and health relevance [7].

Quinoa seeds and leaves contain many secondary metabolites, such as polyphenols, terpenoids, steroids, and nitrogen-containing compounds. Polyphenols, including phenolic acids, flavonoids, and tannins, constitute bioactive secondary metabolites that are reported to play significant roles, notably for antimicrobial, antioxidant, anti-inflammatory, antitumor, and anticarcinogenic activities [6,34]. Most recent studies on quinoa leaves primarily report the total content of polyphenols and flavonoids. Notably, the contents of these secondary metabolites are higher in young leaves compared to quinoa seeds [35,36]. Gawlik-Dziki et al. reported that, in the leaves of *Chenopodium quinoa* variety Faro, the predominant phenolic acids were ferulic, sinapic, chlorogenic, and gallic acids, while rutin, kaempferol, and isorhamnetin were the most abundant flavonoids [13].

Considering that phenolic acids and flavonoids, based on kaempferol and quercetin aglycones, are the most abundant compounds in quinoa, with previously reported biological activities, the present HPTLC study investigated major representative compounds from these groups. This approach aimed to determine not only their Rf values but also the colors they developed with NP reagent–PEG, allowing for the interpretation of the HPTLC fingerprints of sweet and bitter quinoa leaf extracts. Characteristic colors of different polyphenols (standard solutions) upon treatment with NP reagent–PEG are shown in Figure 2. The yellow/orange spots correspond to quercetin and its derivatives hyperoside, isoquercitroside, and rutin; the green and blue spots correspond to kaempferol and chlorogenic acid, respectively [31].

According to previous studies on the extraction of polyphenols, hydromethanolic solvents are particularly suited to extract these compounds. And so, in this study, the mixture methanol–water 80:20 (v/v) was retained. The HPTLC fingerprints of polyphenols from sweet and bitter quinoa leaf extracts (Table 1) indicate (Figure 3) an abundance of glycosides based on the flavonols quercetin and kaempferol (yellow and green spots, respectively) and of phenolic acids (blue spots). These compounds have been extensively studied, mostly in vitro, for their chemopreventive and anticarcinogenic effects [6,13]. The red spots probably correspond to chlorophylls [37]. Interestingly, Gawlik-Dziki et al. associated the presence of high levels of the flavonols kaempferol, isorhamnetin, and rutin in a quinoa leaf extract (young leaves at 90 days; ethanol 50%, *v*/*v*) with an inhibition of prostate cancer cell proliferation, motility, and competence for gap junctional communication [13,31].

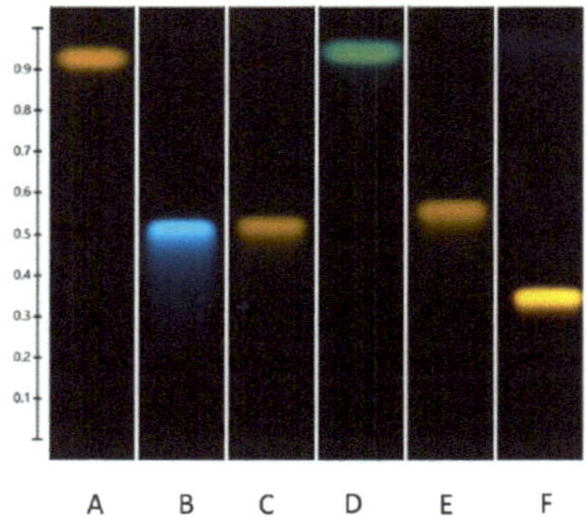

Figure 2. HPTLC chromatogram of standard phenolic compounds: quercetin, 3.2 µg (**A**); chlorogenic acid, 4.8 µg (**B**); hyperoside, 2.0 µg (**C**); kaempferol, 4.0 µg (**D**); isoquercitroside, 4.0 µg (**E**); rutin, 6.4 µg (**F**). Mobile phase: formic acid–water–methyl ethyl ketone–ethyl acetate (10:10:30:50, *v*/*v*/*v*/*v*); treatment with NP reagent–PEG and visualization under UV_{365nm}.

The HPTLC profiles of INIAP-Tunkahuan (Figure 3; tracks 1, 2, 3) and INIAP-Pata de Venado (Figure 3; tracks 4, 5, 6) leaves from Pichincha vary according to phenological stages. As exemplified by peak profiles of INIAP-Tunkahuan (Figure 3), the later vegetative stages (60 and 80 days vs. 40 days, respectively) lead (i) to increased peak areas for kaempferol derivatives, by 8 and 23% (green peak 1) and by 26 and 29% (green peak 2), and quercetin derivatives, by 16 and 17% (yellow peak 1) and by 31 and 35% (yellow peak 2); and (ii) to decreased peak areas for phenolic acids, by 41 and 55% (blue peak 1) and by 18 and 70% (blue peak 2). These results are consistent with those of Buitrago et al. [14], who recorded 1.3 times higher contents of flavonoids (90 days versus 30 and 180 days) in quinoa leaf ethanolic extracts (ethanol 96% *v*/*v*). Similarly, Baldeon et al. confirmed that the total polyphenol content in quinoa leaves varies with the phenological stage. Interestingly, they reported that the highest levels of total phenolic compounds in four Peruvian quinoa cultivars (Blanca Junin, Pasankalla, Salcedo INIA, and Blanca Criolla) were observed during the second phenological stage, 42 days after sowing. Another study on young Peruvian quinoa leaves from three cultivars (Titicaca, Puno, and Vikinga) demonstrated that the contents of ferulic acid, isoquercetin, and rutin significantly varied depending on plant density, harvesting moment, and cultivar [38,39].

The influence of phenological stages on leaf flavonoid content and/or profile was also shown for plant species from many different families, e.g., Moraceae (*Morus alba* L. [40]), Passifloraceae (*Passiflora alata*, Dryander [41]), and Portulacaceae (*Portulaca oleracea* L. [42]).

The HPTLC fingerprints of sweet quinoa leaves appear quite similar at days 60 and 80, indicating that there are no significant effects of variety (INIAP-Tunkahuan or INIAP-Pata de Venado) or cultivation place (Pichincha or Chimborazo); these harvesting times correspond to the end of the vegetative stage (cf. Figure 1), when leaves are most abundant and locally consumed as vegetables [13]. In both sites, over all the cultivations of samples, the environmental conditions were quite similar (cf. Section 2.1), precluding any of the well-known meteorological impacts on polyphenol profiles [43]. The results of this study differ from those reported in the seeds of two Peruvian species (*Chenopodium quinoa*, six ecotypes, and *Chenopodium pallidicaule* Aellen, three varieties), in which the contents of the most relevant flavonoids, quercetin and kaempferol glycosides, as well as phenolic acids, differed according to both species, ecotypes, and varieties [44].

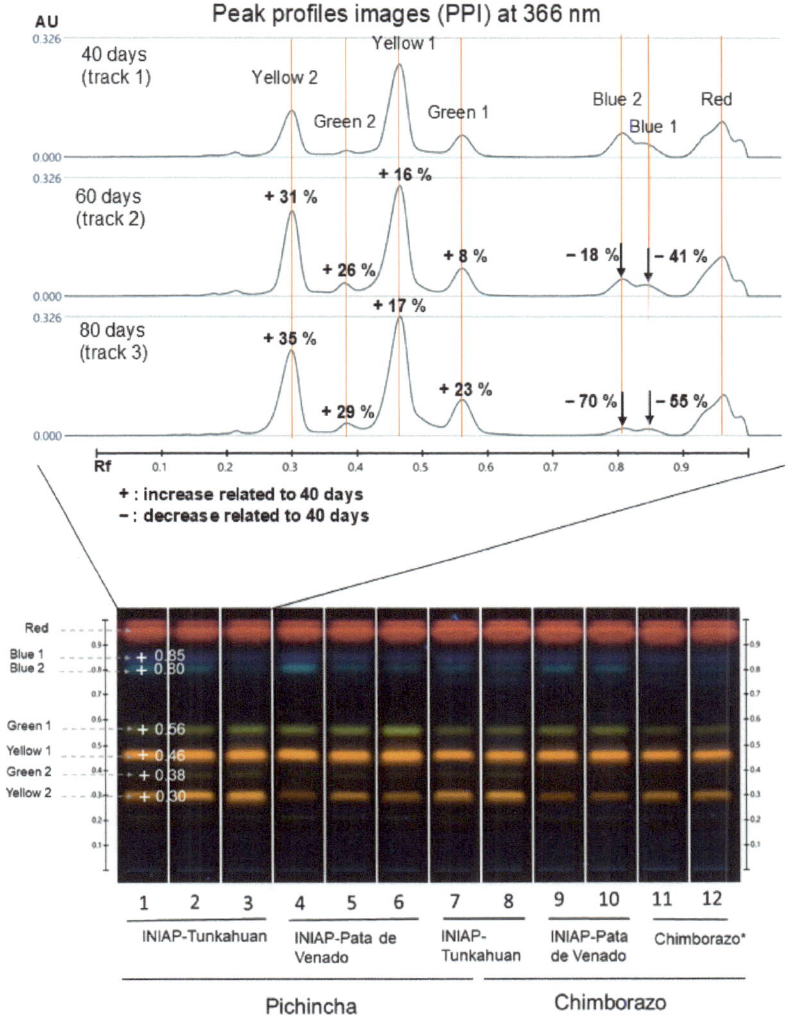

Figure 3. HPTLC fingerprints of polyphenols in Ecuadorian quinoa leaf extracts (methanol–water, 80:20, *w*/*w*; sample–solvent ratio, 1:20 *w*/*v*; application volumes, 4 µL). Mobile phase: formic acid–water–methyl ethyl ketone–ethyl acetate (10:10:30:50, *v*/*v*/*v*/*v*). Derivatization with NP and PEG; examination under UV_{365nm}. Quinoa leaves from Pichincha: INIAP-Tunkahuan (tracks 1, 2, 3), INIAP-Pata de Venado (tracks 4, 5, 6). Quinoa leaves from Chimborazo: INIAP-Tunkahuan (tracks 7, 8), INIAP-Pata de Venado (tracks 9, 10), Chimborazo* genotype (tracks 11, 12).

The chromatographic profiles of Chimborazo bitter quinoa leaves (Figure 3; tracks 11, 12) were comparable to those of the sweet varieties. The endemic Chimborazo bitter genotype has the particularity of developing morphotypes bearing leaves with three clearly differentiated colors, yellow, green, and red [45], and is then considered a *"mestiza"* genotype. Figure 4 shows the polyphenol profiles of extracts obtained from leaves separately collected according to color (these samples were obtained in a specific experiment where quinoa plants of these three colors were separately cultivated); these can be compared with samples composed by mixing the three types of leaves (Figure 3; tracks 11, 12). Interestingly, regardless of the morphotype color, their polyphenolic fingerprints appear quite similar; the color difference could be attributed to carotenoids, which would be interesting to further investigate.

Figure 4. HPTLC fingerprints of polyphenols in yellow, green, and red leaves of the Chimborazo bitter quinoa variety extracts (methanol–water; 80:20, *w*/*w*; sample–solvent ratio, 1:20 *w*/*v*; application volumes, 4 µL). Mobile phase: formic acid–water–methyl ethyl ketone–ethyl acetate (10:10:30:50, *v*/*v*/*v*/*v*). Derivatization with NP and PEG; examination under UV_{365nm}. Samples as per Table 1.

3.2. *DPPH• Assay*

Phenolic compounds are a diverse and widely occurring group of phytochemicals in plant foods, reputed for their beneficial effects on health. They are of considerable interest due to their antioxidant and enzyme-inhibiting properties, as well as their ability to scavenge free radicals and reactive oxygen species. Flavonol glycosides isolated from *Chenopodium quinoa* seeds have demonstrated antioxidant activity in the DPPH• assay. Zhu et al. reported the isolation of six flavonol glycosides from quinoa seeds, all of which exhibited antioxidant activity in this test. Notably, the two identified quercetin 3-glycosides showed significantly stronger activity than the four kaempferol 3-glycosides present [13,34,46]. Also, Gawlik-Dziki et al. reported that hydroethanolic extracts of quinoa leaves (50% *v*/*v* ethanol), which contained considerable amounts of ferulic, sinapic, chlorogenic, and gallic acids, as well as rutin, kaempferol, and isorhamnetin, were able to prevent lipid oxidation. However, their reducing power and antioxidant activity, measured using an $ABTS^{+\bullet}$ decolorization assay, were significantly lower [13].

The direct HPTLC-DPPH• assay was applied here to assess the free radical scavenging activity of quinoa leaf extracts, with the distinct advantages of easiness and rapidity [33,40]. The intensely violet DPPH• is reduced into the yellow 2,2-diphenyl-1-picrylhydrazine (DPPH-H), upon reaction with a hydrogen atom donor. The quinoa compounds with free radical scavenging properties were separated on the HPTLC plate, appearing as yellow zones against the violet DPPH• background; the intensity of reagent decolorization is proportional to a compound's free radical scavenging potency and amount [40]. Figure 5a profiles the compounds that contribute to the free radical scavenging activity of sweet and bitter quinoa leaves, compared with rutin. Ninety seconds after derivatization, the same three yellow zones (Rf 0.32, 0.48, and 0.85) appear in all quinoa leaf samples, regardless of harvest time (40, 60, and 80 days), variety, and place of cultivation (Pichincha and Chimborazo). Baldion et al. assessed the influence of the phenological profile on the antioxidant capacity of leaves from four Peruvian quinoa varieties (Blanca Junín, Blanca Criolla, Pasankalla, and Salcedo). They reported the highest ABTS+• radical scavenging capacity in Pasankalla (1492 µmol TEAC/g) at the second stage, 44 days after sowing. During the same stage, Blanca Junín (1127 µmol TEAC/g), Salcedo (1042 µmol TEAC/g), and Blanca Criolla (946 µmol TEAC/g) also reached their maximum values. This study observed that the highest ABTS+• radical scavenging capacities coincided with the highest total phenolic content at the same stage, a trend also reported previously in Peruvian Andean medicinal leaves, where correlations between ABTS+• and total phenolic content were established. Additionally, studies by Gawlik-Dziki et al. (2013) on quinoa leaves indicated strong relationships between the concentration of phenolic compounds and antioxidant activity, including ABTS+• antiradical capacity [13,38].

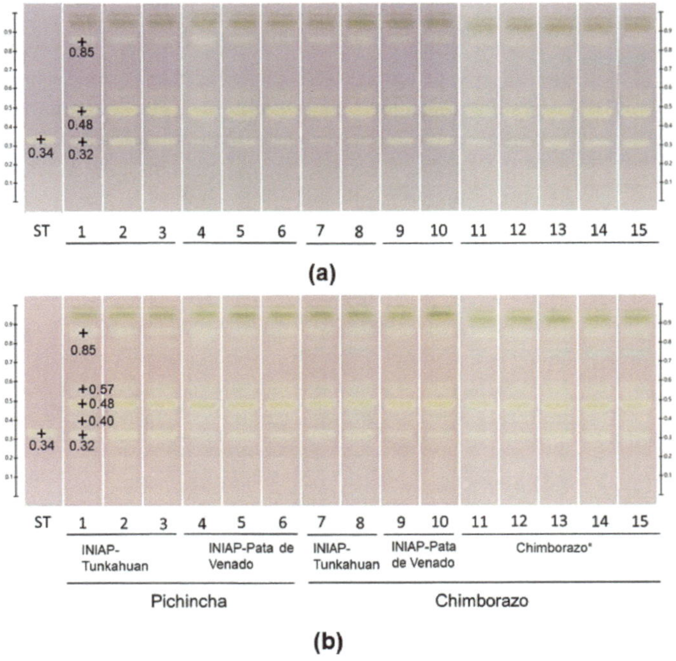

Figure 5. Free radical scavenging activity of Ecuadorian quinoa leaf extracts (methanol–water, 80:20, *w*/*w*; sample–solvent ratio, 1:20 *w/v*; application volumes, 6 µL). Mobile phase: formic acid–water–methyl ethyl ketone–ethyl acetate (10:10:30:50, *v*/*v*/*v*/*v*). Derivatization with DPPH• for 90 s (**a**) and 100 min (**b**) and examination under visible light. Rutin 0.50 µg (track, ST). Quinoa leaves from Pichincha: INIAP-Tunkahuan (tracks 1, 2, 3), INIAP-Pata de Venado (tracks 4, 5, 6); quinoa leaves from Chimborazo: INIAP-Tunkahuan (tracks 7, 8), INIAP-Pata de Venado (tracks 9, 10), Chimborazo variety (tracks 11, 12) and Chimborazo* genotype yellow, green, and red colors (tracks 13, 14, 15). Samples as per Table 1.

Upon derivatization, the plate was maintained at RT in the dark for 30 min (Figure S1, Supplementary Materials) and 100 min (Figure 5b); with time, the intensity of the discolored bands increased, with a higher number of yellowish zones appearing. Recent studies on HPTLC bioautography by Agatonovic-Kustrin et al. [33] and Islam et al. [47] indicate that the rates of reaction between DPPH• and substrates vary widely, according to their antioxidant potency and amounts, and that times between 30 and 60 min are needed to complete the reaction.

Additionally, regardless of the color of Chimborazo bitter quinoa leaves, as shown in Figure 5a,b, they show similar fingerprints of free radical scavenging activity (composite samples in tracks 11, 12; yellow, green, and red leaves in tracks 13, 14, 15, respectively).

3.3. *α-Amylase Inhibitory Activity*

The HPTLC bioautography with "micro droplet" spraying was optimized in our laboratory for the screening of compounds with α-amylase inhibitory activity [31] to considerably reduce the reagent needs (from 40–200 mL to 2–3 mL) and plate smearing usually associated with dipping methods [33,48,49]. A clear and defined blue zone (Rf 0.38; marked with a solid red line; Figure 6a) highlighting the position of α-amylase inhibitor(s) was observed in all samples, regardless of the harvest time, variety, and place of cultivation. Slight blue areas (Rf 0.53 or 0.55; marked with a red dotted line), possibly due to a low effectiveness or concentration of α-amylase inhibitor(s), were observed in some samples, especially in samples with a harvest time of 80 days, regardless of the variety or cultivation site. To clearly see these blue zones, the applied sample volumes were increased (Figure 6b). A previous study suggested that the prominent blue area at the lower edge of the plates may indicate incomplete removal of migrated formic acid, leading to enzyme denaturation and preventing starch degradation [28]. Hence, only the zone above formic acid can be inspected for active compounds. In an in vitro assay, Hemalatha et al. reported a strong inhibition of α-amylase from hydromethanolic quinoa bran extracts (80% v/v methanol), an activity attributed to phenolic acids (vanillic, ferulic, and chlorogenic acids) and flavonoids (the flavonols quercetin, kaempferol, myricetin, rutin; the flavones luteolin and apigenin; the flavanone naringenin; the flavan-3-ol catechin; and the isoflavone daidzein); moreover, the inhibitory activity of quinoa milled fractions ranged in the order of bran > hulls > whole seed > dehulled seed > milled seed. A significant correlation was also noted between α-amylase inhibitory activity and total phenolic compounds [5].

On the other hand, Coronado-Olano et al. reported both α-amylase and α-glucosidase inhibitory activities in three commercial varieties of *Chenopodium quinoa* (Salcedo INIA and Negra Collana) and two commercial varieties of *Chenopodium pallidicaule* Aellen locally known as *"cañihua"* (INIA 406-ILLPA and Cupi). Salcedo INIA, Negra Collana, and Cupi showed a significantly higher inhibition of α-amylase compared to INI 406-ILLPA. Furthermore, in the varieties of quinoa and cañihua, a significant correlation was observed between α-amylase inhibition and total phenolics, total flavonoids, DPPH• radical scavenging activity, gallic acid, and chlorogenic acid [50].

On our HPTLC chromatograms, at least two blue zones were observed in most of the samples marked with a solid red line (Figure 6b). Interestingly, a third but faint blue zone in the samples of the INIAP-Tunkahuan variety, 80 days from both Pichincha and Chimborazo, was observed (marked with a red dotted line). This may be attributed to the higher phenolic compound content in quinoa leaves harvested at 80 days compared to those collected at 40 and 60 days after sowing. These qualitative results that indicate an inhibition of α-amylase by quinoa leaf extracts should be further confirmed through in vivo studies to validate a potentially interesting antidiabetic activity.

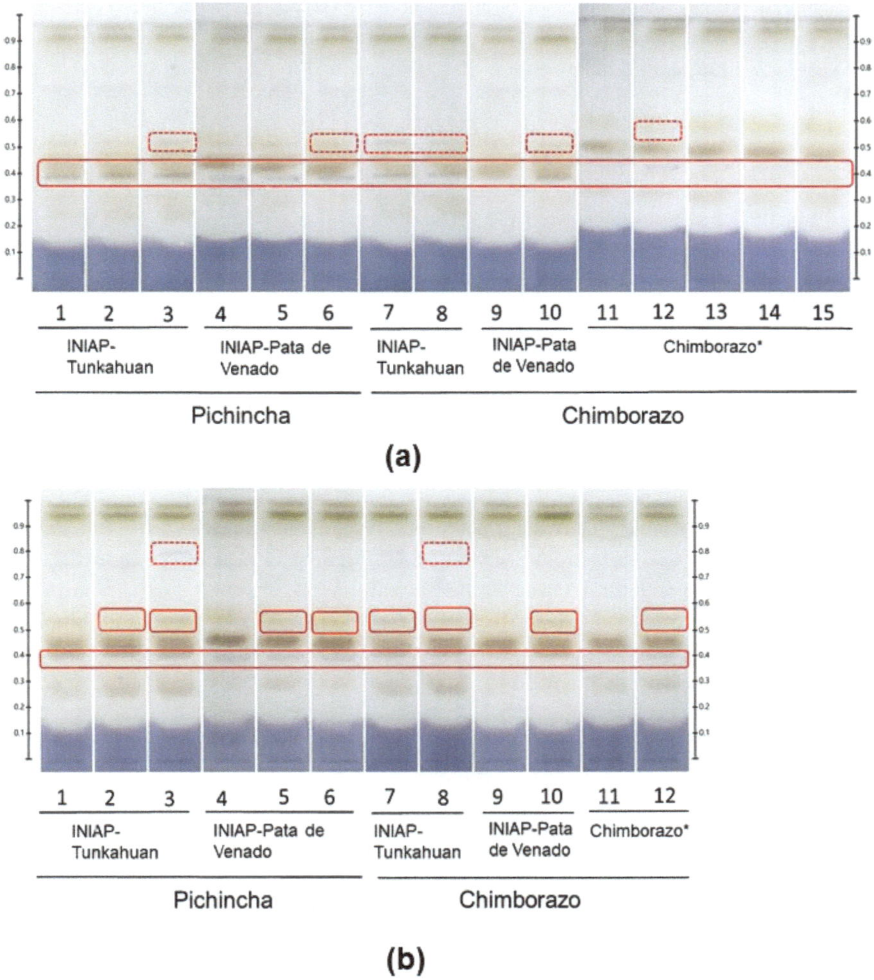

Figure 6. α-Amylase inhibitory activity of Ecuadorian quinoa leaf extracts [methanol–water; 80:20, *w*/*w*; sample–solvent ratio, 1:20 *w*/*v*; application volumes, 8 µL (**a**) and 14 µL (**b**)]. Mobile phase: formic acid–water–methyl ethyl ketone–ethyl acetate (10:10:30:50, *v*/*v*/*v*/*v*). Derivatization with α-amylase, starch, and iodine; examination under visible light. Quinoa leaves from Pichincha: INIAP-Tunkahuan (tracks 1, 2, 3), INIAP-Pata de Venado (tracks 4, 5, 6); quinoa leaves from Chimborazo: INIAP-Tunkahuan (tracks 7, 8), INIAP-Pata de Venado (tracks 9, 10), Chimborazo variety (tracks 11, 12), and Chimborazo* genotype, yellow, green and red colors (tracks 13, 14, 15). Samples as per Table 1. The red marks indicate α-amylase inhibition zones. The large blue smear at the bottom of the plate (Rf 0.0 to 0.2) is due to inhibition of α-amylase by the formic acid of the mobile phase. Samples as per Table 1.

As shown in Figure S2, the development with a more polar mobile phase (formic acid–water–methyl ethyl ketone–ethyl acetate; 10:20:40:30, *v*/*v*/*v*/*v*) allows a better resolution of α-amylase inhibitory bands and a reduction in the formic acid band height at the bottom of the plate.

The major advantage of bioautography is its ability to quickly screen many samples for various bioactivities, such as antioxidant and enzyme inhibition, while using minimal reagent quantities (3 and 2 mL). This method ensures a uniform background and high detection sensitivity. The optimized fast screening technique enables the rapid localization of bioactive compounds within complex plant matrices and colored extracts, as demonstrated with quinoa leaves. This is particularly beneficial because it eliminates the need for sample pretreatment before analysis, streamlining the process significantly.

3.4. Comparison of HPTLC Fingerprints of Quinoa Leaves Versus Seeds

Due to their radical scavenging properties, polyphenols such as flavonoids and phenolic acids are considered major contributors to the antioxidant activity of quinoa leaves [13], but previous reports [51,52] indicate that there is not necessarily a correlation between polyphenol content and DPPH• scavenging. For quinoa leaves, however, there is a clear correlation, as the radical scavenging bands correspond to glycosides of quercetin, kaempferol, and phenolic acids (Figure 3, in all tracks, yellow, green, and blue spots, respectively).

Several in vitro studies indicate a strong inhibition of α-amylase and α-glucosidase by quinoa seed extracts [5,6], and Graf et al. reported that quinoa seeds possess in vivo antidiabetic properties, notably attributed to leached phytoecdysteroids and flavonoids [4]. By contrast, little information is available so far on the eventual antidiabetic properties of quinoa leaves. Interestingly, our HPTLC–bioautographies show that the α-amylase inhibitors we evidence in leaves (blue zones in all tracks of Figure 6, Rf: 0.40) do not correspond to the major polyphenol (yellow bands in Figure 3, Rf: 0.30 and 0.46).

Given the demonstrated chemopreventive interest of seeds [53], their profiles were compared with those obtained from leaves; as the bitter quinoa seeds, Chimborazo variety, were not available, only the sweet varieties could be compared (Figure 7; IT, tracks a, b, c and PV, tracks d, e, f). These HPTLC profiles indicate much less intense bands in seed extracts, for the detection of polyphenols, free radical quenchers, and α-amylase inhibitors.

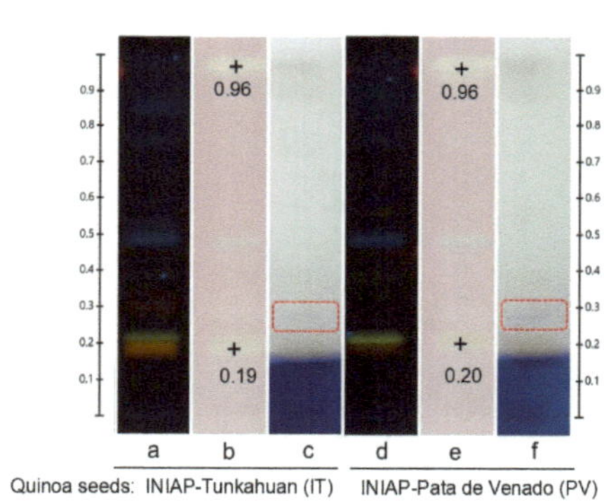

Figure 7. HPTLC fingerprints of polyphenols (derivatization with NP and PEG; examination under UV_{365nm}): tracks a (IT) and d (PV); free radical scavenging activity (derivatization with DPPH• for 100 min and examination under visible light): tracks b (IT) and e (PV); α-amylase inhibition (derivatization with α-amylase, starch, and iodine; examination under visible light): tracks c (IT) and f (PV) of Ecuadorian quinoa seed extracts (methanol–water; 80:20, *w*/*w*; sample–solvent ratio, 1:40 *w*/*v*, and application volumes, 8, 12, and 16 µL, respectively). Mobile phase: formic acid–water–methyl ethyl ketone–ethyl acetate (10:10:30:50, *v*/*v*/*v*/*v*). Samples as per Table 1. The red marks indicate α-amylase inhibition zones. The large blue smear at the bottom of the plate (Rf 0.0 to 0.2) is due to inhibition of α-amylase by the formic acid of the mobile phase. Samples as per Table 1.

These results indicate a major interest in further investigating quinoa leaves, which could be an asset for producers to develop new products with high added value.

4. Conclusions

Quinoa leaves harvested after 80 days of cultivation consistently exhibited higher levels of polyphenols and α-amylase inhibition, regardless of production location or variety.

Moreover, the content of free radical scavenging compounds appeared independently of variety, cultivation time, and production site.

The HPTLC–bioautography methods described in the present study will be a useful tool for the quality control of raw materials in the future production of health-promoting, high-value products based on quinoa leaves. Reliable, fast, and easy to perform, they provide in situ information about the biological activity of separated compounds. Compared to the often-praised quinoa seeds, the much less consumed quinoa leaves present a high content of polyphenols with free radical scavenging activity, and an interesting α-amylase inhibitory property; these biological activities indicate a remarkable potential of quinoa leaves, which may be relevant for the management of diabetes but also for the chemoprevention and/or treatment of pathologies related to oxidative stress. Therefore, further research is essential to explore the potential of quinoa leaves as a dietary vegetable, in the development of functional foods, or for the extraction of bioactive compounds destined for the food and/or pharmaceutical industries. Such initiatives could significantly benefit quinoa smallholders in the Andean regions, promoting economic empowerment and sustainability.

Supplementary Materials: The following supporting information can be downloaded at https://www.mdpi.com/article/10.3390/nutraceuticals5010001/s1, Figure S1: Free radical scavenging activity of Ecuadorian quinoa leaf extracts; Figure S2: α-Amylase inhibitory activity of Ecuadorian quinoa leaf extracts.

Author Contributions: Conceptualization, V.T., P.D. and A.N.; methodology, V.T., C.P., D.B. and E.V.; software, V.T. and C.P.; validation, P.D. and A.N.; formal analysis, V.T. and C.P.; investigation, V.T.; resources, D.B. and E.V.; data curation, V.T.; writing—original draft preparation, V.T.; writing—review and editing, V.T., E.V., P.D. and A.N.; visualization, V.T.; supervision, P.D. and A.N.; project administration, V.T.; funding acquisition, V.T., P.D. and A.N. All authors have read and agreed to the published version of the manuscript.

Funding: This research was funded by the Académie De Recherche Et d'Enseignement Supérieur (ARES) in Belgium through the project "*Appui Institutionnel auprès de l'Université d'Ecuador*".

Data Availability Statement: The original contributions presented in this study are included in the article/Supplementary Materials. Further inquiries can be directed to the corresponding author.

Acknowledgments: V. Taco gratefully acknowledges the fellowship granted to her by ARES. We would like to thank the Camag® company for the lending of a Visualizer system and their technical support and Maquita Cushunchic (MCCH) in Ecuador for supplying the quinoa leaf samples for this study. We thank Raúl López, Marie-Lou Gauthier, Diana Iza, and Benjamín Martínez for their help and technical support.

Conflicts of Interest: The authors declare no conflicts of interest.

References

1. Abderrahim, F.; Huanatico, E.; Segura, R.; Arribas, S.; Gonzalez, M.C.; Condezo-Hoyos, L. Physical Features, Phenolic Compounds, Betalains and Total Antioxidant Capacity of Coloured Quinoa Seeds (*Chenopodium quinoa* Willd.) from Peruvian Altiplano. *Food Chem.* **2015**, *183*, 83–90. [CrossRef] [PubMed]
2. Derbali, W.; Manaa, A.; Spengler, B.; Goussi, R.; Abideen, Z.; Ghezellou, P.; Abdelly, C.; Forreiter, C.; Koyro, H.-W. Comparative Proteomic Approach to Study the Salinity Effect on the Growth of Two Contrasting Quinoa Genotypes. *Plant Physiol. Biochem.* **2021**, *163*, 215–229. [CrossRef]
3. Nowak, V.; Du, J.; Charrondiere, U.R. Assessment of the Nutritional Composition of Quinoa (*Chenopodium quinoa* Willd.). *Food Chem.* **2016**, *193*, 47–54. [CrossRef]
4. Graf, B.L.; Poulev, A.; Kuhn, P.; Grace, M.H.; Lila, M.A.; Raskin, I. Quinoa Seeds Leach Phytoecdysteroids and Other Compounds with Anti-Diabetic Properties. *Food Chem.* **2014**, *163*, 178–185. [CrossRef] [PubMed]

5. Hemalatha, P.; Bomzan, D.P.; Sathyendra Rao, B.V.; Sreerama, Y.N. Distribution of Phenolic Antioxidants in Whole and Milled Fractions of Quinoa and Their Inhibitory Effects on Alpha-Amylase and Alpha-Glucosidase Activities. *Food Chem.* **2016**, *199*, 330–338. [CrossRef] [PubMed]
6. Lin, H.; Li, W.; Lai, D.; Zhou, L. Quinoa Secondary Metabolites and Their Biological Activities or Functions. *Molecules* **2019**, *24*, 2512. [CrossRef]
7. Vilcacundo, R.; Hernández-Ledesma, B. Nutritional and Biological Value of Quinoa (*Chenopodium quinoa* Willd.). *Curr. Opin. Food Sci.* **2017**, *14*, 1–6. [CrossRef]
8. Köseoğlu Yılmaz, P.; Ertaş, A.; Akdeniz, M.; Koçyiğit Avcı, M.; Kolak, U. Chemical compositions by LC-MS/MS and GC-MS and biological activities of *Chenopodium album* subsp. album var. microphyllum. *Ind. Crops Prod.* **2019**, *141*, 111755. [CrossRef]
9. Alandia, G.; Rodriguez, J.P.; Jacobsen, S.-E.; Bazile, D.; Condori, B. Global Expansion of Quinoa and Challenges for the Andean Region. *Glob. Food Secur.* **2020**, *26*, 100429. [CrossRef]
10. Brady, K.; Ho, C.-T.; Rosen, R.T.; Sang, S.; Karwe, M.V. Effects of Processing on the Nutraceutical Profile of Quinoa. *Food Chem.* **2007**, *100*, 1209–1216. [CrossRef]
11. Ruiz, K.B.; Biondi, S.; Oses, R.; Acuña-Rodríguez, I.S.; Antognoni, F.; Martinez-Mosqueira, E.A.; Coulibaly, A.; Canahua-Murillo, A.; Pinto, M.; Zurita-Silva, A.; et al. Quinoa Biodiversity and Sustainability for Food Security under Climate Change. A Review. *Agron. Sustain. Dev.* **2014**, *34*, 349–359. [CrossRef]
12. Andreotti, F.; Bazile, D.; Biaggi, C.; Callo-Concha, D.; Jacquet, J.; Jemal, O.M.; King, O.I.; Mbosso, C.; Padulosi, S.; Speelman, E.N.; et al. When Neglected Species Gain Global Interest: Lessons Learned from Quinoa's Boom and Bust for Teff and Minor Millet. *Glob. Food Secur.* **2022**, *32*, 100613. [CrossRef]
13. Gawlik-Dziki, U.; Swieca, M.; Sulkowski, M.; Dziki, D.; Baraniak, B.; Czyz, J. Antioxidant and Anti-cancer Activities of *Chenopodium quinoa* Leaves Extracts-in Vitro Study. *Food Chem. Toxicol.* **2013**, *57*, 154–160. [CrossRef] [PubMed]
14. Buitrago, D.; Buitrago-Villanueva, I.; Barbosa-Cornelio, R.; Coy-Barrera, E. Comparative Examination of Antioxidant Capacity and Fingerprinting of Unfractionated Extracts from Different Plant Parts of Quinoa (*Chenopodium quinoa*) Grown under Greenhouse Conditions. *Antioxidants* **2019**, *8*, 238. [CrossRef] [PubMed]
15. Brahmi, F.; Hauchard, D.; Guendouze, N.; Madami, K.; Kiendrebeogo, M.; Kamagaju, L.; Stévigny, C.; Chibane, M.; Duez, P. Phenolic Composition, in Vitro Antioxidant Effects and Tyrosinase Inhibitory Activity of Three Algerian Mentha Species. *Ind. Crops Prod.* **2015**, *74*, 722–730.
16. Espin, S.; Gonzalez-Manzano, S.; Taco, V.; Poveda, C.; Ayuda-Duran, B.; Gonzalez-Paramas, A.M.; Santos-Buelga, C. Phenolic Composition and Antioxidant Capacity of Yellow and Purple-Red Ecuadorian Cultivars of Tree Tomato (*Solanum betaceum* Cav.). *Food Chem.* **2016**, *194*, 1073–1080. [CrossRef] [PubMed]
17. Saso, L.; Firuzi, O. Pharmacological Applications Antioxidants Lights and Shadows. *Curr. Drug Targets* **2014**, *15*, 1177–1199. [CrossRef]
18. Halliwell, B.; Gutteridge, J. *Free Radicals in Biology and Medicine*, 3rd ed.; Oxford University Press: Oxford, UK, 1999.
19. Hirose, Y.; Fujita, T.; Ishii, T.; Ueno, N. Antioxidative Properties and Flavonoid Composition of *Chenopodium quinoa* Seeds Cultivated in Japan. *Food Chem.* **2010**, *119*, 1300–1306. [CrossRef]
20. Bai, L.; Li, X.; He, L.; Zheng, Y.; Lu, H.; Li, J.; Zhong, L.; Tong, R.; Jiang, Z.; Shi, J. Antidiabetic Potential of Flavonoids from Traditional Chinese Medicine. A Review. *Am. J. Chin. Med.* **2019**, *47*, 933–957. [CrossRef]
21. Herrera, T.; Navarro Del Hierro, J.; Fornari, T.; Reglero, G.; Martin, D. Inhibitory Effect of Quinoa and Fenugreek Extracts on Pancreatic Lipase and Alpha-Amylase under in Vitro Traditional Conditions or Intestinal Simulated Conditions. *Food Chem.* **2019**, *270*, 509–517. [CrossRef] [PubMed]
22. Stojkovic, D.; Smiljkovic, M.; Ciric, A.; Glamoclija, J.; Griensven, L.; Ferreira, I.C.F.R.; Sokovic, M. An Insight into Antidiabetic Properties of Six Medicinal and Edible Mushrooms: Inhibition of α-Amylase and α-Glucosidase Linked to Type-2 Diabetes. *S. Afr. J. Bot.* **2019**, *120*, 100–103. [CrossRef]
23. Mahmood, N. A Review of α-Amylase Inhibitors on Weight Loss and Glycemic Control in Pathological State Such as Obesity and Diabetes. *Comp. Clin. Pathol.* **2016**, *25*, 1253–1264. [CrossRef]
24. López-Fernández, M.P.; Burrieza, H.P.; Rizzo, A.J.; Martínez-Tosar, L.J.; Maldonado, S. Cellular and Molecular Aspects of Quinoa Leaf Senescence. *Plant Sci.* **2015**, *238*, 178–187. [CrossRef] [PubMed]
25. Legerska, B.; Chmelova, D.; Ondrejovic, M.; Miertus, S. The TLC-Bioautography as a Tool for Rapid Enzyme Inhibitors Detection—A Review. *Crit. Rev. Anal. Chem.* **2022**, *52*, 275–293. [CrossRef] [PubMed]
26. Simoes-Pires, C.A.; Hmicha, B.; Marston, A.; Hostettmann, K. A TLC Bioautographic Method for the Detection of Alpha- and Beta-Glucosidase Inhibitors in Plant Extracts. *Phytochem. Anal.* **2009**, *20*, 511–515. [CrossRef] [PubMed]
27. Dewanjee, S.; Gangopadhyay, M.; Bhattacharya, N.; Khanra, R.; Dua, T.K. Bioautography and Its Scope in the Field of Natural Product Chemistry. *J. Pharm. Anal.* **2015**, *5*, 75–84. [CrossRef] [PubMed]

28. Taco, V.; Palmieri, C.; Duez, P.; Nachtergael, A. Optimized High-Performance Thin-Layer Chromatography-bioautography Screening of Ecuadorian *Chenopodium quinoa* Willd. Leaf Extracts for Inhibition of α-Amylase. *JPC—J. Planar Chromatogr.—Mod. TLC* **2021**, *34*, 561–567. [CrossRef]
29. Sosa-Zuniga, V.; Brito, V.; Fuentes, F.; Steinfort, U. Phenological Growth Stages of Quinoa (*Chenopodium quinoa* Willd) Based on the BBCH Scale. *Ann. Appl. Biol.* **2017**, *171*, 117–124. [CrossRef]
30. Yzarra, W.; López, F.; Dediós, N.; Medina, E.; Gómez, H.; Rosas, R.; Montañez, L.; Escalante, J. Manual de observaciones fenológicas. In *Manual-Guía*, 4th ed.; Senamhi, Ed.; Servicio Nacional de Meteorología e Hidrología del Perú: Lima, Peru, 2017; p. 27.
31. Liu, X.; Ahlgren, S.; Korthout, H.; Salome-Abarca, L.F.; Bayona, L.M.; Verpoorte, R.; Choi, Y.H. Broad Range Chemical Profiling of Natural Deep Eutectic Solvent Extracts Using a High Performance Thin Layer Chromatography-Based Method. *J. Chromatogr. A* **2018**, *1532*, 198–207. [CrossRef] [PubMed]
32. Europe, C. High-Performance Thin-Layer Chromatography of Herbal Drugs and Herbal Drug Preparations (2.8.25). In *European Pharmacopoeia*; Council of Europe: Strasbourg, France, 2019.
33. Agatonovic-Kustrin, S.; Morton, D.W.; Adam, A.; Mizaton, H.H.; Zakaria, H. High-Performance Thin-Layer Chromatographic Methods in the Evaluation of the Antioxidant and Anti-Hyperglycemic Activity of Myrmecodia Platytyrea as a Promising Opportunity in Diabetes Treatment. *J. Chromatogr. A* **2017**, *1530*, 192–196. [CrossRef]
34. Tang, Y.; Li, X.; Zhang, B.; Chen, P.; Liu, R.; Tsao, R. Characterisation of Phenolics, Betanins and Antioxidant Activities in Seeds of Three *Chenopodium quinoa* Willd. Genotypes. *Food Chem.* **2015**, *166*, 380–388. [CrossRef] [PubMed]
35. Villacrés, E.; Quelal, M.; Galarza, S.; Iza, D.; Silva, E. Nutritional Value and Bioactive Compounds of Leaves and Grains from Quinoa (*Chenopodium quinoa* Willd.). *Plants* **2022**, *11*, 213. [CrossRef]
36. Pathan, S.; Siddiqui, R. Nutritional Composition and Bioactive Components in Quinoa (*Chenopodium quinoa* Willd.) Greens: A Review. *Nutrients* **2022**, *14*, 558. [CrossRef] [PubMed]
37. Hynstova, V.; Sterbova, D.; Klejdus, B.; Hedbavny, J.; Huska, D.; Adam, V. Separation, Identification and Quantification of Carotenoids and Chlorophylls in Dietary Supplements Containing Chlorella Vulgaris and Spirulina Platensis Using High Performance Thin Layer Chromatography. *J. Pharm. Biomed. Anal.* **2018**, *148*, 108–118. [CrossRef]
38. Baldeón, E.; Sánchez-Pizarro, A.; Repo-Carrasco-Valencia, R.; Barat, J.; Grau, R.; Cadenillas-Martínez, A.; Comettant-Rabanal, R. Nutritional Value and Chemical Composition of Quinoa Leaves (*Chenopodium quinoa* Wild.) at Different Phenological Stages. *Manglar* **2023**, *20*, 379–387. [CrossRef]
39. Stoleru, V.; Vitanescu, M.; Teliban, G.; Cojocaru, A.; Vlase, L.; Gheldiu, A.; Mangalagiu, I.; Amăriucăi-Mantu, D.; Burducea, M.; Zheljazkov, V.; et al. Phytosterol and Polyphenol Contents and Quinoa Leave Yields Variation in Relationships to Variety, Density and Harvesting Date. *Agronomy* **2022**, *12*, 2397. [CrossRef]
40. Hu, L.; Wang, C.; Guo, X.; Chen, D.; Zhou, W.; Chen, X.; Zhang, Q. Flavonoid Levels and Antioxidant Capacity of Mulberry Leaves: Effects of Growth Period and Drying Methods. *Front. Plant Sci.* **2021**, *12*, 684974. [CrossRef] [PubMed]
41. Ming, L.C.; Maia-Almeida, C.I.; Monte Conceição, D.; Yuthi Yuhara, T.; Ortiz Mayo Marques, M.; Mendes Moreschi, S.R.; Leonel, S.; Bonon, A.J.; Castro Tavares, R.; Fidelis, R.R.; et al. Phytomass and Flavonoid Production in Different Organs and Phenological Stages of *Passiflora Alata* Dryander. *J. Med. Plants Res.* **2012**, *6*, 5695–5700.
42. Saffaryazdi, A.; Ganjeali, A.; Farhoosh, R.; Cheniany, M. Variation in Phenolic Compounds, Alpha-Linolenic Acid and Linoleic Acid Contents and Antioxidant Activity of Purslane (*Portulaca oleracea* L.) during Phenological Growth Stages. *Physiol. Mol. Biol. Plants* **2020**, *26*, 1519–1529. [CrossRef]
43. Rodrigues, A.S.; Pérez-Gregorio, M.R.; García-Falcón, M.S.; Simal-Gándara, J.; Almeida, D.P.F. Effect of Meteorological Conditions on Antioxidant Flavonoids in Portuguese Cultivars of White and Red Onions. *Food Chem.* **2011**, *124*, 303–308. [CrossRef]
44. Repo-Carrasco-Valencia, R.; Hellström, J.; Pihlava, J.; Mattila, P. Flavonoids and Other Phenolic Compounds in Andean Indigenous Grains: Quinoa (*Chenopodium quinoa*), kañiwa (*Chenopodium pallidicaule*) and kiwicha (*Amaranthus caudatus*). *Food Chem.* **2010**, *120*, 128–133. [CrossRef]
45. Gandarillas, H.; Nieto, C.; Castillo, R. *Razas de Quinua en Ecuador*; INIAP, Estación Experimental Santa Catalina, Programa de Cultivos Andinos: Quito, Ecuador, 1989.
46. Abugoch James, L. Chapter 1 Quinoa (*Chenopodium quinoa* Willd.): Composition, Chemistry, Nutritional, and Functional Properties. *Adv. Food Nutr. Res.* **2009**, *58*, 1–31. [PubMed]
47. Islam, M.K.; Sostaric, T.; Lim, L.Y.; Hammer, K.; Locher, C. Development and Validation of an HPTLC–DPPH Assay and Its Application to the Analysis of Honey. *JPC—J. Planar Chromatogr.—Mod. TLC* **2020**, *33*, 301–311. [CrossRef]
48. Camara, M.A.; Tian, M.; Yang, L.; Wang, S. Application of Thin-Layer Chromatography in Enzyme Activity and Inhibitors Studies of Glucose-6-Phosphate Dehydrogenase. *J. Planar Chromatogr.* **2015**, *28*, 333–336. [CrossRef]
49. Parveen, R.; Khan, N.; Zahiruddin, S.; Ibrahim, M.; Anjum, V.; Parveen, B.; Khan, M.A. TLC-Bioautographic Evaluation for High-Throughput Screening and Identification of Free Radical Scavenging and Antidiabetic Compounds from Traditional Unani Medicinal Plant: Citrullus Colocynthis Schrad. *J. AOAC Int.* **2020**, *103*, 669–677. [CrossRef]

50. Coronado-Olano, J.; Repo-Carrasco-Valencia, R.; Reategui, O.; Toscano, E.; Valdez, E.; Zimic, M.; Best, I. Inhibitory Activity against α-amylase and α-glucosidase by Phenolic Compounds of Quinoa (*Chenopodium quinoa* Willd.) and Cañihua (*Chenopodium pallidicaule* Aellen) from the Andean region of Peru. *Pharmacogn. J.* **2021**, *13*, 896–901. [CrossRef]
51. Brahmi, F.; Guedouze, N.; Hauchard, D.; Okusa, P.; Kamagaju, L.; Madani, K.; Duez, P. Phenolic Profile and Biological Activities of *Micromeria graeca* (L.) Benth. ex Rchb. *Int. J. Food Prop.* **2017**, *20*, 2070–2083.
52. Guendouze-Bouchefa, N.; Madani, K.; Chibane, M.; Boulekbache-Makhlouf, L.; Hauchard, D.; Kiendrebeogo, M.; Stévigny, C.; Ndjolo Okusa, P.; Duez, P. Phenolic Compounds, Antioxidant and Antibacterial Activities of Three Ericaceae from Algeria. *Ind. Crops Prod.* **2015**, *70*, 459–466. [CrossRef]
53. Vilcacundo, R.; Miralles, B.; Carrillo, W.; Hernandez-Ledesma, B. In Vitro Chemopreventive Properties of Peptides Released from Quinoa (*Chenopodium quinoa* Willd.) Protein under Simulated Gastrointestinal Digestion. *Food Res. Int.* **2018**, *105*, 403–411. [CrossRef]

nutraceuticals

Article

Obesity Prevention Effects of Avocado (*Persea americana*) Seed Powder in High-Fat Diet-Induced Obesity in Rats

Shoeshoe Mokhele [1,2], Oluwaseyi Aboyade [1] and David R. Katerere [1,*]

[1] Department of Pharmaceutical Sciences, Tshwane University of Technology, 175 Nelson Mandela Drive, Pretoria 0001, South Africa; shoeshoe.mokhele@smu.ac.za (S.M.); oluwaseyi.aboyade@gmail.com (O.A.)

[2] Department of Pharmaceutical Sciences, School of Pharmacy, Sefako Makgatho Health Sciences University, Molotlegi Street, Pretoria 0204, South Africa

[*] Correspondence: katereredr@tut.ac.za

Abstract: Avocado seed contains 64% of the phenolic compounds of the whole fruit. This makes avocado seed a potential candidate for the development of treatments for different illnesses, including obesity (the major risk factor for metabolic disorders). The aim of this study was to investigate the effects of avocado seed powder on high-fat diet-induced obesity in rats. Sprague Dawley rats (16 rats) were fed a high-fat diet for 10 weeks. After 10 weeks, the rats were assigned into two groups of eight animals each and were fed either a high-fat diet (HFD; control group) or a high-fat diet containing avocado seed powder (HFD-A; treatment group) for 6 weeks. Animals were weighed weekly, and weekly weight gain was determined. Animals in the treatment (avocado seed) group showed significantly lower body weight gain (7.8 ± 9.63 g) than animals in the control group (33.9 ± 10.84 g) at the end of this study. The treatment group presented with lower triglycerides than the control, with LDL and HDL comparable to the control group. Avocado seed powder showed potential to reduce obesity in rats fed a high-fat diet. Avocado seed can therefore be investigated further as a potential anti-obesity nutraceutical.

Keywords: *Persea americana*; avocado seed; high-fat diet; obesity reversal; anti-obesity; weight gain; weight loss

Citation: Mokhele, S.; Aboyade, O.; Katerere, D.R. Obesity Prevention Effects of Avocado (*Persea americana*) Seed Powder in High-Fat Diet-Induced Obesity in Rats. *Nutraceuticals* **2024**, *4*, 417–429. https://doi.org/10.3390/nutraceuticals4030025

Academic Editors: Ivan Cruz-Chamorro and Guillermo Santos Sánchez

Received: 10 July 2024
Revised: 2 September 2024
Accepted: 3 September 2024
Published: 9 September 2024

1. Introduction

Obesity may lead to metabolic syndrome, a cluster of metabolic disorders that includes insulin resistance, hyperglycemia, dyslipidemia and hypertension [1]. There is an increased risk of developing type 2 diabetes, cardiovascular diseases [2] and certain cancers [3] in obese individuals. The aforementioned diseases are major contributors to the increased morbidity and premature mortality recorded worldwide [4]. According to the World Health Organization (WHO), more than 1 billion people were obese worldwide in 2022 and the prevalence of obesity will continue to rise [5]. Urgent interventions are therefore needed to combat obesity and related disorders. Among the interventions against obesity and its co-morbidities, is the use of anti-obesity agents and nutraceuticals. Plant materials are potential sources of therapeutic agents and nutraceuticals. They contain abundant bioactive compounds that play major roles in the prevention and treatment of diseases [6]. In addition to their nutritive value, nutraceuticals possess medicinal properties and have been used in the management of different diseases such as cancers, arthritis and metabolic disorders [7]. Different nutraceuticals possessing anti-obesity effects have been reported [8]. With increasing interest towards the use of functional foods for disease management, avocado fruit and seed are potential leads to the discovery of anti-obesity nutraceuticals.

Avocado (*Persea americana* Mill) has been reported to possess medicinal properties [9–11]. It is from the Lauraceae family, and it is cultivated in tropical and sub-tropical areas in Mexico, Brazil, India and South Africa [12]. Avocado fruit has an olive-green peel with a smooth, fatty, thick and almost creamy-textured pulp. Avocado seed makes up 13% of the fruit and contains 64% of the phenolic compounds contributing to 57% of the antioxidant activity of the whole fruit [13]. The seed contains higher antioxidant activity, and higher procyanidin content than the pulp [14].

Phytochemicals found in avocado seed include flavonoids, alkaloids, saponins, glycosides and tannins [15–17]. The phenolic compounds in avocado seed include protocatechuic acid, kaempferide, vanillic acid, chlorogenic acid, syringic acid, rutin and kaempferol [13]. Phenolic compounds reduce adipogenesis, inhibit digestive enzymes and possess anti-inflammatory and antioxidant effects [18,19]. These properties play a major role in obesity management. The seed also contains nutrients such as carbohydrates, proteins and fats [16,20] and minerals such as magnesium, potassium, sodium, zinc and chromium, in addition to vitamin C [12,16,21]. These trace elements affect different pathways of glucose and lipid metabolism [22].

Avocado seed extracts have been found to have anti-hypertensive properties, lower total cholesterol levels and possess hypoglycemic activities [12,15,23]. Even though the seeds contain beneficial phytochemicals, they are usually disposed during processing and consumption, thus contributing to environmental waste. The use of avocado seed may contribute towards the management of obesity and its comorbidities, while also reducing organic waste burden. This study therefore aimed to investigate the obesity prevention effects of avocado seed powder on obesity induced by a high-fat diet in rats.

2. Materials and Methods

Hass avocado (*Persea americana Mill*) fruits were purchased from a local market in Makhado, Limpopo province, South Africa. Sprague Dawley rats were sourced from South African Vaccines Producers, Johannesburg, while standard animal pellets were bought from Epol, South Africa. Palm stearin was obtained from commercial food provider Ingreto in Johannesburg, South Africa.

2.1. *Animal Feed Preparation*

The avocado seeds were cut into small pieces with a knife and dried in an oven at 30 °C for 72 h. Dried seeds were milled into powder and stored in the fridge at 4 °C until further use. Standard commercial Epol pellets were ground to a powder together with molasses (constituting 8% of the final feed) as a binding agent. Palm stearin (20% of the total final feeds) was added to the chow to make a high-fat diet (HFD). Powdered avocado seed (20% of the total feed) was included in the HFD mix to make a high-fat diet with avocado seed powder for treatment. The ingredients were all mixed and pelleted using a pelleting machine. The formulated pellets were HFD and HFD with avocado seeds (HFD-A). The pellets were spread on an open surface for approximately 1 h to air cool after production. They were then weighed, packaged in 5 kg plastic bags, sealed, labeled and stored in a refrigerator (−4 °C) until use.

2.1.1. Nutritional Analysis of Food Pellets

Nutritional analysis of the formulated pellets was conducted at the South African Grain Laboratory (SAGL) using validated methods. The pellets were analyzed for protein content, fat content, fiber content, moisture content and ashes [24]. Protein content was determined using an AACC international approved method of analysis (the combustion method). Fat was extracted from the samples with petroleum ether using Soxhlet extraction apparatus, based on AACC method 30-25.01. To determine the moisture content, the feed sample weighing 2 g was dried in an oven at 105 °C for 5 h and the dried feed was reweighed. The moisture content was calculated as the difference in weights.

2.1.2. Elemental Analysis of Food Pellets

Elemental analysis was performed using the method by [25]. The pellets were crushed into a powder with a mortar and pestle. The powdered samples were acid digested to decompose the organic and inorganic components of the samples and release the elements from the sample matrix [26]. The solutions were then cooled and diluted with distilled water to make a sample concentration of 33.33 mg/mL. Then, the concentrations of boron (B), sodium (Na), magnesium (Mg), potassium (K), calcium (Ca), manganese (Mn), iron (Fe), copper (Cu), molybdenum (Mo), selenium (Se), arsenic (As) and chromium (Cr) were determined with an inductively coupled plasma-optical emission spectrometer (ICP-OES). Where readings were above the linear dynamic range, the samples were diluted further. Standards of the elements were obtained from SMM instruments; Johannesburg, South Africa (SA), Industrial analytical; Johannesburg, South Africa (SA) and Teknolab Sorbent AB; Kungsbacka, Sweden, and were used at the concentration of 50 µg/mL.

2.2. *In Vivo Assays*

Ethical approval to conduct this study was obtained from the Animal Ethics Committee of the University of Pretoria (V038-15) and the Animal Research Ethics Committee of Tshwane University of Technology (AREC2015/10/005(2)). This study was conducted following the South African Medical Research Council's guidelines on ethics for medical research (use of animals in research and training [27]), and the South African National Standard (the care and use of animals for scientific purposes [28]). Other than the researcher, the animals were under the care of a certified animal technician with indirect supervision of a veterinarian. The animals were observed for any signs of distress such as behavioral changes, tremors, salivation, diarrhea, lethargy, sleep, convulsions or coma throughout this study. At the end of this study, the animals were euthanized under anesthesia.

2.2.1. Animal Model

Eight-to-twelve-week-old male Sprague Dawley rats sourced from South African Vaccine Producers (Johannesburg; South Africa) were used as animal models in this study. The rats were kept at the University of Pretoria Biomedical Research Centre in conventional cages (one rat per cage with sawdust bedding, facial tissues, egg containers, shredded paper as nesting material, wooden sticks as gnawing material and 15 cm PVC pipe for hiding). The room was maintained at a temperature of ±22 °C, controlled relative humidity (50–60%) and a light/dark cycle of 12 h [29]. The animals received a high-fat diet (HFD) to induce obesity for 10 weeks, after which they were randomly assigned to 2 groups of 8 animals per group. Rats in one group (n = 8) continued with a HFD (control group) for 6 more weeks, while the treatment group (n = 8) received a high-fat diet with avocado seed (HFD-A; treatment group) for 6 weeks. The animals were weighed at the beginning of the study (week 0), then weekly and the weight for each animal was recorded. The weight gain for each animal was calculated as the difference between its weekly weight and its weight at week 0 of the treatment (Equation (1)), while percentage weight gain was calculated using Equation (2). Food intake was monitored by weighing the amount of food that was given and the amount that remained after every 48 h to determine food consumption using Equation (3).

$$\text{Weekly weigh gain} = WweekX - Wweek0 \tag{1}$$

where W = weight, X = week number and 0 = initial

$$\%\text{weight gain} = \frac{\text{Weekly weight gain}}{\text{Initial weight}} \times 100 \tag{2}$$

$$\text{Daily food intake} = \frac{Mass\ food\ in - Mass\ of\ food\ left}{2\ days} \tag{3}$$

The animals were euthanized at the end of week 16 (thus, 10 weeks of obesity induction and 6 weeks of treatment). The animals were sacrificed by cardiac puncture under isoflurane anesthesia [30]. After the animal was confirmed dead, the kidneys, liver, heart, adipose tissue and brain were isolated and weighed. Blood samples were collected and sent to the pathology lab for liver function tests (ALT, AST and ALP) and cholesterol analysis (total cholesterol, high-density lipoproteins, low-density lipoproteins and triglycerides).

2.2.2. Histopathology

Histopathology was conducted on 3 randomly selected rat samples from each group by IDEXX laboratory (Pretoria North, South Africa). Organ specimens were collected and fixed in 10% buffered formalin. The organs were cut, and the tissues were processed in an automated histological tissue processor overnight. Histological sections of 5 μm were cut on a microtome and the resultant slides were stained with hematoxylin and eosin dyes in an automated histological stainer. The slides were examined, and morphological findings recorded.

2.3. Phytochemical Analysis

To determine the phytochemicals responsible for the obesity prevention effects of avocado seed powder observed in this study, avocado seed powder methanol extract was subjected to analysis by liquid chromatography coupled with mass spectroscopy (LC-MS), as described before [31].

2.4. Data Analysis

Results were reported as mean ± standard deviation (n = 8). Data collected were analyzed using Stata software version 14 (StataCorp LLC: College Station, TX, USA). A generalized least squares (GLS) regression model was used to determine significant statistical differences between the treatment groups and the control group. This method was utilized because it improved statistical efficiency and minimized the chances of reporting erroneous inferences. The results were considered significantly different at $p < 0.05$.

3. Results

3.1. Nutritional and Elemental Analysis

Results of the nutritional composition of the formulated pellets are shown in Table 1. Although HFD showed a higher protein than HFD-A, HFD-A showed a higher fat content than HFD. For elemental composition, a solution obtained from the acid digestion of HFD-A showed a higher content of Mg, K, Ca, Mn and Fe than a solution from HFD (Table 2).

Table 1. Nutritive value of pellets in %.

Analysis	HFD	HFD-A
Moisture (%)	9.30	8.10
Protein (%)	16.09	13.70
Ash (%)	5.46	4.82
Fat (%)	19.50	24.50
Fiber (%)	4.60	4.70

Table 2. Elemental content of the solution obtained from acid digestion of the pellets in mg/L.

Trace Elements (mg/L)	B	Cu	Se	Mo	As	Cr	Na	Mg	K	Ca	Mn	Fe
HFD	0.22	0.60	0.23	<−0.17	0.01	<−0.08	34.01	17.08	69.03	42.75	0.52	1.89
HFD-A	0.25	0.34	0.16	<−0.18	0.01	<−0.07	30.26	22.52	>105.0	57.08	0.76	2.52

3.2. In Vivo Study

3.2.1. Animal Food Intake and Effects of Avocado Seed Powder on Animal Weights

The food intake by animals in both groups was comparable, as animals in HFD consumed an average 22.18 ± 0.66 g/day while animals in HFD-A consumed an average of 28.04 ± 3.11 g/day. The difference in food intake by the treatment group was not significant ($p > 0.05$) when compared with food intake by the control group. The weekly weights of the animals for the treatment period are presented in Table 3, with weight gain/loss presented in Figure 1.

Table 3. Weekly weights of the animals across the treatment period (mean ± SD; n = 8).

Week	HFD (Control) Group Weight (g)	HFD-A (Treatment) Group Weight (g)
0	345.88 ± 9.16	306.75 ± 4.06
1	340.00 ± 11.48	288.38 ± 6.65
2	350.00 ± 11.77	288 ± 6.59
3	355.88 ± 10.37	286.88 ± 7.28
4	362.38 ± 12.93	289.88 ± 9.51
5	363.63 ± 15.84	293.25 ± 9.22
6	379.75 ± 12.67	314.5 ± 9.72

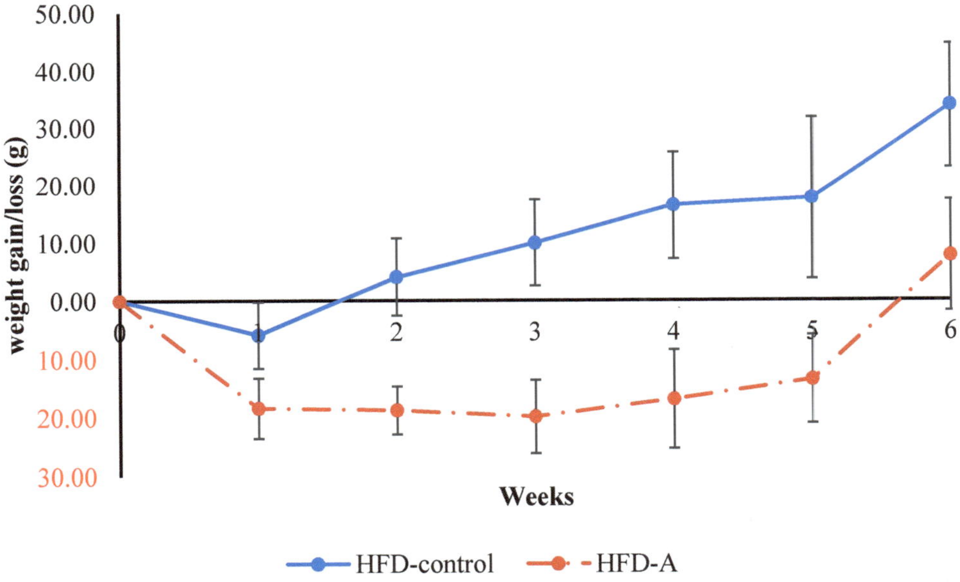

Figure 1. Weight gain per group during treatment period of this study (error bars represent the standard deviation of mean (n = 8)).

There was a significant weight loss in the first week of treatment for the animals in the HFD-A group (Figure 1). The animals in this group did not gain weight until week 3, after which they showed minimal weight gain compared with the control group ($p < 0.05$) until the end of this study. HFD-A gained 7.75 ± 9.63 g, while HFD gained 33.88 ± 10.84 g by the end of this study. Percentage weight gain for animals was 9.81% and 2.53% for HFD and HFD-A, respectively, by the end of this study. A significant difference in average animal weights between the two groups was observed throughout this study.

3.2.2. Effects of Avocado Seed Powder on Animal Organ Weight and Biochemical Parameters

There was no significant difference in the relative heart weight between the two groups (Figure 2). HFD-A presented with a significantly higher brain weight than the control. Adipose tissue was significantly lower in HFD-A compared with the control.

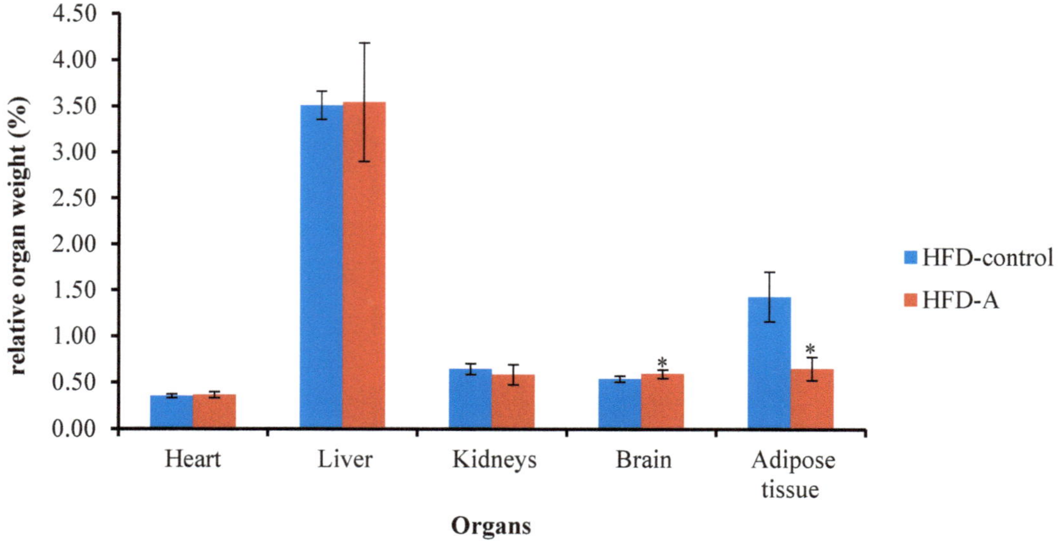

Figure 2. Percentage organ weight relative to the body weight of the animals. * Statistical significance ($p < 0.05$) when treatment group is compared with HFD-control.

Regarding biochemical parameters (Table 4), the treatment group showed triglyceride levels significantly lower ($p < 0.05$) than that of the control group, with comparable HDL and LDL. Both groups presented comparable total cholesterol levels. The ALP in HFD-A group was significantly higher than in the control group.

Table 4. Biochemical parameters of the animals at the end of this study (mean ± SD; n = 8).

Analysis	HFD	HFD-A
Total cholesterol (mmol/L)	2.38 ± 0.13	2.23 ± 0.32
Triglycerides (mmol/L)	0.49 ± 0.10	0.34 ± 0.05 *
HDL (mmol/L)	0.98 ± 0.05	1.00 ± 0.08
LDL (mmol/L)	0.59 ± 0.08	0.69 ± 0.11
ALT (U/L)	65.75 ± 14.67	70.75 ± 12.93
ALP (U/L)	130.00 ± 22.81	175.88 ± 28.40 *
AST (U/L)	83.63 ± 18.95	74.29 ± 10.13

* Statistical significance ($p < 0.05$) when HFD-A is compared with HFD.

3.2.3. Post-Mortem Results

All the animals were in good health with no abnormalities detected in both groups. The livers of some animals in both groups were histologically within normal limits, while some revealed mild vascular changes with varying distribution patterns in the hepatocyte cytoplasm. Macroscopic pathology was not detected with all other tissues.

3.3. Phytochemical Analysis

Figure 3 presents LC-MS chromatogram for the avocado seed powder extracts. Further analysis of peaks 309 and 351 fragments showed the presence of cinnamic acid. Fragments of peak 371 showed the presence of ferulic acid, neochlorogenic acid and flavone. Peak 702 fragments showed the presence of isoferulic acid-3-O-glucoronide and rothindin. From the LC-MS chromatograms, it was established that the main constituents of avocado seed powder extracts were ferulic acid, neochlorogenic acid and flavone.

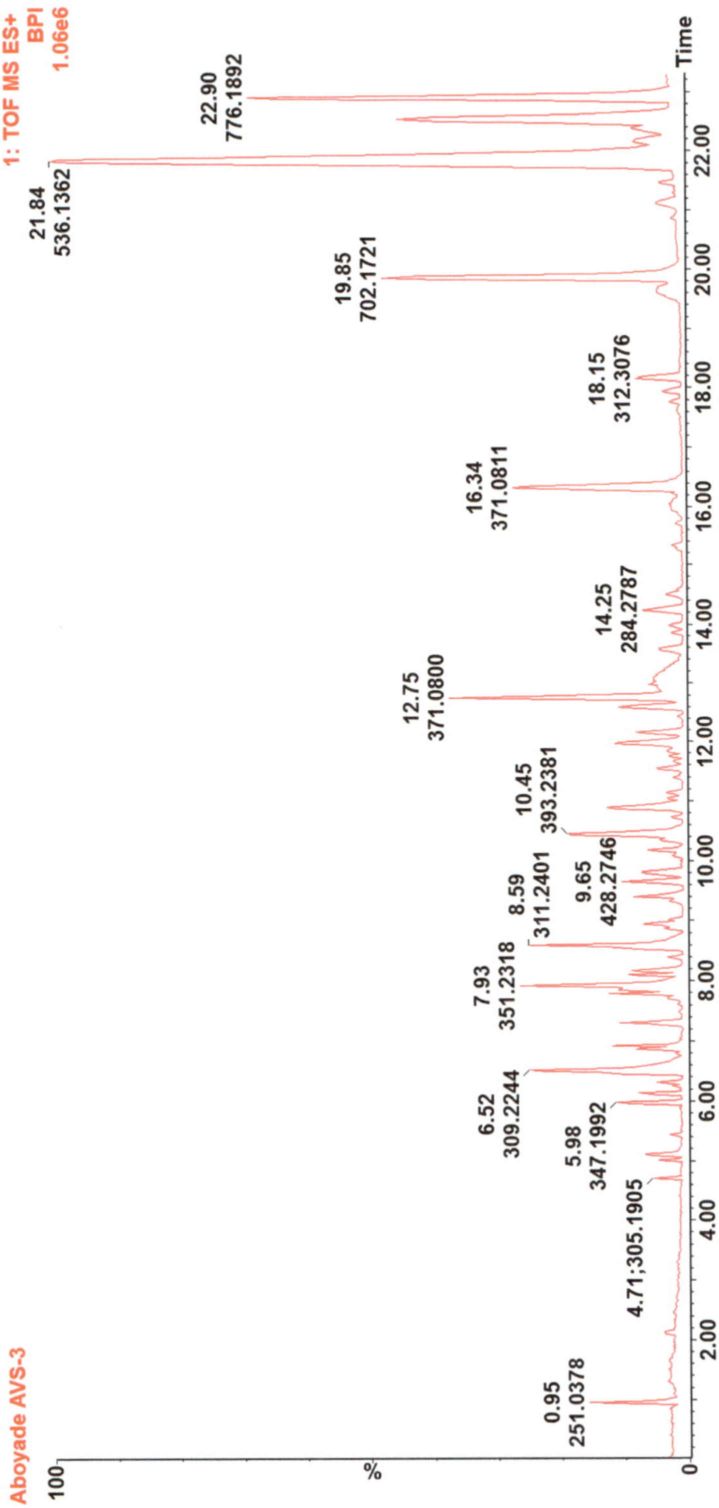

Figure 3. LC-MS chromatogram for avocado seed.

4. Discussion

Rats are considered the best models for studying human metabolic syndrome as they have similar metabolic patterns as humans [32]. The animals maintained their good health throughout this study and none of them became sick or died from the treatment.

Food intake determines calorie intake, which is one of the risk factors of obesity development. High food intake might lead to a higher calorie intake than energy expenditure, leading to the accumulation of body fat and obesity development. Sprague Dawley rats have shown the ability to become obese when placed on a HFD [33]. HFD has also been used to induce obesity in Zucker rats and mice [34–36].

The treatment group gained significantly less weight than the control group in this study, showing the potential of avocado seed to reduce weight gain. A previous study showed that avocado fruits (not avocado seed) resulted in drastic weight loss and reduction in BMI in individuals who received avocados daily for 6 months in an obesity reduction study [9]. Avocado seed oil may be responsible for the reduction of weight observed in animals fed with HFD-A (shown by the high crude fat content of HFD-A compared with HFD). Avocado seed oil is full of monounsaturated fatty acids, beneficial for the management of obesity and other inflammatory disorders. These fatty acids increase the secretion of adiponectin and reduce the plasma levels of pro-inflammatory cytokines [37].

Animals in the treatment group showed significantly less adipose tissue when compared with the control in this study. Similar results were previously obtained with *Cosmos caudatus* Kunth leaf [32]. Animals on a high-fat diet might present with high-fat deposits regardless of their weight [38]. Since the accumulation of adipose tissue in obesity is due to an increase in the number of adipocytes (hyperplasia), which follows increased adipocyte differentiation and size (hypertrophy), a decrease in adipose tissue mass might be associated with a reduction in adipocyte differentiation [35]. This is shown by a decrease in the mRNA level of differentiation biomarkers such as PPARγ, C/EBPα and adiponectin in weight reduction [35]. Therefore, avocado seed powder might possibly inhibit one or more of the processes involved in adipocyte differentiation and proliferation [39].

Trace elements such as zinc, iron, calcium, copper, manganese and selenium, play a major role in the regulation of metabolic pathways, including lipid regulation [40], and act as co-factors for enzymes. Disturbances in their homeostasis have been noted in obesity [41,42] and may lead to oxidative stress, reduction in immunity and increased expression of inflammatory mediators. Their supplementation to the required concentration may, therefore, play a role in ameliorating obesity. Avocado seed powder showed the presence of Fe, K, Ca, Mn and Mg in this study. Talabi et al. (2016) reported avocado seed to contain calcium, phosphorus, potassium and sodium [20]. While the other study reported the presence of sodium, potassium, calcium, magnesium, iron, manganese and zinc in avocado seed [43]. According to Bouglé et al. (2009), trace elements decrease oxidative and inflammatory stress, thereby preventing obesity and its co-morbidities. Calcium suppresses 1-α-25-dihydroxycholecalciferol, leading to increased adipocyte differentiation and lipogenesis from fatty acids and inhibits lipolysis [42]. The role of magnesium and calcium in obesity modulation has been demonstrated with deep sea water (DSW), for instance. DSW resulted in weight reduction in mice, decreased levels of adiponectin and PPARγ and decreased levels of circulating adipogenic proteins in the DSW group compared with the tap water group and this was attributed to the presence of Mg and Ca [44]. In addition to regulating insulin production and action, magnesium is a co-factor for different enzymatic reactions involved in glucose modulation [45].

In this study, HFD-A showed a potential to decrease triglycerides, with LDL, HDL and total cholesterol comparable to animals fed with HFD. Obesity is associated with various biochemical abnormalities such as high triglyceride levels and low HDL [46]. High-fat-containing meals increase the level of cholesterol and also alter lipid profiles [47]. Higher levels of cholesterol and triglycerides were previously obtained from rodents on a HFD (21% fat content of the food) than with the control (5% fat content) [48]. A good treatment would therefore be able to decrease total cholesterol, triglyceride and LDL while increasing the level of HDL. Avocado oil has been found to alleviate non-alcoholic fatty liver disease, hyperglycemia and dyslipidemia in rats fed with a high-fat/high-fructose diet by improving mitochondrial function and reducing the levels of reactive oxygen species, pro-inflammatory factors and lipid peroxidation [10].

Though the mechanism of obesity prevention was not established from this study, phytochemicals such as apigenin (flavone), hesperetin (flavanone), cyanidin (anthocyanins) and curcumin (phenolic acid) among others have been found to combat obesity in different ways [49]. These phytochemicals have been reported to reduce the digestion of fat (probably through lipase inhibition), hence reduce lipid absorption from the gastro-intestinal tract, and also effect anti-inflammatory actions [49]. Polyphenols also cause a decrease in lipogenesis and increase lipolysis [17,39,50,51]. The flavonoids and saponins in avocado are associated with inhibition in the inflammatory pathways involved in obesity development and the pathophysiology of its comorbidities [51,52]. These phytochemicals may be responsible for the observed anti-obesity effects of avocado seed powder, as the LC-MS analysis indicated the presence of cinnamic acid, ferulic acid, neochlorogenic acid, isoferulic acid-3-O-glucoronide and rothindin in this study. Ferulic acid and chlorogenic acid have been reported to elicit their anti-obesity effects through the inhibition of pancreatic lipase [53]. As obesity is associated with inflammation, agents with anti-inflammatory effects would be beneficial. The reduction of inflammatory cytokines has been reported as one of the mechanisms the phytochemicals follow in obesity management. This mechanism was reported for cinnamic acid [54], ferulic acid [55] and neochlorogenic acid [56], in preventing or reversing obesity. It was reported that the anti-obesity activity of ferulic acid was due to its potential to inhibit inflammatory markers such as tumor necrosis factor-α (TNF-α) and monocyte chemoattractant protein-1 (MCH-1) [55]. This was also observed in a study by Tian et al., 2022, where ferulic acid reduced the production of the inflammation cytokines TLR-4, NF-κB, TNF-α, IL-1β and IL-6 and reduced weight gain in mice fed a high-fat diet [57]. In addition, ferulic acid reduced leptin levels and increased ghrelin levels in mice fed with HFD [55].

Avocado seed has been reported to contain catechins, hydroxybenzoic acids, hydroxycinnamic acids, flavonols and procyanidins [13]. Another study showed the presence of 5-O-caffeoylquinic acid, quercetin 3-O-glucoside, quercetin 3-O-galactoside, quercetin 3-O-rutinoside, quercetin-3,4′-diglucoside and quercetin 3-O-arabinoside [58]. These phytochemicals have been reported to play a significant role in the treatment of obesity and its co-morbidities [59]. Hence, the potential anti-obesity properties shown by avocado seed powder in this study may be attributed to these phytochemicals. The phytochemical composition of plants is affected by genetic, environmental and physical factors [60]. Parameters for extraction methods such as temperature, duration of extraction, solvent-to-solid ratio and storage conditions affect the phytochemical content [61]. These may explain the differences in the phytochemical composition of the avocado seed from this study and the other studies.

Although animals in the treatment group showed increased brain weight, both the post-mortem and histology examination did not confirm injury to any other organ and did not reveal any lesions due to toxic effects in any of the organs in both groups. As a major organ of metabolism, the liver is more prone to drug toxicity [62]. Liver injury can increase the permeability of hepatocytes, as they lose their integrity, accompanied by the leakage of the enzymes into the bloodstream [63]. The increase in liver enzymes in the blood might therefore be directly proportional to the extent of the injury. The results of the liver enzymes showed that the animals on HFD-A had higher ALP than those on HFD. Apart from its use in hepatic function analysis, ALP is also elevated in bone disease [64]. An increase in ALP levels of the treatment group may therefore require further investigations. An increase in ALT and AST levels on animals fed avocado seeds has been reported previously [65]. However, another study found no significant changes in liver enzymes associated with avocado seed [66]. The changes in liver enzymes might indicate possible liver damage in this group. However, the histopathology results did not show any detectable liver injury or injury to any other organ. The vascularization observed in this study might be due to microvesicular lipidosis, which has been reported before in mice on HFD [67]. Vascularization might also show hepatocellular damage under oxidative stress [68]. Since other findings (clinical and anatomical) did not point towards any clinically significant hepatic changes, the importance of the observed changes was uncertain due to the degree of the hepatic lesions (varying between normal, minimal/slight and mild).

In this study, avocado seed showed the potential to reduce weight gain in rats fed with a high-fat diet. Avocado seed should be investigated further for its mechanisms of action and safety.

5. Conclusions

Avocado seed powder contains different phytochemicals and trace elements with potential for use in management of obesity. In this study, avocado seed powder has shown potential to prevent obesity and reduce triglycerides. However, the long-term effects of the avocado seed powder ingestion still needs to be investigated. Further studies should be conducted over a long period to determine the mechanisms of action of avocado seed in managing obesity and its safety in humans. Until that is done, we would not recommend it's chronic use in supplements.

Author Contributions: Conceptualization, D.R.K.; methodology, S.M., O.A. and D.R.K.; validation, O.A. and D.R.K.; formal analysis, S.M.; investigation, S.M.; resources, S.M., O.A. and D.R.K.; data curation, S.M., O.A. and D.R.K.; writing—original draft preparation, S.M.; writing—review and editing, S.M., O.A. and D.R.K.; visualization, S.M.; supervision, O.A. and D.R.K.; project administration, S.M., O.A. and D.R.K.; funding acquisition, O.A. and S.M. All authors have read and agreed to the published version of the manuscript.

Funding: This study was financially supported by the South African Medical Research Council (MRC) career development award and the National Research Foundation (NRF) grant number: 99073, Tshwane University of Technology (TUT) full-time postgraduate program 2015 and National Research Foundation (NRF) grant number: 111184.

Institutional Review Board Statement: Ethical approval to conduct this study was obtained from the Animal Ethics Committee of the University of Pretoria (V038-15) and the Animal Research Ethics Committee of Tshwane University of Technology (AREC2015/10/005(2)). This study was conducted following the South African Medical Research Council's guidelines on ethics for medical research: use of animals in research and training [27], and the South African National Standard: the care and use of animals for scientific purposes [28].

Informed Consent Statement: Not applicable.

Data Availability Statement: The datasets for this study are not available online; however, they can be found upon request by interested researchers to the corresponding author.

Acknowledgments: Authors would like to acknowledge the following: animal research staff members of the University of Pretoria for animal lodging and care; Tshepo Sekele, Victor Rametse, Lefa Tswaledi and Peter Ohwofasa for assisting with animal care; Princess Ramokolo and Tholang Mokhele for statistical analysis; Tshwane University of Technology (TUT), National Research Foundation (NRF) and South African Medical Research council (SAMRC) for funding.

Conflicts of Interest: The authors declare no conflicts of interest.

References

1. Klein, S.; Gastaldelli, A.; Yki-Järvinen, H.; Scherer, P.E. Why does obesity cause diabetes? *Cell Metab.* **2022**, *34*, 11–20. [CrossRef] [PubMed]
2. Pi-Sunyer, X. The Medical Risks of Obesity. *Postgrad. Med.* **2009**, *121*, 21–33. [CrossRef]
3. Tzenios, N. Obesity as a Risk Factor for Different Types of Cancer. *EPRA Int. J. Res. Dev.* **2023**, *8*, 97–100. [CrossRef]
4. Popkin, B.M.; Ng, S.W. The nutrition transition to a stage of high obesity and noncommunicable disease prevalence dominated by ultra-processed foods is not inevitable. *Obes. Rev.* **2021**, *23*, e13366. [CrossRef] [PubMed]
5. WHO. *World Obesity Day 2022—Accelerating Action to Stop Obesity*; WHO: Geneva, Switzerland, 2022.

6. Iwu, M.M. Introduction: Therapeutic Agents from Ethnomedicine. In *Advances in Phytomedicine*; Elsevier: Amsterdam, The Netherlands, 2002; pp. 1–22.
7. Ruchi, S. Role of nutraceuticals in health care: A review. *Int. J. Green Pharm.* **2017**, *11*. [CrossRef]
8. Nijhawan, P.; Behl, T. Nutraceuticals in the management of obesity. *Obes. Med.* **2019**, *17*, 100168. [CrossRef]
9. Devalaraja, S.; Jain, S.; Yadav, H. Exotic fruits as therapeutic complements for diabetes, obesity and metabolic syndrome. *Food Res. Int.* **2011**, *44*, 1856–1865. [CrossRef]
10. García-Berumen, C.I.; Vargas-Vargas, M.A.; Ortiz-Avila, O.; Piña–Zentella, R.M.; Ramos-Gómez, M.; Figueroa–García, M.d.C.; Mejía-Zepeda, R.; Rodríguez–Orozco, A.R.; Saavedra–Molina, A.; Cortés-Rojo, C. Avocado oil alleviates non-alcoholic fatty liver disease by improving mitochondrial function, oxidative stress and inflammation in rats fed a high fat–High fructose diet. *Front. Pharmacol.* **2022**, *13*, 1089130. [CrossRef]
11. Ochoa-Zarzosa, A.; Báez-Magaña, M.; Guzmán-Rodríguez, J.J.; Flores-Alvarez, L.J.; Lara-Márquez, M.; Zavala-Guerrero, B.; Salgado-Garciglia, R.; López-Gómez, R.; López-Meza, J.E. Bioactive molecules from native Mexican avocado fruit (*Persea americana* var. drymifolia): A review. *Plant Foods Hum. Nutr.* **2021**, *76*, 133–142. [CrossRef]
12. Alhassan, A.J.; Sule, M.S.; Atiku, M.K.; Wudil, A.M.; Abubakar, H.; Mohammed, S.A. Effects of aqueous avocado pear (*Persea americana*) seed extract on alloxan induced diabetes rats. *Greener J. Med. Sci.* **2012**, *2*, 005–011. [CrossRef]
13. Pahua-Ramos, M.E.; Ortiz-Moreno, A.; Chamorro-Cevallos, G.; Hernández-Navarro, M.D.; Garduño-Siciliano, L.; Necoechea-Mondragón, H.; Hernández-Ortega, M. Hypolipidemic effect of avocado (*Persea americana* Mill) seed in a hypercholesterolemic mouse model. *Plant Foods Hum. Nutr.* **2012**, *67*, 10–16. [CrossRef] [PubMed]
14. Wang, W.; Bostic, T.R.; Gu, L. Antioxidant capacities, procyanidins and pigments in avocados of different strains and cultivars. *Food Chem.* **2010**, *122*, 1193–1198. [CrossRef]
15. Ezejiofor, A.N.; Okorie, A.; Orisakwe, O.E. Hypoglycaemic and tissue-protective effects of the aqueous extract of *Persea americana* seeds on alloxan-induced albino rats. *Malays. J. Med. Sci.* **2013**, *20*, 31–39. [PubMed]
16. Henry, L.N.; Mtaita, U.Y.; Kimaro, C.C. Nutritional efficacy of avocado seeds. *Glob. J. Food Sci. Technol.* **2015**, *3*, 192–196.
17. Leite, J.J.G.; Brito, H.S.; Cordeiro, R.A.; Brilhante, R.S.N.; Sidrim, J.J.C.; Bertini, L.M.; de Morais, S.M.; Rocha, M.F.G. Chemical composition, toxicity and larvicidal and antifungal activities of *Persea americana* (avocado) seed extracts. *Rev. Soc. Bras. Med. Trop.* **2009**, *42*, 110–113. [CrossRef]
18. Sebastián, D.F.-S.; Alaniz-Monreal, S.; Rabadán-Chávez, G.; Vázquez-Manjarrez, N.; Hernández-Ortega, M.; Gutiérrez-Salmeán, G. Anthocyanins: Potential Therapeutic Approaches towards Obesity and Diabetes Mellitus Type 2. *Molecules* **2023**, *28*, 1237. [CrossRef]
19. Aloo, S.-O.; Ofosu, F.K.; Kim, N.-H.; Kilonzi, S.M.; Oh, D.-H. Insights on Dietary Polyphenols as Agents against Metabolic Disorders: Obesity as a Target Disease. *Antioxidants* **2023**, *12*, 416. [CrossRef]
20. Talabi, J.Y.; Osukoya, O.A.; Ajayi, O.O.; Adegoke, G.O. Nutritional and antinutritional compositions of processed Avocado (*Persea americana* Mill) seeds. *Asian J. Plant Sci. Res.* **2016**, *6*, 6–12.
21. Dreher, M.L.; Davenport, A.J. Hass avocado composition and potential health effects. *Crit. Rev. Food Sci. Nutr.* **2013**, *53*, 738–750. [CrossRef]
22. González-Domínguez, Á.; Domínguez-Riscart, J.; Millán-Martínez, M.; Lechuga-Sancho, A.M.; González-Domínguez, R. Exploring the association between circulating trace elements, metabolic risk factors, and the adherence to a Mediterranean diet among children and adolescents with obesity. *Front. Public Health* **2023**, *10*, 1016819. [CrossRef]
23. Imafidon, K.; Amaechina, F. Effects of aqueous seed extract of *Persea americana* Mill. (avocado) on blood pressure and lipid profile in hypertensive rats. *Adv. Biol. Res.* **2010**, *4*, 116–121.
24. Agoreyo, B.; Obansa, E.; Obanor, E. Comparative nutritional and phytochemical analyses of two varieties of Solanum melongena. *Sci. World J.* **2012**, *7*, 5–8.
25. Dolan, S.P.; Capar, S.G. Multi-element analysis of food by microwave digestion and inductively coupled plasma-atomic emission spectrometry. *J. Food Compos. Anal.* **2002**, *15*, 593–615. [CrossRef]
26. Mester, Z.; Sturgeon, R.E. *Sample Preparation for Trace Element Analysis*; Elsevier: Amsterdam, The Netherlands, 2003; Volume 41.
27. SAMRC. *Guidelines on Ethics for Medical Research: Use of Animals in Research and Training 2004*; South African Medical Research Council: Cape Town, South Africa, 2004; pp. 11–16.
28. South African National Standard. *Care and Management of Laboratory Animals—Rodents (Mice, Rats and Hamsters) in the Care and Use of Animals for Scientific Purposes*; SABS: Pretoria, South Africa, 2008; pp. 180–187.
29. Olaokun, O.O.; McGaw, L.J.; van Rensburg, I.J.; Eloff, J.N.; Naidoo, V. Antidiabetic activity of the ethyl acetate fraction of *Ficus lutea* (Moraceae) leaf extract: Comparison of an in vitro assay with an in vivo obese mouse model. *BMC Complement. Altern. Med.* **2016**, *16*, 110. [CrossRef]
30. Parasuraman, S.; Raveendran, R.; Kesavan, R. Blood sample collection in small laboratory animals. *J. Pharmacol. Pharmacother.* **2010**, *1*, 87–93. [CrossRef]
31. Mokhele, M.; Tswaledi, D.; Aboyade, O.; Shai, J.; Katerere, D. Investigation of Aloe ferox leaf powder on anti-diabesity activity. *S. Afr. J. Bot.* **2020**, *128*, 174–181. [CrossRef]
32. Rahman, H.A.; Sahib, N.G.; Saari, N.; Abas, F.; Ismail, A.; Mumtaz, M.W.; Hamid, A.A. Anti-obesity effect of ethanolic extract from *Cosmos caudatus* Kunth leaf in lean rats fed a high fat diet. *BMC Complement. Altern. Med.* **2017**, *17*, 122. [CrossRef]

33. Nilsson, C.; Raun, K.; Yan, F.-F.; O Larsen, M.; Tang-Christensen, M. Laboratory animals as surrogate models of human obesity. *Acta Pharmacol. Sin.* **2012**, *33*, 173–181. [CrossRef]
34. Kalupahana, N.S.; Claycombe, K.; Newman, S.J.; Stewart, T.; Siriwardhana, N.; Matthan, N.; Lichtenstein, A.H.; Moustaid-Moussa, N. Eicosapentaenoic acid prevents and reverses insulin resistance in high-fat diet-induced obese mice via modulation of adipose tissue inflammation. *J. Nutr.* **2010**, *140*, 1915–1922. [CrossRef]
35. Moon, M.-H.; Jeong, J.-K.; Lee, J.-H.; Park, Y.-G.; Lee, Y.-J.; Seol, J.-W.; Park, S.-Y. Antiobesity activity of a sphingosine 1-phosphate analogue FTY720 observed in adipocytes and obese mouse model. *Exp. Mol. Med.* **2012**, *44*, 603–614. [CrossRef] [PubMed]
36. Tsuda, T.; Horio, F.; Uchida, K.; Aoki, H.; Osawa, T. Dietary cyanidin 3-O-β-D-glucoside-rich purple corn color prevents obesity and ameliorates hyperglycemia in mice. *J. Nutr.* **2003**, *133*, 2125–2130. [CrossRef]
37. Hamouda, A.F. Study on the effect of avocado on apoptosis, oxidative stress and injuries induced by diethyl nitrosamine in rat liver. *J. Pharm. Pharmacol.* **2015**, *3*, 243–252. [CrossRef]
38. Cha, M.C.; Chou, C.J.; Boozer, C.N. High-fat diet feeding reduces the diurnal variation of plasma leptin concentration in rats. *Metabolism* **2000**, *49*, 503–507. [CrossRef]
39. Hasani-Ranjbar, S.; Jouyandeh, Z.; Abdollahi, M. A systematic review of anti-obesity medicinal plants—An update. *J. Diabetes Metab. Disord.* **2013**, *12*, 28. [CrossRef]
40. Li, C.-P.; Song, Y.-X.; Lin, Z.-J.; Ma, M.-L.; He, L.-P. Essential trace elements in patients with dyslipidemia: A meta-analysis. *Curr. Med. Chem.* **2024**, *31*, 3604–3623. [CrossRef]
41. Nimrouzi, M.; Abolghasemi, J.; Sharifi, M.H.; Nasiri, K.; Akbari, A. Thyme oxymel by improving of inflammation, oxidative stress, dyslipidemia and homeostasis of some trace elements ameliorates obesity induced by high-fructose/fat diet in male rat. *Biomed. Pharmacother.* **2020**, *126*, 110079. [CrossRef]
42. Bouglé, D.; Bouhallab, S.; Bureau, F.; Zunquin, G. Effects of trace elements and calcium on diabetes and obesity, and their complications: Protective effect of dairy products–A mini-review. *Dairy Sci. Technol.* **2009**, *89*, 213–218. [CrossRef]
43. Alagbaoso, C.A.; Tokunbo, I.I.; Osakwe, O.S. Comparative study of antioxidant activity and mineral composition of methanol extract of seeds of ripe and unripe avocado pear (*Persea americana*, Mill.). *NISEB J.* **2015**, *15*, 123–127.
44. Hwang, H.S.; Kim, H.A.; Lee, S.H.; Yun, J.W. Anti-obesity and antidiabetic effects of deep sea water on ob/ob mice. *Mar. Biotechnol.* **2009**, *11*, 531–539. [CrossRef] [PubMed]
45. Barbagallo, M.; Dominguez, L.J. Magnesium and type 2 diabetes. *World J. Diabetes* **2015**, *6*, 1152. [CrossRef] [PubMed]
46. El-Awdan, S.A.; Jaleel, G.A.A.; Saleh, D.O. Grape seed extract attenuates hyperglycaemia-induced in rats by streptozotocin. *Bull. Fac. Pharm. Cairo Univ.* **2013**, *51*, 203–209. [CrossRef]
47. Angéloco, N.; Deminice, R.; Leme ID, A.; Lataro, R.C.; Jordão, A.A. Bioelectrical impedance analysis and anthropometry for the determination of body composition in rats: Effects of high-fat and high-sucrose diets. *Rev. Nutr.* **2012**, *25*, 331–339. [CrossRef]
48. Chadli, F.K.; Nazih, H.; Krempf, M.; Nguyen, P.; Ouguerram, K. Omega 3 fatty acids promote macrophage reverse cholesterol transport in hamster fed high fat diet. *PLoS ONE* **2013**, *8*, e61109. [CrossRef]
49. Carrera-Quintanar, L.; Lopez Roa, R.I.; Quintero-Fabián, S.; Sánchez-Sánchez, M.A.; Vizmanos, B.; Ortuño-Sahagún, D. Phytochemicals that influence gut microbiota as prophylactics and for the treatment of obesity and in-flammatory diseases. *Mediat. Inflamm.* **2018**, *2018*, 9734845. [CrossRef]
50. Chandrasekaran, C.V.; Vijayalakshmi, M.A.; Prakash, K.; Bansal, V.S.; Meenakshi, J.; Amit, A. Review article: Herbal approach for obesity management. *Am. J. Plant Sci.* **2012**, *3*, 1003–1014. [CrossRef]
51. Mohamed, G.A.; Ibrahim, S.R.; Elkhayat, E.S.; El Dine, R.S. Natural anti-obesity agents. *Bull. Fac. Pharm. Cairo Univ.* **2014**, *52*, 269–284. [CrossRef]
52. Wintola, O.A.; Afolayan, A.J. Phytochemical constituents and antioxidant activities of the whole leaf extract of Aloe ferox Mill. *Pharmacogn. Mag.* **2011**, *7*, 325–333. [CrossRef]
53. Kasabri, V.; Al-Hallaq, E.K.; Bustanji, Y.K.; Abdul-Razzak, K.K.; Abaza, I.F.; Afifi, F.U. Antiobesity and antihyperglycaemic effects of *Adiantum capillus-veneris* extracts: In vitro and in vivo evaluations. *Pharm. Biol.* **2016**, *55*, 164–172. [CrossRef] [PubMed]
54. Lee, A.G.; Kang, S.; Im, S.; Pak, Y.K. Cinnamic Acid Attenuates Peripheral and Hypothalamic Inflammation in High-Fat Diet-Induced Obese Mice. *Pharmaceutics* **2022**, *14*, 1675. [CrossRef] [PubMed]
55. de Melo, T.; Lima, P.; Carvalho, K.; Fontenele, T.; Solon, F.; Tomé, A.; de Lemos, T.; Fonseca, S.d.C.; Santos, F.; Rao, V.; et al. Ferulic acid lowers body weight and visceral fat accumulation via modulation of enzymatic, hormonal and inflammatory changes in a mouse model of high-fat diet-induced obesity. *Braz. J. Med. Biol. Res.* **2017**, *50*, e5630. [CrossRef] [PubMed]
56. Singdam, P.; Naowaboot, J.; Senggunprai, L.; Boonloh, K.; Hipkaeo, W.; Pannangpetch, P. The mechanisms of neochlorogenic acid (3-caffeoylquinic acid) in improving glucose and lipid metabolism in rats with insulin resistance induced by a high fat-high fructose diet. *Trends Sci.* **2023**, *20*, 6455. [CrossRef]
57. Tian, B.; Geng, Y.; Wang, P.; Cai, M.; Neng, J.; Hu, J.; Xia, D.; Cao, W.; Yang, K.; Sun, P. Ferulic acid improves intestinal barrier function through altering gut microbiota composition in high-fat diet-induced mice. *Eur. J. Nutr.* **2022**, *61*, 3767–3783. [CrossRef] [PubMed]
58. Kosińska, A.; Karamac, M.; Estrella, I.; Hernández, T.; Bartolomé, B.; Dykes, G.A. Phenolic compound profiles and antioxidant capacity of *Persea americana* Mill. peels and seeds of two varieties. *J. Agric. Food Chem.* **2012**, *60*, 4613–4619. [CrossRef] [PubMed]
59. Zhang, Y.; Huang, C. Targeting adipocyte apoptosis: A novel strategy for obesity therapy. *Biochem. Biophys. Res. Commun.* **2012**, *417*, 1–4. [CrossRef]

60. Li, H.; Tsao, R.; Deng, Z. Factors affecting the antioxidant potential and health benefits of plant foods. *Can. J. Plant Sci.* **2012**, *92*, 1101–1111. [CrossRef]
61. Michiels, J.A.; Kevers, C.; Pincemail, J.; Defraigne, J.O.; Dommes, J. Extraction conditions can greatly influence antioxidant capacity assays in plant food matrices. *Food Chem.* **2012**, *130*, 986–993. [CrossRef]
62. Hardisty, J.F.; Brix, A.E. Comparative hepatic toxicity: Prechronic/chronic liver toxicity in rodents. *Toxicol. Pathol.* **2005**, *33*, 35–40. [CrossRef]
63. Wintola, O.; Sunmonu, T.; Afolayan, A. Toxicological evaluation of aqueous extract of Aloe ferox Mill. in loperamide-induced constipated rats. *Hum. Exp. Toxicol.* **2010**, *30*, 425–431. [CrossRef]
64. Kunutsor, S.K.; Bakker, S.J.L.; Kootstra-Ros, J.E.; Gansevoort, R.T.; Gregson, J.; Dullaart, R.P.F. Serum alkaline phosphatase and eisk of incident cardiovascular disease: Interrelationship with high sensitivity C-reactive protein. *PLoS ONE* **2015**, *10*, e0132822. [CrossRef]
65. Zakariya, U.A.; Umar, U.A.; Dambazau, S.M.; Sulaiman, A. Comparative Hepatotoxic Effects of Aqueous and Phenolic Extracts of Avocado (*Persea americana*) Seed in Wistar Albino Rats. *Int. J. Biochem. Res. Rev.* **2016**, *10*, 1–6. [CrossRef]
66. Ozolua, R.; Anaka, O.; Okpo, S.; Idogun, S. Acute and sub-acute toxicological assessment of the aqueous seed extract of *Persea americana* Mill (Lauraceae) in rats. *Afr. J. Tradit. Complement. Altern. Med.* **2009**, *6*, 573–578. [CrossRef] [PubMed]
67. Kanfi, Y.; Peshti, V.; Gil, R.; Naiman, S.; Nahum, L.; Levin, E.; Kronfeld-Schor, N.; Cohen, H.Y. SIRT6 protects against pathological damage caused by diet-induced obesity. *Aging Cell* **2010**, *9*, 162–173. [CrossRef] [PubMed]
68. Bhushan, B.; Saxena, P.N.; Saxena, N. Biochemical and histological changes in rat liver caused by cypermethrin and be-ta-cyfluthrin. *Arch. Ind. Hyg. Toxicol.* **2013**, *64*, 57–67.

nutraceuticals

Review

Momordica balsamina L.: A Plant with Multiple Therapeutic and Nutritional Potential—A Review

Marème Thiaw [1,2], Issa Samb [1,*], Manon Genva [2], Mohamed Lamine Gaye [3] and Marie-Laure Fauconnier [2,*]

1 Department of Chemistry, Training and Research Unit for Applied Sciences and Information and Communication Technology, University Alioune Diop of Bambey (UADB), Bambey 30, Senegal; mareme.thiaw@doct.uliege.be
2 Laboratory of Chemistry of Natural Molecules, Gembloux Agro-Bio Tech, University of Liege, Passage des Deportes, 2-5030 Gembloux, Belgium; m.genva@uliege.be
3 Department of Chemistry, Faculty of Sciences and Technology (FST), University Cheikh Anta Diop Dakar (UCAD), Dakar 5005, Senegal; mlgayeastou@yahoo.fr
* Correspondence: issa.samb@uadb.edu.sn (I.S.); marie-laure.fauconnier@uliege.be (M.-L.F.); Tel.: +221-77-568-34-78 (I.S.); +32-81-62-22-89 (M.-L.F.)

Abstract: This review seeks to deepen our comprehension of the African plant *Momordica balsamina* L. by elucidating its therapeutically important molecules and nutrient composition. Commonly referred to as the balsam apple, this plant species is extensively harnessed for its diverse therapeutic potential across its various organs, including leaves, fruits, roots, and stems. Numerous bioactive molecules have been isolated or identified within this plant, notably encompassing polyphenols, flavonoids, terpenes, and carotenoids. These compounds exhibit a wide array of biological activities, ranging from antioxidative, anti-inflammatory, anti-diabetic and anti-carcinogenic to anti-malarial properties, among others. Furthermore, the leaves of *Momordica balsamina* L. stand out for their abundant micronutrients, proteins, and amino acids. This investigation aims to shed light not only on the botanical characteristics of the *Momordica balsamina* plant and its potential applications in traditional medicine but also on its chemical composition, biological functionalities, and physicochemical attributes, thus accentuating its nutritional advantages. Nonetheless, an intriguing avenue presents itself for the exploration of strategies to conserve this species, delve deeper into its potential within the cosmetics industry, and innovate methodologies for the synthesis or biosynthesis of these bioactive molecules.

Keywords: *Momordica balsamina* L.; phytotherapy; phytochemicals; biological properties; nutritional values; cosmetic uses

Citation: Thiaw, M.; Samb, I.; Genva, M.; Gaye, M.L.; Fauconnier, M.-L. *Momordica balsamina* L.: A Plant with Multiple Therapeutic and Nutritional Potential—A Review. *Nutraceuticals* **2023**, *3*, 556–573. https://doi.org/10.3390/nutraceuticals3040040

Academic Editors: Ivan Cruz-Chamorro and Guillermo Santos Sánchez

Received: 15 September 2023
Revised: 9 November 2023
Accepted: 15 November 2023
Published: 17 November 2023

1. Introduction

For millennia, humanity has drawn upon environmental resources to ensure survival and well-being. Even in contemporary times, plants remain invaluable sources of sustenance and medicinal properties [1,2]. Nevertheless, a scarcity of ethnobotanical and chemical investigations has obscured the full extent of their pharmacopeia, nutritional value, and therapeutic potential.

In the present era, uncovering the botanical intricacies and therapeutic merits of plants presents a formidable challenge. Indeed, plants are fundamental and essential components in pharmaceutical research and production [1]. Consequently, delving into their chemical compositions and nutritional profiles emerges as a critical pursuit to enhance their application in medicine. Notably, plant-based derivatives constitute over 50% of the global medicinal repertoire [3]. These plants are frequently encountered in herbalists' domains, traditional healing practices, marketplaces, or their regions of origin. The leaves, barks, roots, and fruits of medicinal plants constitute the most utilized components for phytotherapeutic purposes, often administered via maceration, infusion, digestion, or decoction [4].

The Cucurbitaceae family, widely employed in traditional medicine, encompasses herbaceous vines with tendrils and, in some cases, shrubs. Certain of these species exhibit widespread distribution across tropical and subtropical regions [4,5]. Cucurbits notably hold a significant position in Africa, offering a range of edibles (such as squash, pumpkin, and melon) and utilitarian items (like gourds and sponges). The fruits of numerous cucurbit species, rich in micronutrients, constitute dietary staples in Africa, sometimes consumed as juices to alleviate micronutrient deficiencies. Plants within the *Momordica* genus are of notable importance for their remedial attributes [6].

Momordica balsamina, commonly known as balsam apple or referred to as Mburbuf or Mborbof in the Wolof language (native to Senegal), belongs to the *Momordica* genus. Flourishing in arid expanses of tropical and subtropical Africa, *M. balsamina* serves a dual role, being both a dietary staple and a fundamental element of traditional African medicine. It finds application in alleviating diverse symptomatic manifestations, including those associated with diabetes and malaria [7].

The primary objective of this article is to provide a comprehensive botanical delineation of *M. balsamina*, followed by an exhaustive exploration of its numerous applications. Subsequently, a detailed synthesis of its chemical composition, biological activities, and nutritional potential is presented. *M. balsamina*, with its numerous applications in the food sector, as a source of medicine, and potentially as a source for new cosmetic ingredients, is well worth this review paper. The uniqueness of this review lies in its extensive coverage, encompassing not only the phytochemical and pharmacological aspects of *M. balsamina* but also its nutritional potential, traditional uses, and cosmetic applications. The literature search was carried out using the following databases: Google Scholar, Science Direct, Pub Chem, ACS, and Scopus. The keywords mainly used were "*Momordica balsamina*", associated with thematic words such as botany, traditional use, biological properties, nutritional value, and cosmetic use.

2. Botanical Description and Geographical Distribution of *Momordica balsamina*

The genus *Momordica* comprises approximately fifty-nine herbaceous or perennial herbaceous species, occasionally manifesting as small shrub climbers. These plants belong to the cucurbit family and are indigenous to tropical and subtropical regions of Africa, Asia, and Australia [8].

M. balsamina (Figure 1) is an annual herbaceous, monoecious, climbing plant with glabrous or lightly hairy stems up to 4 to 5 m long, equipped with tendrils for attachment to nearby vegetation. It grows at altitudes of 0 to 1293 m [3,4,9].

The leaves of *M. balsamina* (see Figure 1a) exhibit a light green hue. They are alternate, waxy, simple, and lobed, featuring three to five distinct lobes reaching up to half of the leaf blade. These leaves can reach a length of 12 cm [4,9,10].

The flowers (see Figure 1d) are pale yellow, unisexual, and solitary. They take on a rounded, trumpet-like form, characterized by a pedicel measuring up to 0.5 cm in length. The receptacle spans 0.5–1 mm, while the sepals are narrow and can grow up to 0.5 cm. Petals range from 0.5 to 1.5 cm in length, surrounding a unicellular inferior ovary [3,4,10].

The fruits of *M. balsamina* (see Figure 1b) adopt a spindle-shaped appearance, with colors shifting from orange to vibrant red. These fruits exhibit about 19 rows of regular or irregular, short, blunt, non-prickly, creamy, or yellowish spines. Upon ripening (when orange), the fruits, which measure 25–60 mm, spontaneously split into three coiled valves, revealing numerous seeds enshrouded by a vivid (see Figure 1c), extremely adhesive scarlet aril. This aril is edible and possesses a sweetness akin to watermelon. The seeds are encased within a crimson pulp. Oval and compressed, the seeds span 9 to 12 mm in length [3,4,10].

M. balsamina is a prevalent vegetable within tropical and subtropical zones [6]. Originating from tropical Africa, it has proliferated across drier areas of South Africa and coastal regions of Australia. Its cultivation has extended to European gardens, as well as to Central American, Arabian, tropical Asian, and Indian regions. In India, it thrives naturally in forests during the rainy season. This species has been introduced to parts of the Neotropics, gaining naturalization in the United States and Pakistan [9,11]. It can be found growing in the wild across southern African regions, including Botswana, Swaziland, Namibia, and South Africa. Within South Africa, it flourishes in the Eastern and Northern Cape, Limbo, and Kwazulu-Natal provinces [3].

Figure 1. Different parts of the *M. balsamina* plant: (**a**) leaves, (**b**) fruits, (**c**) seeds, and (**d**) flower (Mareme THIAW, Ngoundiane (Thies-Senegal) 2022).

3. Chemical Composition of *Momordica balsamina*

M. balsamina is revered in traditional medicine for its diverse chemical composition, contributing to its efficacy in treating a multitude of diseases. Each part of the plant is utilized in accordance with its distinct chemical composition and the specific ailment to be treated.

The plant's leaves, fruits, stems, seeds, and bark collectively contain a spectrum of compounds, including resins, alkaloids, flavonoids, glycosides, steroids, terpenes, cardiac glycosides, and saponins, among others [3,12–14].

Indeed, the aerial part of *M. balsamina* has been found to contain a variety of compounds, including terpenoids [12–14], alkaloids, saponins [13,14], tannins [12,13], flavonoids, anthocyanins, mucilages, and reducing compounds [12]. Additionally, the presence of glycosides and glycoside saponins has also been noted [14].

The fruit of *M. balsamina* notably contains a significant quantity of saponins, steroid rings, and carbohydrates. Alkaloids, tannins, flavonoids, glycosides, steroids, and terpenes are also present, albeit in smaller amounts [3].

Based on findings from ethnopharmacology surveys, phytochemical screenings, and extract fractionation via chromatography, a range of biological tests can be conducted. These include assessments in vitro and in vivo of antioxidant, antibacterial, anti-diabetic, anti-inflammatory, anti-malarial, antiviral, and anti-HIV properties, among others (Table 1) [15–18]. The observed biological activities can be attributed to the presence of identified or isolated metabolites within *M. balsamina* extracts, including flavonoids, terpenes, phenolic acids, carotenoids, and more (refer to Table 1).

Numerous phenolic acids, such as quinic acid and various chlorogenic acids, have been identified within the plant [16,17]. Unique compounds, including pseudolaroside A acid and isocitric feruloyl acid, have been reported for the first time in relation to *M. balsamina* [19]. Common flavonoids, including kaempferol, quercetin, and isorhamnetin, have been identified, each exhibiting distinct glycoside forms [16,17]. Notably, certain flavonoids demonstrate in vitro and in vivo an array of biological activities, such as anti-inflammatory, antioxidant, anti-diabetic, and anti-malarial effects [16,17,20].

Cucurbitane triterpenoids, the main constituents isolated from this species encompassing balsaminols, balsaminosides, balsaminagenins, karavilagenins, and cucurbalsaminols, have been extensively investigated by Ramalhete et al. (2010), (2011), and (2022) for their diverse bioactivities, including anti-diabetic, anti-malarial, antiplasmodial, anti-cancer, antibacterial, and P-glycoprotein (P-gp)-inhibitory effects [18,20,21]. Carotenoids, including zeaxanthin, lutein, and β-carotene, renowned for their antioxidative, anti-inflammatory, and anti-aging properties, have also been identified within the plant [17,20]. Biosynthetic pathways for cucurbitane have been established by Ramalhete et al. (2022) [18]. It proposed biosynthetic routes for the new cucurbalsaminane skeleton starting from a tetracyclic cucurbitane skeleton, with an α,β-unsaturated carbonyl at C-7.

Balsamin, a ribosome-inactivating protein (RIP), was isolated from the seeds of the plant *M. balsamina* [11]. It allows the in vitro inhibition of HIV-1 replication at the translation stage [11,22]. Ajji et al. (2018) [22] have shown that balsamin also possesses anti-tumor, antibacterial, and DNase-like activity. Moreover, a broad-spectrum antibacterial activity has been documented [18].

The chemical diversity of the plant has made it effective in treating various health problems, and its potential in traditional medicine has been substantiated by numerous biological tests, highlighting the significance of *M. balsamina* as a valuable resource in both traditional medicine and modern pharmacology for a wide range of health issues.

Table 1. Metabolites of *M. balsamina* and their biological activities.

Class of Compounds	Compounds	Biological Activity	Essay/Conclusion	References
	Balsamin	Antiviral activity	In vitro. Inhibits HIV-1 replication in T-cell lines and human primary $CD4^+$ T cells during the translation of viral proteins with an IC_{50} of approximately 10 nM after three days of treatment.	[23]
		Antibacterial	In vitro Inhibition the growth in a dose-dependent manner of pathogens *Staphylococcus epidermidis* (MICs = 1.56 μg/mL) and *Staphylococcus aureus* (MICs = 6.25 μg/mL).	[24]
Flavonoids	Kaempferol	Depigmenting	In vitro. Tyrosine inhibition in B16 melanoma cells with IC_{50} 171.40 μM.	[25]

Table 1. *Cont.*

Class of Compounds	Compounds	Biological Activity	Essay/Conclusion	References
Flavonoids	Kaempferol	Anti-inflammatory	In vitro. Potent inhibitory activity relative to Nitric Oxide (NO) production, induced by Lipopolysaccharides (LPS) in RAW 264.7 cells (IC_{50} 15.40 μM) without cytotoxicity, and inhibited NF-κB-mediated luciferase activities (IC_{50} 90.30 μM).	[25]
		Managing cancer-associated ailments	In vitro. Effectively inhibit (IC_{50} 43 μmol/L) multiple cancer-associated pathways in triple-negative breast cancer cells (TNBC cells) simultaneously.	[26]
	Quercetin	Anti-carcinogenic/ Antioxidant	In vitro. Scavenge oxygen-free radicals H_2O_2 (IC_{50} 5 μM) and O_2^- (IC_{50} 9 μM); inhibit lipid peroxidation (IC_{50} 60 μM); and quench 8 ohdg formation via UV light irradiation (IC_{50} 0.8 μM) and Fenton reaction (IC_{50} 80 μM).	[27]
		Antioxidant/ Anti-inflammatory	In vivo Preserve the function of the liver in acute alcoholic injury by upregulating the expression of Interleukin 10 and Oxygenase-1 and thus inhibiting NLRP3 inflammasome activation and inflammatory factor secretion.	[28]
	Isorhamnétine	Anti-inflammatory	In vitro. At 30 and 60 mg/kg, inhibits the inflammatory response to lipopolysaccharides (LPS) in RAW 264.7 cells and in an acute lung injury (ALI) model. It significantly suppresses the overproduction of pro-inflammatory cytokines and neutrophil migration and reduces the histopathological changes and lung edema induced by LPS.	[29]
Phenolic Acids	Quinic acid	Anti-inflammatory	In vitro. At 0.10 μg/mL, inhibits vascular cell adhesion molecule-1 (VCAM-1) expression via the suppression of mitogen-activated Protein (MAP) kinase and NF-κb signaling pathways in TNF-α-stimulated vascular smooth muscle cells (VSMCs).	[30]
		Antiviral	In vitro. Inhibits HBV (Hepatitis B Virus) DNA replication and Hepatitis B surface antigen (HBsAg) production in HepG2.2.15 cells infected with duck Hepatitis B virus.	[31]
Triterpenoids	Karavilagenin C	Anti-malarial	In vitro. Anti-malarial activity against the chloroquine-sensitive (3D7) (IC_{50} 10.40 μM) and the chloroquine-resistant (Dd2) (IC_{50} 11.20 μM) strains of Plasmodium *falciparum*.	[21]
		Antibacterial	In vitro. At 3 μM inhibition, the activity of bacterial efflux pumps of methicillin-resistant *Staphylococcus aureus* (MRSA) COL_{oxa} by increasing the intracellular accumulation of ethidium bromide.	[32]

Table 1. *Cont.*

Class of Compounds	Compounds	Biological Activity	Essay/Conclusion	References
Triterpenoids	Karavilagenin C	P-glycoprotein modulation activity	In vivo. Inhibition of P-glycoprotein in a mouse lymphoma cell transfected with the human ABCB1 gene. Fluorescence activity ratio (FAR) 42.10 at 2 μM.	[18]
	Balsaminagenin B	Antibacterial	In vitro. Inhibition of the activity of bacterial efflux pumps of *Enterococcus faecalis* by increasing the intracellular accumulation of ethidium bromide at 30 μM.	[18]
	Cucurbalsamin one B	P-glycoprotein modulation activity	In vivo. Inhibition of P-glycoprotein in a mouse lymphoma cell transfected with the human ABCB1 gene. Fluorescence activity ratio (FAR) 76.90 at 2 μM.	[18]
	Balsaminoside A	Anti-malarial	In vitro. Inhibition of *P. falciparum* 3D7 (IC_{50} 4.60 μM) and Dd2 (IC_{50} 4.00 μM stem cell growth.	[20]
	karavilagenin E	Anti-malarial	In vitro. Inhibition of *P. falciparum* 3D7 (IC_{50} 4.70 μM) and Dd2 (IC_{50} 8.20 μM) stem cell growth.	[20]
	Balsaminol groups, Balsaminoside and Balsamina-genine groups, Karavilagenin groups, Cucurbalsaminol groups	Anti-diabetic, Anti-malarial, Antiviral, Anticancer, antibacterial, P-glycoprotein (P-gp) inhibitors, and others.		[18,20,21]

4. Nutritional Value of *Momordica balsamina*

M. balsamina stands as a plant rich in vital nutrients and holds a significant place in African cuisine [33]. The leaves of *M. balsamina* deemed a significant green-leaf vegetable, possess multifaceted nutritional potential. It offers a source of nutrients (Figure 2) that complements major dietary components [9]. Table 2 highlights the nutritional composition, encompassing protein, fiber, fat, and caloric value of *M. balsamina* leaves. Notably, the protein content of this wild vegetable is remarkably high (287.70 ± 1.80 g/kg or 28.77%), surpassing that of numerous other vegetables [9]. This elevated protein content positions it as a promising protein supplement or meat substitute, especially for resource-scarce rural communities [9]. In line with recommended daily protein allowances, a mere 3 g of *M. balsamina* leaves could significantly contribute to daily protein requirements.

With its relatively low crude fiber content (37.20 ± 7.90 g/kg), *M. balsamina* leaves contribute substantially to dietary fiber intake, an essential component in human nutrition [9]. Characteristic of most green leafy vegetables, *M. balsamina* boasts a modest caloric value (1892.20 kcal/kg), making it a favorable dietary addition for those aiming to manage caloric intake [34].

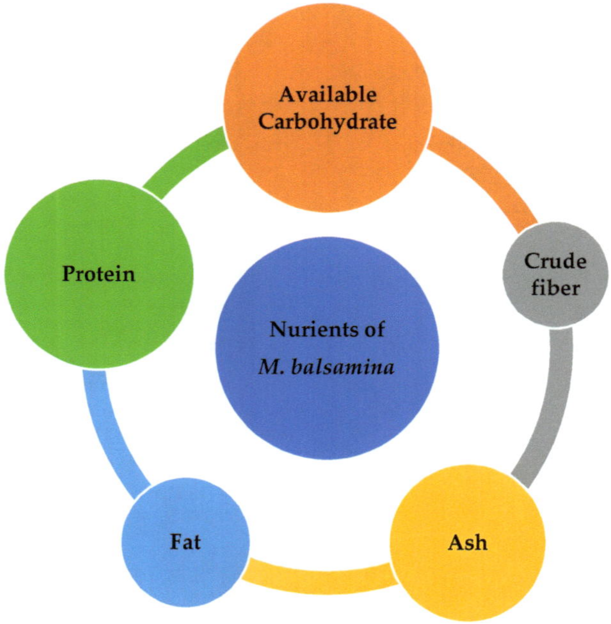

Figure 2. Nutrients of *M. balsamina* leaves.

Its impressively high ash content (127.00 ± 17.00 g/100 g) establishes *M. balsamina* as a dependable source of essential minerals (Table 3) [9]. Notably, potassium and calcium dominate the mineral composition (27.05 and 22.20 g/kg, respectively). Even a modest consumption of 4 g for children and 6 g for adults of *M. balsamina* can fulfill a part of daily calcium needs, along with potassium requirements ranging from 1.50 g to 17.00 g [9]. These results are not in agreement with studies by Karumi et al. (2004) [35] who report higher concentrations of Fe, Zn, and Mn at 28.00 ± 0.52 g/kg, 15.91 ± 0.13 g/kg, and 30.27 ± 0.32 g/kg, respectively (n = 5). In contrast, copper, potassium, and calcium show lower levels at 6.85 ± 0.22 g/kg, 0.98 ± 0.11 g/kg, and 2.45 ± 0.44 g/kg, respectively. The elevated potassium content suggests its potential as a dietary addition to mitigate hypertension, stroke, cardiac disorders, kidney damage, and osteoporosis, particularly in conjunction with low-sodium diets [9]. In contrast, the levels of sodium, magnesium, zinc, and iron remain relatively low, underscoring the need for alternative sources of these minerals. A high potassium content is a valuable source for managing hypertension and other cardiovascular conditions [2].

While *M. balsamina* leaves are found to be low in vegetable lipids, aligning with the characteristic low-fat nature of leafy greens, the results substantiate its role in promoting health and preventing obesity [34].

Nutritional data on *M. balsamina* fruits reveal that they are among the most diverse and promising vegetable crops. They contain a range of essential minerals, with potassium (43.67 mg/100 g) and phosphorus (9.67 mg/100 g) being the most significant. Additionally, these fruits contain calcium (2.02 mg/100 g), sodium (0.53 mg/100 g), iron (1.24 mg/g), and zinc (0.38 mg/100 g). Consequently, incorporating *M. balsamina* fruits into the human diet is an intriguing possibility [36].

Table 4 delves into the amino acid composition of *M. balsamina* leaves, revealing a preponderance of non-essential amino acids, primarily glutamic acid and aspartic acid, comprising 12.38 and 8.21 mg/100 g of protein, respectively [34]. However, it is noteworthy that while these proteins, including balsamin, are rich in certain amino acids, their profile lacks balanced proportions of the eight essential amino acids, including isoleucine, leucine, lysine, methionine, phenylalanine, threonine, valine, and histidine.

In the final analysis, nutritional data reveal that *M. balsamina* offers a wealth of nutrients, making it an essential dietary component with the potential to enhance dietary diversity, support health management, and provide nutrient supplementation for diverse populations. However, it is important to note that its amino acid profile does not achieve an optimal balance of the eight essential amino acids. The nutritional composition of the *M. balsamina* may be variable, probably dependent on farming. This highlights the importance of incorporating this plant into a diet that includes other food sources to achieve a well-balanced nutritional intake.

Table 2. Nutritional composition of *M. balsamina* leaves.

Nutrients	Composition (g/kg) [9,34]	Recommended Daily Intake [37]
Protein	287.70 ± 1.80	1.05–0.85 g/kg body weight/day (Child) 0.80 g/kg body weight/day (Adult)
Fat	53.70 ± 8.60	
Crude fiber	37.20 ± 7.90	
Ash	127.00 ± 17.00	
Moisture	710.00 ± 0.90	
Available Carbohydrate	390.50 ± 2.00	
Calorific Value (Kcal/kg)	1892.20	

Age range: children (1 to 18); adults (19 to 70); mean ± standard deviation of three replicates.

Table 3. Mineral composition of *M. balsamina* leaves and fruits.

Minerals	Leaves (g/kg) [9]	Fruits (mg/100 g) [36]	Recommended Daily Intake [37]
Calcium	22.20 ± 0.50	2.02	210–800 mg/day (Child); 1000–1300 mg/day (Adult)
Magnesium	3.82 ± 0.06	-	80–410 mg/day (Child); 310–420 mg/day (Adult)
Phosphorus	3.24 ± 0.01	9.67	460–1250 mg/day (Child); 700 mg/day (Adult)
Potassium	27.05 ± 0.27	43.67	0.4–4.5 g/day (Child); 4.7 g/day (Adult)
Sodium	0.06 ± 0.02	0.53	1.5–2.3 g/day (Child); 2.3 g/day (Adult)
Iron	0.14 ± 0.01	1.24	7–15 mg/day (Child); 8–18 mg/day (Adult)
Manganese	0.15 ± 0.00	-	0.6–2.2 mg/day (Child); 1.8–2.3 mg/day (Adult)
Zinc	0.39 ± 0.01	0.38	3–8 mg/day (Child); 8–11 mg/day (Adult)

Age range: children (1 to 18); adults (19 to 70); mean ± standard deviation of three replicates.

Table 4. Amino acid composition of *M. balsamina* leaves.

Amino Acid	Concentration (g/100 g Protein) [34]	Recommended Daily Intake in mg/day Body Weight for +12 Years [38]
	Essential amino acids	
Isoleucine	2.94	10
Leucine	8.38	14
Lysine	3.94	12
Methionine	0.90	
Phenylalanine	3.94	
Threonine	3.13	7
Valine	4.11	10
Histidine	2.50	8–12

Table 4. *Cont.*

Amino Acid	Concentration (g/100 g Protein) [34]	Recommended Daily Intake in mg/day Body Weight for +12 Years [38]
	Non-essential Amino Acids	
Cysteine	0.56	
Total Sulfur	1.46	
Tyrosine	2.62	
Total aromatic	6.46	
Alanine	4.16	
Arginine	4.87	
Aspartic acid	8.21	
Glutamic acid	12.38	
Glycine	4.66	
Proline	3.21	
Serine	4.00	

5. Biological Properties of *Momordica balsamina*

M. balsamina leaves have been employed as a vegetable, consumed as a decoction, or incorporated into various recipes in conjunction with other plants [39]. The leaves, fruits, seeds, and bark of this plant offer substantial medicinal benefits, including anti-HIV, anti-plasmodial, anti-diarrheal, antiseptic, antibacterial, antiviral, anti-inflammatory, antimicrobial, hypoglycemic, antioxidant, analgesic, hepato-protective properties, and so many others (Figure 3) [3,40,41].

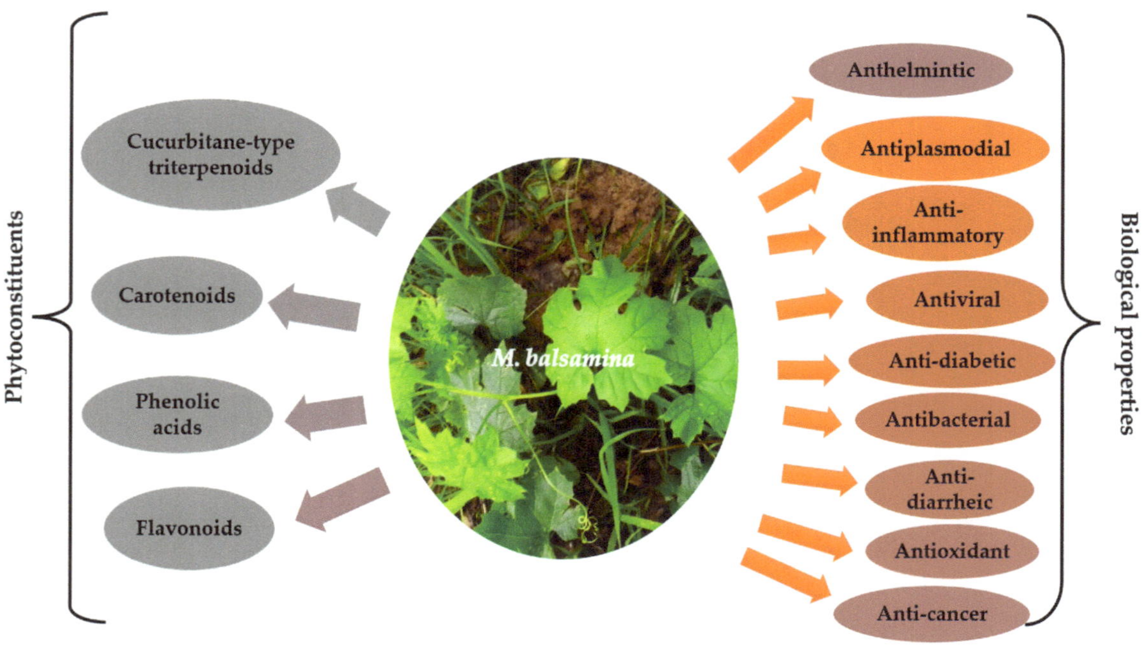

Figure 3. Biological properties and phytoconstituents of *Momordica balsamina*.

Indeed, the methanolic extract of the *M. balsamina* aerial part exhibits antiplasmodial activities that are highly promising with very low toxicity. This extract showed in vitro activity on three strains of *Plasmodium falciparum* (*P. falciparum*) (F32-Tanzania, FcB1-Colombia, and FcM29-Cameroon) and in vivo activity on mice infected with *Plasmodium vinckei*, which were treated either intraperitoneally or orally in a 4-day suppression test. This could potentially explain its widespread use in traditional malaria treatment in Niger [42]. In a study by Ramalhete et al. (2010) [20], the anti-malarial activity of new triterpenoids isolated from *M. balsamina* was demonstrated. These triterpenes and, notably, certain derivatives of karavilagenin exhibited substantial in vitro activity against *P. falciparum* strains 3D7 and Dd2, with low cytotoxicity. Furthermore, Clarkson et al. (2004) [43] demonstrated the in vitro activity of the aqueous extract of leaves and stems of *M. balsamina* against *P. falciparum D10*.

The antibacterial activity of *M. balsamina* was studied in vitro by Abdulhamid et al. (2023) [44], using methanol leaf extracts against *Staphylococcus aureus* and *Escherichia coli* over a 24 h period. They reported that methanol extracts from *M. balsamina* exhibited sensitivity (measured by the zone of inhibition) against the tested bacteria, suggesting that *M. balsamina* leaves could serve as a natural antibacterial alternative for infections caused by these bacteria. This discovery aligns with previous research by Otimenyin et al. (2008) [45] and Jigam et al. (2004) [46], who also demonstrated in vitro antibacterial activity against various bacteria, including *Pseudomonas aeruginosa*, *Salmonella typhi*, *Bacillus subtilis*, *Proteus mirabilis*, and *Klebsiella pneumoniae*, using extracts from *M. balsamina*. Additionally, a protein isolated from *M. balsamina* known as balsamin has shown in vitro antibacterial properties against bacteria such as *Staphylococcus aureus*, *Salmonella enterica*, *Staphylococcus epidermidis*, and *Escherichia coli*, suggesting its potential as a nutraceutical [24]. The ethyl acetate extract of the aerial parts of the *M. balsamina* plant also exhibited remarkable antibacterial activity in vitro against the *E. faecalis* strain [47]. Thus, crude extracts of *M. balsamina* possess significant potential as antimicrobial agents, potentially explaining their traditional use in treating diabetic wounds [45]. *M. balsamina* leaves could serve as a natural antibacterial alternative for infections caused by these bacteria and can be utilized in the treatment of infectious diseases.

The *M. balsamina* plant exhibits remarkable anti-HIV potential. In fact, the plant protein MoMo30 (30 kDa protein), isolated from *M. balsamina*, effectively inhibits HIV-1 in vitro at nanomolar levels while exhibiting minimal cellular toxicity at inhibitory concentrations [48]. This finding is in line with the work of Coleman et al. (2022) [49]. They found that the aqueous extract of *M. balsamina* leaves inhibits HeLa-CD4+-LTR-βgal cells with HIV-$1_{NL4\text{-}3}$ by more than 50% at concentrations of 0.02 mg/mL and above, with no toxicity in the 0–0.5 mg/mL range [49]. Other antiviral properties of the plant have been demonstrated. For instance, in vitro tests against the growth of Newcastle disease virus on a cell line of chicken embryo fibroblast (CEF) using aqueous extracts of fruit pulp and leaves have shown promising potential in the management of Newcastle disease [50]. Ampitan et al. (2023) [51] demonstrated the in vivo efficacy of *M. balsamina* extract against avian paramyxovirus-1 infection in broilers. These results indicate that the plant can prevent the virus from attaching to host cells, thus confirming its antiviral potential against this virus [50].

Khumalo et al. (2023) [52] conducted a study demonstrating that oral administration of methanolic leaf extracts of *M. balsamina* to pre-diabetic rats, induced by an experimental high-fat, high-carbohydrate diet, attenuated diabetes-associated renal function abnormalities. This treatment reduced damage and restored renal function. Similar research by Siboto et al. (2018) [53], using diabetic rats induced by streptozotocin (STZ), suggests that *M. balsamina* may have beneficial effects on processes associated with renal disorders in STZ-induced diabetic rats [52,53]. The in vivo study conducted by Sani et al. (2019) [54] on diabetic rats, induced by an injection of alloxan monohydrate, further confirms this property of the plant. Kgopa et al. (2020) [55] revealed that relatively non-polar extracts from *M. balsamina* fruits, such as hexane and ethyl acetate extracts, can increase glucose uptake by RIN-m5F β-cells In vitro. They argued that *M. balsamina* fruit extracts may exert their anti-diabetic effects not only via increased insulin sensitivity and inhibition of intestinal glucose absorption but also via the stimulation of insulin synthesis and secretion [55].

The results of research by A.T. Samaila's [56] revealed that *M. balsamina* leaf and root extracts have the potential to generate drugs and compounds with antifungal properties, effectively targeting pathogenic fungi.

Mabasa et al. (2012) [16] demonstrated in vitro anti-inflammatory activity in *M. balsamina* leaves on RAW 264.7 cell lines. They confirmed that *M. balsamina* leaves contain nontoxic secondary metabolites that may play a pivotal role in human health as anti-inflammatory agents [16]. Ndhlala et al. (2011) [57] showed the in vitro anti-inflammatory capacity of the aqueous extract of *M. balsamina* via the inhibition of cyclo-oxygenase 1 and cyclo-oxygenase 2.

Biological studies have indicated that *M. balsamina* possesses anticancer properties. In fact, a significant number of triterpenoids tested in vitro by Silva (2017) [58] demonstrated robust P-glycoprotein (P-gp) modulation activity with a Fluorescence Activity Report (FAR) > 10 and synergistic interactions with the antitumor drug doxorubicin. These findings support their potential as multidrug resistance (MDR) reversers.

This plant possesses significant antioxidant potential. The in vitro antiradical capacity, as demonstrated by Odhav et al. (2007) [59] on methanol extracts of its leaves, exhibits strong inhibition of free radicals.

Studies in vitro conducted by Okpara et al. (2017) [60] on the methanolic extract of *M. balsamina* fruit also demonstrated its potential effectiveness in managing diarrhea. The methanolic extract, when administered at moderate doses (400 mg/kg and 800 mg/kg), significantly reduced the gastrointestinal transit of activated charcoal in mice. Furthermore, oral administration of the extract (at doses of 200 mg/kg, 400 mg/kg, and 800 mg/kg) provided dose-dependent protection to Wistar rats against castor oil-induced diarrhea (2 mL/rat) and significantly reduced castor oil-induced enteropooling in Wistar rats.

The aqueous leaf extract of *M. balsamina* demonstrated a promising effect on acetic acid-induced twisting/stretching in Wistar rats in a dose-dependent manner. This extract was found to have significant antinociceptive (analgesic) effects, with a reduction of up to 71.3% observed at a dose of 400 mg/kg body weight [61].

M. balsamina is a plant with significant anti-malarial potential. Triterpenoids, isolated from *M. balsamina*, possess a broad range of potent biological activities, including hepatoprotective, anti-malarial, anti-inflammatory, cardiovascular, anti-diabetic, and antiparasitic effects [20].

Continued investigations and exploration of the bioactive compounds within *M. balsamina* hold immense potential for the development of novel drugs and therapeutic approaches across various medical disciplines. The identification and isolation of its bioactive compounds have opened new avenues for drug development and therapeutic interventions spanning diverse medical fields. The wide-ranging pharmacological properties of *M. balsamina*, substantiated by scientific research, not only underscore its significance in traditional medicine but also illuminate its potential in modern pharmaceutical research. This plant's diverse bioactive compounds have demonstrated their ability to address various health conditions, making it a valuable resource for both traditional and modern medicine. However, it should be noted that some biological studies have limitations either due to the high concentrations required to produce a biological effect or the low content of specific active ingredients in the plant. This is the case with balsamin, a bioactive protein found in the seeds of the plant but in low concentrations.

6. Traditional Uses of *Momordica balsamina*

M. balsamina is an intriguing plant renowned for its potent medicinal properties. It finds application in treating a variety of conditions, such as hypertension, digestive disorders, malaria, diabetes, and measles prevention [4].

Furthermore, *M. balsamina* is recommended for addressing dermatological ailments like pimples, skin spots, and edema. It serves as a remedy for wounds, eczema, allergies, and skin infections while also functioning as a skin moisturizer. Its effectiveness in addressing dermatological issues may be associated with its scientifically proven anti-inflammatory and antibacterial properties.

The plant assumes the role of an anthelmintic when employing its fruits, seeds, and leaves. Leaves are used to combat fever and excessive uterine bleeding, while the whole plant is utilized to manage syphilis, rheumatism, hepatitis, and diverse skin conditions. It has applications as an abortifacient, aphrodisiac, galactogen, and diabetic treatment [4,62]. *M. balsamina* is used for its anti-diabetic properties. Indeed, diabetic patients consume leaf decoctions orally to regulate blood sugar levels. The plant is employed in various forms, such as infusions, poultices, and herbal teas, to address issues like liver problems, stomach cramps, and ulcers [4].

The utilization of *M. balsamina* in traditional medicine is both diverse and extensive, spanning various countries and cultures. This plant has found its way into the pharmacopeias of different regions, where its different parts are harnessed to address a multitude of health concerns. Below is an overview of its traditional medicinal applications (refer to Figure 4).

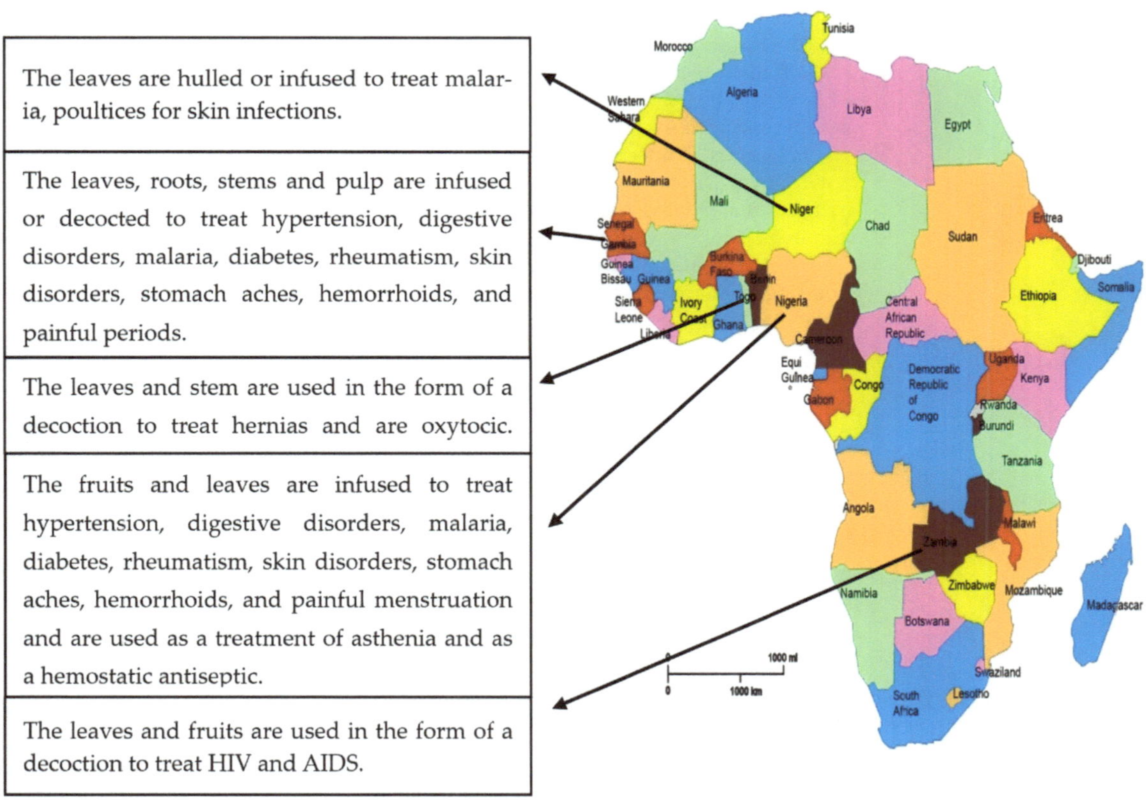

Figure 4. The utilization of *M. balsamina* in traditional medicine in Africa [2,3,42,63]. Data presented in the figure refer to traditional uses of the plants based on ethnobotanical studies.

In Senegal, it is employed for its anti-inflammatory, anthelminthic, and antibacterial properties for treating dermatosis, stomachaches, rheumatism, hemorrhoids, painful menstruation, and other ailments. An aqueous extract of its leaves alleviates menstrual discomfort in young girls. It also finds anti-malarial and anti-diabetic properties, often in the forms of decoction or infusion. The Wolof people (Senegal) utilize the fruits for purgative and deworming purposes [2,3]. This use highlights its anthelmintic properties.

In Nigeria, *M. balsamina* addresses asthenia and digestive issues, while the Fulani people use it as a deworming agent and tranquilizer [63]. It is even integrated into prescriptions for mental health concerns. Additionally, the maceration of the entire plant acts as a galactagogue and is employed for chest massages to alleviate intercostal pain [3].

In Togo, a decoction of *M. balsamina*'s aerial parts, along with honey and *Zanthoxyloides fagara* root bark, functions as an oxytocic agent. Togolese people frequently utilize the leaves and stems. Oral administration of calcined aerial parts powder is employed to treat hernias [63].

In different regions, the plant's fruits and leaves are used as hemostatic antiseptics for wound treatment. In Zambia, the use of *M. balsamina* is associated with its antiviral potential. Indeed, a decoction of the entire plant is used to treat conditions like syphilis, human immunodeficiency virus (HIV), and acquired immunodeficiency syndrome (AIDS) [3].

In Niger, crushed *M. balsamina* leaves are used as poultices for skin infections. Pounded leaves are recommended for hemorrhoids, diabetes, and fever. In combination with *Combretum micranthum*, it is utilized to treat malaria. Some women mix it with henna for abortion [63].

The Hausa people (Mali) use the whole plant for embalming, drinking, and bathing to ward off malevolent spirits [63].

In Syria, an infusion of fruit or leaf powder is used as an antiseptic and for treating asthenia and hemostasis [3].

In Indonesia, bitter melon is not only known as a vegetable but also traditionally used as a laxative, a fever remedy, and an appetite stimulant. Additionally, bitter melon leaves are used to alleviate menstrual discomfort, heal burns, treat skin diseases, and act as a vermifuge [64].

The traditional medicinal applications of *M. balsamina* are both diverse and extensive, reflecting its adaptability and effectiveness in addressing a multitude of health concerns. These uses have been passed down through generations and are integrated into the traditional healing practices of various countries and cultures. *M. balsamina* has found a place in the pharmacopeias of different regions, where its different parts are harnessed to improve the well-being of individuals and communities. Its role as a natural remedy for a wide range of ailments highlights its importance in traditional medicine.

In the realm of traditional medicine, *M. balsamina* has long held a revered position for its therapeutic effects on a multitude of ailments. Scientific studies have now provided empirical evidence to validate these traditional uses, further emphasizing their importance in preserving and promoting traditional healing practices. Moreover, in the context of modern pharmaceutical research, *M. balsamina* emerges as a promising candidate. Its potential as an anti-diabetic agent, as well as its antimicrobial, anticancer, HIV-inhibiting, anti-inflammatory, anti-malarial properties, and many others, showcase the plant's phytoconstituents offering a wealth of opportunities for drug discovery and innovation. This interdisciplinary approach, bridging the wisdom of traditional medicine with the rigor of scientific research, paves the way for the advancement of healthcare and the enhancement of global well-being. It is essential to recognize that many traditional uses of *M. balsamina* rely on its biological or pharmacological properties. Nevertheless, it should be noted that the effectiveness of *M. balsamina* for certain health issues in traditional uses still requires further investigation regarding its biological or pharmacological properties.

7. Cosmetic Use of *Momordica balsamina*

In the realm of cosmetics, *M. balsamina* finds extensive application in addressing dermatological concerns. Notably, the juice extracted from the leaves of *M. balsamina* has been employed as a natural soap [62]. This unique formulation, often referred to as "pimples natural soap," harnesses the properties of *M. balsamina* to combat pimples, skin spots, and edema. When mixed with water, it produces a mildly soapy solution. The plant's soothing attributes extend to the treatment of wounds, eczema, pimples, allergies, and skin infections. Additionally, *M. balsamina* serves as a highly effective skin moisturizer [10,62].

The utilization of *M. balsamina* for dermatological issues can indeed be linked to its demonstrated anti-inflammatory [16] and antibacterial [44–46] properties, as supported by biological studies. These properties make the plant a valuable natural remedy for skin-related problems. Incorporating *M. balsamina* into dermatological treatments or skincare routines holds promise as a natural and holistic approach to addressing skin issues.

The use of *M. balsamina* in natural soap formulations reflects its role in promoting skin health and well-being. However, there have been limited studies conducted on its cosmetic applications. Therefore, it would be intriguing to conduct further research into the cosmetic uses of *M. balsamina*. Indeed, the cosmetic potential of *M. balsamina* is an intriguing area that warrants further exploration via research. While some traditional uses of *M. balsamina* in skincare and cosmetics have been documented, there is still much to discover about its specific cosmetic applications and the potential benefits it can offer in modern skincare and beauty products.

8. Other Uses of *Momordica balsamina*

The versatility of the *M. balsamina* plant transcends beyond therapy and encompasses various other domains. The whole plant extract demonstrates insecticidal properties, making it a valuable tool in pest control. The seeds of the plant are often utilized as arrow poison. The plant contributes to mitigating micronutrient deficiencies within the soil, aiding in soil health and fertility enhancement [3]. Extracts from the leaves of *M. balsamina* serve as effective metal cleaners [62]. Additionally, the ethanolic extract of *M. balsamina* leaves has been studied for its potential to safeguard copper from corrosion within acidic environments [19]. In the Hausa regions of Nigeria and the Niger Republic, the leaves are incorporated into green vegetable soups consumed by nursing mothers. It is believed that these soups aid in post-labor blood regeneration and the purification of breast milk [64]. Among farmers, the plant is harnessed to augment milk production in cows [63]. The multifaceted applications of *M. balsamina* highlight its valuable contributions across diverse sectors, ranging from cosmetics and therapy to agriculture and beyond.

Its anthelmintic properties enable it to combat gastrointestinal disorders (diarrhea) in animals [65]. Breeders in Benin use *M. balsamina* leaves and stems in combination with potash in the form of maceration to combat diarrhea in animals [66].

The multifaceted applications of *M. balsamina* highlight its valuable contributions across diverse sectors beyond traditional medicine. They underscore the importance of understanding and preserving traditional knowledge of the plant's uses in various communities.

9. Adverse Effects of *Momordica balsamina*

Toxicity and cytotoxicity studies were conducted on crude extracts of *M. balsamina*. Oral administration of the methanolic extract of *M. balsamina* to albino mice for one week revealed that doses of 30 and 40 mg/kg do not induce any toxicity. However, when the dosage ranges from 50 mg/kg to 150 mg/kg, instances of mouse fatalities are observed due to the toxicity of the methanolic extract, with an average death rate of 20% in most cases. An intoxication syndrome becomes evident in mice starting from 50 mg/kg, followed by occasional cases of death, with the highest average death rate reaching 40% [63]. In contrast, the larval cytotoxicity test conducted by Dehou et al. [16] using the *Artemia salina* model indicates that the hydro-ethanoic and dichlomethanic extracts do not exhibit toxicity, with LC_{50} values of 2.25 mg/mL and 5.20 g/mL, respectively.

Mabasa et al. (2021) [16] have demonstrated that *M. balsamina* leaves have no cytotoxic activity against human colorectal adenocarcinoma cell lines HT29 and Caco2.

These findings suggest that while certain extracts of *M. balsamina* may exhibit toxicity at high doses, others, such as the hydro-ethanoic and dichlomethanic extracts, appear to be safer. The specific components and concentrations of the extracts play a crucial role in determining their toxicity levels. The toxicity and cytotoxicity studies on crude extracts of

M. balsamina provide important insights into the safety and potential risks associated with the use of these extracts.

Furthermore, some compounds isolated from *M. balsamina*, including balsaminaepoxide, balsaminatriol, balsaminal, balsaminol G, karavilagenin A, karavilagenin B, karavilagenin C, and kuguacin B, have demonstrated interactions with the anti-cancer drug doxorubicin, as reported by Ramalhete et al., (2016) [67]. In fact, the interaction of these non-cytotoxic compounds with doxorubicin was assessed in vitro in a combination chemotherapy model using ABCB1-transfected mouse T lymphoma cells (L5178Y-MDR). These compounds displayed a synergistic interaction with the anti-cancer drug doxorubicin [58,67].

These findings highlight the complexity of *M. balsamina*'s effects on different models and cell lines. While some extracts and compounds may exhibit toxicity at high doses, others appear to be safe. The specific components and concentrations of the extracts play a significant role in determining their toxicity levels. Additionally, the interactions with anti-cancer drugs suggest potential applications in combination chemotherapy. These studies emphasize the importance of careful dosage and safety assessments when using *M. balsamina* for various purposes.

10. Conclusions and Future Directions

In summary, this review underscores the extensive utilization of *M. balsamina* within traditional African medicine. The breadth of its traditional applications finds validation in its diverse array of biological properties and intricate chemical compositions. Consequently, we advocate for the incorporation of *M. balsamina* in phytotherapeutic practices. It is imperative, however, to acknowledge the potential toxicity associated with many plants employed in traditional medicine, and therefore, meticulous attention must be dedicated to ensuring proper usage, adherence to prescribed doses, and mitigation of potential toxic effects. Simultaneously, it becomes essential to safeguard a sustainable and ample reservoir of active compounds, thereby averting the depletion of vital natural resources. In this context, comprehensive investigations into the innovative methodologies and strategies for the synthesis or biosynthesis of these naturally occurring bioactive constituents. However, it is important to acknowledge that the development of new drugs, whether derived from natural sources like *M. balsamina* or synthetic compounds, is a complex and lengthy process. The transition from laboratory and preclinical studies to clinical trials is a critical step in drug development. Clinical studies are necessary to evaluate the safety and efficacy of *M. balsamina*-based treatments in humans. Considering that most biological tests are conducted on crude extracts of *M. balsamina* or even on isolated compounds in experimental studies, it is now crucial to progress toward clinical studies for the development of new drugs.

Overall, *M. balsamina*'s nutritional richness suggests a pivotal role in enhancing dietary diversity, managing health, and supplementing nutrients across populations. *M. balsamina*, rich in vital micronutrients and macronutrients, holds significance in African cuisine as a versatile plant. Its leaves, a valuable green vegetable, offer high protein content, making it a potential protein supplement, particularly for resource-scarce communities. With considerable dietary fiber and modest caloric value, it aids nutritional intake and suits calorie management.

Furthermore, given its prevalent application in addressing dermatological concerns, particularly in Senegal, there exists a compelling rationale to embark on comprehensive studies exploring its potential within the cosmetic sphere. As the intricate interplay between tradition, science, and sustainability unfolds, embracing the multifaceted potentials of *M. balsamina* beckons us toward a realm of holistic health and well-being.

Author Contributions: Conceptualization, M.T., I.S. and M.-L.F.; Methodology, M.T.; Investigation, I.S. and M.-L.F.; Writing and original draft preparation, M.T.; Writing, Review and Editing, I.S., M.-L.F., M.L.G. and M.G.; Project Administration, I.S. and M.-L.F. All authors have read and agreed to the published version of the manuscript.

Funding: Marème Thiaw thanks the Academy of Research and Higher Education (ARES) for the B-MOB mobility grant as part of the 2022–2027 five-year development cooperation program 2022-MAA-522; and the University of Liege for the impulse grant.

Institutional Review Board Statement: Not applicable.

Informed Consent Statement: Not applicable.

Data Availability Statement: Not applicable.

Acknowledgments: We would like to thank the Center for Partnership and Development Cooperation (PACODEL), the Academy of Research and Higher Education (ARES), and the University of Liege (ULiege).

Conflicts of Interest: The authors declare no conflict of interest.

References

1. Rikisahedew, J.J.; Naidoo, Y. Phyto-histochemical evaluation and characterization of the foliar structures of *Tagetes minuta* L. (Asteraceae). *S. Afr. J. Bot.* **2018**, *115*, 328–329. [CrossRef]
2. Balde, S.; Ayessou, N.C.; Gueye, M.; Ndiaye, B.; Sow, A.; Cissé, O.I.K.; Cissé, M.; Diop, C.G. Ethnobotanical investigations of *Momordica charantia* Linn (Cucurbitaceae) in Senegal. *Int. J. Biol. Chem. Sci.* **2019**, *13*, 466–474. [CrossRef]
3. Thakur, G.; Bag, M.; Sanodiya, B.S.; Bhadauriya, P.; Debnath, M.; Prasad, G.B.; Bisen, P.S. *Momordica balsamina*: A medicinal and neutraceutical plant for health care management. *Curr. Pharm. Biotechnol.* **2009**, *10*, 667–682. [CrossRef]
4. Omokhua-Uyi, A.G.; Staden, J.V. Phytomedicinal relevance of South African Cucurbitaceae species and their safety assessment: A review. *J. Ethnopharmacol.* **2020**, *259*, 112967. [CrossRef]
5. Paris, H.S.; Tadmor, Y.; Schaffer, A.A. Cucurbitaceae, melons, squash, cucumber. In *Encyclopedia of Applied Plant Sciences*, 2nd ed.; Thomas, B., Murray, B.G., Murphy, D.J., Eds.; Academic Press: Oxford, UK, 2017; Volume 3, pp. 209–217. [CrossRef]
6. Bouquet, A. Plantes médicinales de la côte d'ivoire. *Off. De La Rech. Sci. Et Tech. D'outre-Mer (ORSTOM) Paris* **1974**, *165*, 38–39. Available online: https://cir.nii.ac.jp/crid/1571135650006923392 (accessed on 1 September 2023).
7. Ramalhete, C.; Mansoor, T.A.; Mulhovo, S.; Molnár, J.; Ferreira, M.J. Cucurbitane-type triterpenoids from the African plant *Momordica balsamina*. *J. Nat. Prod.* **2009**, *72*, 2009–2013. [CrossRef] [PubMed]
8. Schaefer, H.; Renner, S.S. A three-genome phylogeny of *Momordica* (Cucurbitaceae) suggests seven returns from dioecy to monoecy and recent long-distance dispersal to Asia. *Mol. Phylogenet. Evol.* **2010**, *54*, 553–560. [CrossRef]
9. Flyman, M.V.; Afolayan, A.J. Proximate and mineral composition of the leaves of *Momordica balsamina* L.: An under-utilized wild vegetable in Botswana. *Int. J. Food Sci. Nutr.* **2007**, *58*, 419–423. [CrossRef]
10. Duenas-Lopez, M.A. *Momordica balsamina (Common Balsam Apple)*; Invasive Species Compendium (CABI Compendium): Wallingford, UK, 2019; p. 34677. [CrossRef]
11. Kaur, I.; Yadav, S.K.; Hariprasad, G.; Gupta, R.C.; Srinivasan, A.; Batra, J.K.; Puri, M. Balsamin, a novel ribosome-inactivating protein from the seeds of Balsam apple *Momordica balsamina*. *Amino Acids* **2012**, *43*, 973–981. [CrossRef] [PubMed]
12. Dehou, R.; Dougnon, V.; Atchade, p.; Attakpa, C.; Legba, B.; Baba-Moussa, L.; Gbaguidi, F. Phytochemical and toxicological study of *Cassia italica*, *Momordica balsamina* and *Ocimum gratissimum*, three plants used against scabies in southern Benin. *Rev. Ivoir. Sci. Technol.* **2018**, *32*, 286–297.
13. Yolidjé, I.; Keita, D.A.; Moussa, I.; Toumane, A.; Maarouhi, I.M.; Saley, K.; Jean-Luc, P.; Much, O.J.T. Phytochemical Characterization and Larvicidal Activity of Crude Plant Extracts from the Traditional Pharmacopoeia of Niger on Larvae of *Anopheles gambiae*. *Eur. Sci. J.* **2019**, *15*, 30–51. [CrossRef]
14. Mu'azu, A.; Bello, M. Reproductive performance of rabbit feed with graded of balsam apple (*Momordica balsamina*). *J. Agric. Sci. Technol.* **2022**, *9*, 17–26.
15. Bot, Y.S.; Mgbojikwe, L.O.; Nwosu, C.; Abimiku, A.; Dadik, J.; Damshak, D. Screening of the fruit pulp extract of *Momordica balsamina* for anti-HIV property. *Afr. J. Biotechnol.* **2007**, *6*, 047–052.
16. Mabasa, X.E.; Mathomu, L.M.; Madala, N.E.; Musie, E.M.; Sigidi, M.T. Molecular spectroscopic (FTIR and UV-Vis) and hyphenated chromatographic (UHPLC-qTOF-MS) analysis and in vitro bioactivities of the *Momordica balsamina* leaf extract. *Biochem. Res. Int.* **2021**, *2021*, 2854217. [CrossRef]
17. Mashiane, P.; Shoko, T.; Manhivi, V.; Slabbert, R.; Sultanbawa, Y.; Sivakumar, D. A Comparison of Bioactive Metabolites, Antinutrients, and Bioactivities of African Pumpkin Leaves (*Momordica balsamina* L.) Cooked by Different Culinary Techniques. *Molecules* **2022**, *27*, 1901. [CrossRef]
18. Ramalhete, C.; Gonçalves, B.M.; Barbosa, F.; Duarte, N.; Ferreira, M.J. *Momordica balsamina*: Phytochemistry and pharmacological potential of a gifted species. *Phytochem. Rev.* **2022**, *21*, 617–646. [CrossRef]
19. Maibulangu, B.M.; Ibrahim, M.B.; Akinola, L.K. Inhibitory effect of African Pumpkin (*Momordica balsamina* Linn.) leaf extract on copper corrosion in acidic media. *J. Appl. Sci. Environ. Manag.* **2017**, *21*, 1067–1071. [CrossRef]
20. Ramalhete, C.; Lopes, D.; Mulhovo, S.; Molnár, J.; Rosário, V.E.; Ferreira, M.U. New antimalarials with a triterpenic scaffold from *Momordica balsamina*. *Bioorg. Med. Chem.* **2010**, *18*, 5254–5260. [CrossRef] [PubMed]

21. Ramalhete, C.; Lopes, D.; Molnár, J.; Mulhovo, S.; Rosário, V.E.; Ferreira, M.U. Karavilagenin C derivatives as antimalarials. *Bioorg. Med. Chem.* **2011**, *19*, 330–338. [CrossRef]
22. Ajji, P.K.; Sonkar, S.P.; Walder, K.; Puri, M. Purification, and functional characterization of recombinant balsamin, a ribosome-inactivating protein from *Momordica balsamina*. *Int. J. Biol. Macromol.* **2018**, *114*, 226–234. [CrossRef] [PubMed]
23. Kaur, I.; Puri, M.; Ahmed, Z.; Blanchet, F.P.; Mangeat, B.; Piguet, V. Inhibition of HIV-1 Replication by Balsamin, a Ribosome Inactivating Protein of *Momordica balsamina*. *PLoS ONE* **2013**, *8*, e73780. [CrossRef] [PubMed]
24. Ajji, P.K.; Walder, K.; Puri, M. Functional Analysis of a Type-I Ribosome Inactivating Protein Balsamin from *Momordica balsamina* with Anti-Microbial and DNase Activity. *Plant Foods Hum. Nutr.* **2016**, *71*, 265–271. [CrossRef]
25. Rho, H.S.; Ghimeray, A.K.; Yoo, D.S.; Ahn, S.M.; Kwon, S.S.; Lee, K.H.; Cho, D.H.; Cho, J.Y. Kaempferol and Kaempferol Rhamnosides with Depigmenting and Anti-Inflammatory Properties. *Molecules* **2011**, *16*, 3338–3344. [CrossRef] [PubMed]
26. Zhu, L.; Xue, L. Kaempferol Suppresses Proliferation and Induces Cell Cycle Arrest, Apoptosis, and DNA Damage in Breast Cancer Cells. *Oncol. Res.* **2019**, *27*, 629–634. [CrossRef] [PubMed]
27. Cai, Q.; Rahn, R.O.; Zhang, R. Dietary flavonoids, quercetin, luteolin and genistein, reduce oxidative DNA damage and lipid peroxidation and quench free radicals. *Cancer Lett.* **1997**, *119*, 99–107. [CrossRef]
28. Liu, S.; Tian, L.; Chai, G.; Wen, B.; Wang, B. Targeting heme oxygenase-1 by quercetin ameliorates alcohol-induced acute liver injury via inhibiting NLRP3 inflammasome activation. *Food Funct.* **2018**, *9*, 4184–4193. [CrossRef]
29. Chi, G.; Zhong, W.; Liu, Y.; Lu, G.; Lu, H.; Wang, D.; Sun, F. Isorhamnetin protects mice from lipopolysaccharide-induced acute lung injury via the inhibition of inflammatory responses. *Inflamm. Res.* **2016**, *65*, 33–41. [CrossRef]
30. Jang, S.A.; Park, D.W.; Kwon, J.E.; Song, H.S.; Park, B.; Jeon, H.; Sohn, E.H.; Koo, H.K.; Kang, S.C. Quinic acid inhibits vascular inflammation in TNF-α-stimulated vascular smooth muscle cells. *Biomed. Pharmacother.* **2017**, *96*, 563–571. [CrossRef]
31. Wang, G.F.; Shi, L.P.; Ren, Y.D.; Liu, Q.F.; Liu, H.F.; Zhang, R.J.; Li, Z.; Zhu, F.H.; He, P.L.; Tang, W.; et al. Anti-hepatitis B virus activity of chlorogenic acid, quinic acid and caffeic acid in vivo and in vitro. *Antivir. Res.* **2009**, *83*, 186–190. [CrossRef] [PubMed]
32. Ramalhete, C.; Spengler, G.; Martins, A.; Martins, M.; Viveiros, M.; Mulhovo, S.; Ferreira, M.J.U.; Amaral, L. Inhibition of efflux pumps in meticillin-resistant Staphylococcus aureus and Enterococcus faecalis resistant strains by triterpenoids from *Momordica balsamina*. *Int. J. Antimicrob. Agents* **2011**, *37*, 70–74. [CrossRef]
33. Avato, P.; Argentieri, M.P. Editorial to the special issue: "Phytochemicals in nutrition and health: Advances and challenges". *Phytochem. Rev.* **2022**, *21*, 313–316. [CrossRef]
34. Hassan, L.G.; Umar, K.J. Nutritional value of Balsam Apple (*Momordica balsamina* L.) leaves. *Pak. J. Nutr.* **2006**, *5*, 522–529. [CrossRef]
35. Karumi, Y.; Onyeyili, P.A.; Ogugbuaja, V.O. Identification of active principles of *M. balsamina* (Balsam Apple) leaf extract. *J. Med. Sci.* **2004**, *4*, 179–182.
36. Chinthan, K.N.; Manjunathagowda, D.C.; Rathod, V.; Devan, S.R.; Anjanappa, M. Southern balsam pear (*Momordica balsamina* L.): Characterization of underutilized cucurbitaceous vegetable. *S. Afr. J. Bot.* **2022**, *145*, 95–98. [CrossRef]
37. Institute of Medicine. *Dietary Reference Intakes: The Essential Guide to Nutrient Requirements*; Meyers, L.D., Hellwig, J.P., Otten, J.J., Eds.; The National Academies Press: Washington, DC, USA, 2006. [CrossRef]
38. WHO-World Health Organization. Energy and Protein Requirements: Report of a Joint FAO/WHO/UNU Expert Consultation [Held in Rome from 5 to 17 October 1981]. Available online: https://apps.who.int/iris/handle/10665/62734 (accessed on 10 August 2023).
39. De Wet, H.; Ramulondi, M.; Ngcobo, Z.N. The use of indigenous medicine for the treatment of hypertension by a rural community in northern Maputaland, South Africa. *S. Afr. J. Bot.* **2016**, *103*, 78–88. [CrossRef]
40. Bhardwaj, N.; Gauttam, V.; Kalia, A.N. Evaluation of antidiabetic activity of *Momordica balsamina* Linn seeds in experimentally-induced diabetes. *J. Chem. Pharm. Res.* **2010**, *2*, 701–707.
41. Souda, S.; George, S.; Mannathoko, N.; Goercke, I.; Chabaesele, K. Antioxidant and Antibacterial Activity of Methanol Extract of *Momordica balsamina*. *IRA-Int. J. Appl. Sci* **2018**, *10*, 7–17. [CrossRef]
42. Benoit-Vical, F.; Grellier, P.; Abdoulaye, A.; Moussa, I.; Ousmane, A.; Berry, A.; Ikhiri, K.; Poupat, C. In vitro and in vivo Antiplasmodial Activity of *Momordica balsamina* Alone or in a Traditional Mixture. *Chimiothérapie* **2006**, *52*, 288–292. [CrossRef]
43. Clarkson, C.; Maharaj, V.J.; Crouch, N.R.; Grace, O.M.; Pillay, P.; Matsabisa, M.G.; Bhagwandin, N.; Smith, P.J.; Folb, P.I. In vitro antiplasmodial activity of medicinal plants native to or naturalised in South Africa. *J. Ethnopharmacol.* **2004**, *92*, 177–191. [CrossRef]
44. Abdulhamid, A.; Jega, S.A.; Sani, I.; Bagudo, A.I.; Abubakar, R. Phytochemical and antibacterial activity of *Momordica balsamina* leaves crude extract and fractions. *Drug Discov.* **2023**, *17*, e11dd1012. [CrossRef]
45. Otimenyin, S.O.; Uguru, M.O.; Ogbonna, A. Antimicrobial and hypoglycemic effects of *Momordica balsamina*. Linn. *J. Nat. Prod.* **2008**, *1*, 3–9. Available online: http://hdl.handle.net/123456789/3130 (accessed on 1 September 2023).
46. Jigam, A.A.; Akanya, H.O.; Adeyemi, D.J. Antimicrobial and antiplasmodial effects of *Momordica balsamina*. *Nig. J. Nat. Prod. Et Med.* **2004**, *8*, 11–12. [CrossRef]
47. Madureira, A.M.; Ramalhete, C.; Mulhovo, S.; Duarte, A.; Ferreira, M.J.U. Antibacterial activity of some African medicinal plants used traditionally against infectious diseases. *Pharm. Biol.* **2012**, *50*, 481–489. [CrossRef]
48. Khan, M.; Diop, A.; Gbodossou, E.; Xiao, P.; Coleman, M.; De Barros, K.; Duong, H.; Bond, V.C.; Floyd, V.; Kondwani, K.; et al. Anti-human immunodeficiency virus-1 activity of MoMo30 protein isolated from the traditional African medicinal plant *Momordica balsamina*. *J. Virol.* **2023**, *20*, 50. [CrossRef]

49. Coleman, M.I.; Khan, M.; Gbodossou, E.; Diop, A.; DeBarros, K.; Duong, H.; Bond, V.C.; Floyd, V.; Kondwani, K.; Montgomery, R.V.; et al. Identification of a Novel Anti-HIV-1 Protein from *Momordica balsamina* Leaf Extract. *Int. J. Environ. Res. Public Health* **2022**, *19*, 15227. [CrossRef]
50. Chollom, S.C.; Olawuyi, A.K.; Danjuma, L.D.; Nanbol, L.D.; Makinde, I.O.; Hashimu, G.A.; Alesa, M.U.; Esilonu, J.T.; Ogundeji, E.B.; Kwatjel, J.S. Antiviral Potential of Aqueous Extracts of Some Parts of *Momordica balsamina* plant against Newcastle Disease Virus. *J. Adv. Pharm. Educ. Res.* **2012**, 3, 82–92.
51. Ampitan, T.; Garba, M.H.; Adekambi, D.I.; Ampitan, A.; Adelakun, K. Prophylactic potency of methanolic extract of *Momordica balsamina* L. against avian paramyxovirus-1 infection in broiler chickens. *Egypt. J. Anim. Prod.* **2023**, *60*, 25–32. [CrossRef]
52. Khumalo, B.; Siboto, A.; Akinnuga, A.M.; Sibiya, N.; Khathi, A.; Ngubane, P.S. Effects of *Momordica balsamina* on markers associated with renal dysfunction in a diet-induced prediabetic rat model. *Res. Squar.* **2023**, 1–20. [CrossRef]
53. Siboto, A.; Sibiya, N.; Khathi, A.; Ngubane, P. The effects of *Momordica balsamina* methanolic extract on kidney function in STZ-induced diabetic rats: Effects on selected metabolic markers. *J. Diabetes Res.* **2018**, *2018*, 7341242. [CrossRef]
54. Sani, A.A.; Idris, S.B.; Yakubu, S. Anti-Diabetic Activity of Methanolic Extract of *Momordica Balsamina* in Rabbits. *Sci. Res. J.* **2019**, *7*, 1–6. [CrossRef]
55. Kgopa, A.H.; Shai, L.J.; Mogale, M.A. *Momordica balsamina* fruit extracts enhance selected aspects of the insulin synthesis/secretion pathway. *Am. J. Biochem. Biotechnol.* **2020**, *16*, 549–560. [CrossRef]
56. Samaila, A.T. Evaluation of Antifungal Efficacy of *Lawsonia inermis*, *Momordica balsamina* and *Ziziphus mauritiana* against Clinically Isolated Candida Species (*Candida albicans, C. krusei, C. tropicalis,* and *C. parapsilosis*). In Proceedings of the BOUESTI International Conference on Education, Science and Technology (ICEST), Ikere-Ekiti, Nigeria, 21–24 November 2022.
57. Ndhlala, A.R.; Finnie, J.F.; Van Staden, J. Plant composition, pharmacological properties and mutagenic evaluation of a commercial Zulu herbal mixture: Imbiza ephuzwato. *J. Ethnopharmacol.* **2011**, *133*, 663–674. [CrossRef]
58. Silva, C.D.; Ramalhete, C.; Spengler, G.; Mulhovo, S.; Molnar, J.; Ferreira, M.J.U. Triterpenes from *Momordica balsamina* (African pumpkin): ABCB1 inhibition and synergistic interaction with doxorubicin in resistant cancer cells. *Phytochemistry* **2022**, *203*, 113354. [CrossRef]
59. Odhav, B.; Beekrum, S.; Akula, U.S.; Baijnath, H. Preliminary assessment of nutritional value of traditional leafy vegetables in KwaZulu-Natal, South Africa. *J. Food Compost. Anal.* **2007**, *20*, 430–435. [CrossRef]
60. Okpara, J.O.; Salihu, A.A.; Akanni, K.N.; Ogbu, K.I.; Ochai, C.Z. Antidiarrhoel properties of the methanolic fruit extract of *Momordica balsamina* (Linneaus). *Int. J. Sci. Appl. Res.* **2017**, *2*, 17–26.
61. Karumi, Y.; Onyeyili, P.; Ogugbuaja, V. Anti-inflammatory and Antinociceptive (Analgesic) Properties of *Momordical balsamina* Linn. (Balsam Apple) Leaves in Rats. *Pak. J. Biol. Sci.* **2003**, *6*, 1515–1518. [CrossRef]
62. Bosch, C.H. *Momordica balsamina* L., Ressources Végétales De L'Afrique Tropicale. Available online: https://plantnet.org/en/ (accessed on 12 July 2023).
63. Mamadou, A.O. Étude Phytochimique Et De L'activité Antipaludique In Vivo Et In Vitro De *Momordica balsamina* Linn. (Cucurbitaceae). Ph.D. Thesis, FMPOS Bamako, Bamako, Mali, 2005.
64. Putri, H. Determination of Total Flavonoid Content of Ethanol Extract of Forest Bitter Melon (*Momordica balsamina*) Leaves by Uv-Vis Spectrophotometric Method. Ph.D. Thesis, Universitas Muhammadiyah Mataram, Kota Mataram, Indonesia, 2023.
65. Mensah, É.J.P.A.; Kindomihou, V.; Moussa, H.; Houndonougbo, F.; Vodouhè, D.S.; Aiyelaagbe, I.; Sinsin, B. Organic production potential of traditional small ruminant farming systems: A synthesis. *J. Anim. Plant Sci.* **2022**, *52*, 9404–9451.
66. Noudèkè, N.D.; Dotché, I.; Ahounou, G.S.; Karim, I.Y.A.; Farougou, S. Inventory of medicinal plants used in the treatment of diseases that limit milk production of cow in Benin. *J. Adv. Vet. Anim. Res.* **2017**, *4*, 1–14. [CrossRef]
67. Ramalhete, C.; Mulhovo, S.; Molnar, J.; Ferreira, M.J.U. Triterpenoids from *Momordica balsamina*: Reversal of ABCB1-mediated multidrug resistance. *Bioorg. Med. Chem.* **2016**, *24*, 5061–5067. [CrossRef]

nutraceuticals

Review

Effectiveness of Commercial Red Clover (*Trifolium pratense* L.) Products for the Treatment of Symptoms in Menopausal Women—A Narrative Review

Mirjana Zukić [1], Irzada Taljić [2] and Ines Banjari [1,*]

[1] Department of Food and Nutrition Research, Faculty of Food Technology, Josip Juraj Strossmayer University of Osijek, F. Kuhača 18, 31 000 Osijek, Croatia; mirjanazukic2@gmail.com
[2] Faculty of Agriculture and Food Sciences, University of Sarajevo, Zmaja od Bosne 8, 71 000 Sarajevo, Bosnia and Herzegovina; i.taljic@ppf.unsa.ba
* Correspondence: ibanjari@ptfos.hr

Abstract: Red clover (*Trifolium pratense* L.) is found in southeast Europe and Anatolia. Its primary traditional medicinal use includes the treatment of various conditions of the upper respiratory tract. In recent years, its isoflavones have become the focus of research aimed at developing treatments to alleviate menopausal symptoms. Reduced levels of circulating estrogen due to reduced ovarian function can cause short-term symptoms such as hot flashes, palpitations, difficulty sleeping, headaches, fatigue, mood disorders and reduced concentration but also long-term chronic conditions, such as cardiovascular disease, accelerated weight and bone mass loss, atrophic vaginitis, osteoporosis, and cognitive impairment. The aim of this narrative review was to analyze the effects of commercially available and standardized red clover extracts on menopausal women. Eight randomized controlled trials on a total of 8769 menopausal women (aged 40 to 65 years) evaluated the effect of red clover isoflavone extract on menopausal symptoms. In all studies, isoflavone extract treatment showed improvement in all menopausal symptoms, including some common comorbidities, namely, hot flashes (1487 women, 25%), blood lipids (1155 women, 19%), atherosclerosis (6938 women, 79%), risk of breast cancer and endometrial cancer (428 women, 5%), osteoporosis and osteopenia (555 women, 6%), and menopause-related cognitive impairment (3530 women, 40%).

Keywords: red clover; *Trifolium pratense* L.; menopause; red clover extracts; bioactive compounds

Citation: Zukić, M.; Taljić, I.; Banjari, I. Effectiveness of Commercial Red Clover (*Trifolium pratense* L.) Products for the Treatment of Symptoms in Menopausal Women—A Narrative Review. *Nutraceuticals* **2024**, *4*, 430–449. https://doi.org/10.3390/nutraceuticals4030026

Academic Editors: Ivan Cruz-Chamorro and Guillermo Santos Sánchez

Received: 17 June 2024
Revised: 30 July 2024
Accepted: 6 August 2024
Published: 9 September 2024

1. Introduction

Menopause is the physiological or iatrogenically induced cessation of menstruation (amenorrhea) due to reduced ovarian function. The decrease in circulating estrogen levels can cause menopausal symptoms, including short-term symptoms such as hot flashes, palpitations, sleep difficulties, headaches, fatigue, mood disorders, and decreased concentration, as well as long-term chronic conditions such as cardiovascular diseases, accelerated bone loss, atrophic vaginitis, osteoporosis, and cognitive impairment [1]. Perimenopause refers to the years before (duration varies widely) and one year after the last menstruation. Perimenopause is usually characterized by an increased frequency of menstruation followed by thinning (*oligomenorrhea*), but any pattern is possible; conception is still possible during perimenopause. The period after menopause is called postmenopause [2]. The term "climacteric" comes from the Greek word "klimakter", meaning change, and encompasses the entire period of gradual ovarian function decline from perimenopause, through menopause, including postmenopause [3]. Premature menopause, also known as premature ovarian failure, is defined as the cessation of menstruation before the age of 40 and can be caused by factors such as smoking, living at high altitudes, and malnutrition [4]. Iatrogenic (artificially induced) menopause can occur due to medical procedures such as oophorectomy, chemotherapy, pelvic radiation, or any other procedure that disrupts blood supply [5].

As the ovaries age, their response to pituitary gonadotropins, follicle-stimulating hormone (FSH), and luteinizing hormone (LH) weakens, initially causing shortened follicular phases (with shorter and irregular cycles), less frequent ovulations, and consequently reduced progesterone production [6]. Eventually, the follicles become unresponsive, resulting in limited estradiol production. Estrogens (now mainly estrone) still circulate; they are produced by peripheral tissues such as adipose tissue and skin from androgens like androstenedione and testosterone [7]. However, the overall estrogen level is much lower. Around menopause, androstenedione levels decrease by half, but the decline in testosterone, which begins gradually during young adulthood, does not accelerate during menopause as the stroma of the postmenopausal ovary and the adrenal glands continue to secrete significant amounts [8].

The diagnosis of physiological menopause is clinically confirmed when menstruation is absent for one year. The aforementioned manifestations or symptoms can be treated (e.g., with hormones or selective serotonin reuptake inhibitors (SSRIs, antidepressants)). In the United States, the average age of physiological menopause is 51 years, while in Croatia, the average age of menopause is 48.8 years [9,10].

The aim of this narrative review was to analyze the effects of commercially available standardized extracts of red clover (*T. pratense* L.) on women's health during menopause. Eight randomized controlled trials were included in this review; all assessed the effect of red clover isoflavone extract on menopausal symptoms [1,8,9,11–16].

2. Botanical Origins

Clover (*Trifolium*) is one of the most important legumes (*Fabaceae*) in temperate and humid regions. It plays a crucial role in livestock nutrition [8] and makes a significant contribution to agricultural and animal production in Europe and America. Red clover (*Trifolium pratense* L.) naturally grows in southeastern Europe and Anatolia, with Anatolia being recognized as its homeland [17].

The Latin name of the genus *Trifolium* comes from the Greek word *trifolion* (three-leaved), a compound of the words *tria* (three) and *folium* (leaf). The species name *pratensis* (meadow) indicates the plant's habitat. Red clover is a herbaceous perennial plant belonging to the legume family (*Fabaceae*). It naturally grows in the regions of Europe and western Asia and is often found as a real green carpet in meadows and pastures. It reproduces through seeds. In agriculture, red clover is considered a significant plant. Clusters of symbiotic bacteria form on its roots, which bind nitrogen from the air and thereby fertilize the soil. In addition to enriching the soil by giving it fertility, it is also important as a fodder plant. It is an excellent source of nectar and can contribute to substantial honey production. Daily intake can reach up to 3 kg per hive, yielding a total of up to 260 kg of honey per hectare [18]. Seven diploid varieties (2n = 2x = 14) have been identified worldwide. It serves as a natural source of valuable isoflavonoids, utilized in commercial products (e.g., Menoflavon, Rimostil, Promensil) in some parts of the world [17].

The stem is upright, highly branched, and sparsely hairy, growing up to 50 cm. The root system is a taproot, well-branched, ranging from 50 to 100 cm in length, with round nodules. The leaves are alternate, elliptical, 2–4 cm long, and 1–1.5 cm wide, with a distinct white triangular spot, arranged in groups of three on a hairy stalk. At the base of the stalk, there are two membranous stipules. The lower leaves are on stalks around 20 cm long, while the upper ones are short-stalked or nearly sessile [19]. The flowers are hermaphroditic, irregular, clustered in head-like inflorescences on a 1–7 cm long, hairy stalk. The inflorescence is compound. The calyx is bell-shaped, covered with hairs, and has five triangular lobes. The corolla resembles a butterfly, red in color, and is twice as long as the calyx. The ovary has an elevated stigma carrying 1–2 embryonic seeds, with 10 stamens, 9 of which are fused into a tube. Blooming occurs from May to autumn [20]. The fruits are rounded pods containing 1–2 elongated-ovate, smooth seeds [18].

It can be sown alone for hay production or mixed with cereals in crop rotations. It can be cultivated in various soil types, pH values, and environmental conditions, providing good yields [21]. Tetraploid *Trifolium pratense* L., obtained through classical hybridization methods, is an important forage plant in Asia, Europe, North America, New Zealand, and Australia. Records show that in the year 2000, clover seed was produced on 25,422 hectares in European Union member states [red clover (*T. pratense* L.) on 11,031 ha; white clover (*T. repens* L.) on 4346 ha; and other clover species on 10,045 hectares]. France had the largest area for red clover production (5732 hectares), followed by Sweden [22].

3. Bioactive Compounds from Red Clover with Beneficial Health Effects

Red clover contains a number of potent bioactive compounds including polyphenols, flavonoids, coumarin derivatives, cyanogenic glycosides, and volatile oils, as well as vitamins and trace minerals [23], which have been identified in parts of the plant in various quantities, as summarized in Table 1.

Table 1. Bioactive compounds and their quantities identified in all parts of *Trifolium pratense* L.

Plant Part	Bioactive Compounds	Quantity (mg/g in Dry Metter dm)	Reference
Roots	formononetin	2.88	[8,23–26]
	biochanin A	1.95	
	maackiain	2.68	
	pseudobaptigenin	2.19	
Leaves	biochanin A	39.3	[25–30]
	formononetin	32.2	
	genistein	23.4	
	daidzein	5.1	
	glycitein, irilone, quercetin, ononin, maackain, orobol, pratensein, pseudoapigenin, prunetin, prunetin-4-glucopyranoside, genistein-7-galactopyranoside, pinitol-6-methoxycyclo-hexanepentaol	<1	
Stems	formononetin	34.9	[18,28,30]
	biochanin A	18.7	
	genistein	2.37	
	daidzein	34.9	
	prunetin, glycitein, pseudoapigenin, pratensein, irilone, orobol, prunetin-4-glucopyranoside, genistein-7-galactopyranoside, pinitol-6-methoxycyclo-hexanepentaol	<1	

Table 1. *Cont.*

Plant Part	Bioactive Compounds	Quantity (mg/g in Dry Metter dm)	Reference
Seeds	biochanin A	0.93	[24,25,28–31]
	formononetin	0.83	
	genistein	0.64	
	daidzein	0.20	
	flavonol quercetin, taxifolin dihydroxyquercetin, soyasaponin I, 22-O-glycosides, 22-O-diglycosides, astragaloside VIII, prunetin, prunetin-4-glucopyranoside, pinitol-6-methoxycyclohexanepentaol	<1	
Flowers	biochanin A	37.4	[28,30]
	formononetin	23.6	
	genistein	34.1	
	daidzein	4.9	
	prunetin, genistein, prunetin-4-glucopyranoside, genistein-7-galactopyranoside, pinitol-6-methoxycyclohexanepentaol	<1	

In phytotherapy, the flower heads of red clover are used for tinctures, which are collected during the flowering season. Red clover flower juice is used for eye diseases [18]. The red clover concentrate (obtained by extraction by continuous boiling in water) and infusion (soaked in cold or hot water without boiling) are still used as expectorants (means of facilitating the expectoration of tracheal secretions), alternatives (improving the state of body fluids), sedatives, and rheumatism drugs. Red clover is occasionally used to induce menstruation or as a fertility tonic [32–34].

Today, the well-known and clinically tested commercial products, available in North America that contain red clover are Promensil, Rimostil, and Trinovin, all produced by Novogen, Ltd. (North Ryde, NSW, Australia). Given the extent of the available clinical evidence for the aforementioned commercial products, the majority of the evidence presented in this review is based on these particular products.

4. Use of Red Clover in the Treatment of Perimenopause and Postmenopause Symptoms

In the 19th and 20th centuries, infusions (flower tea) or tinctures (ethanol extract) were used to treat upper respiratory tract diseases such as cough, asthma, bronchitis, laryngitis, and tuberculosis. They were also employed as mild sedatives, antispasmodics (for whooping cough, measles), and remedies for rheumatism. Creams and ointments were used for the treatment of burns, wounds, gout, and fungal infections. Southeastern Cherokee Indians used flower tea or aerial parts to treat fever, Bright's disease (chronic inflammation of the kidneys, nephritis), leukorrhea (vaginal discharge), and as a "blood medicine". The Southern Ute Indian tribe, native to Colorado, used red clover syrup as an abortifacient, and they rolled red clover leaves into cigarettes, using them to treat asthma [35–38].

Commercial products containing red clover were introduced to the market in the early 20th century for the treatment of various health conditions. In the year 1900, the Wm. S. Merrell Chemical Company (Cincinnati, OH, USA) produced a trifolium extract for treating syphilis, scrofula, chronic rheumatism, and various skin issues. In the year 1920, Harry Hoxsey and Norman Baker introduced Hoxsey as a cancer remedy. For the same application, in 1934, Flora, Inc. (Lynden, WA, USA) and Flora Manufacturing & Distributing, Ltd. (Burnaby, BC, Canada) produced Flor-Essence or "Indian tea".

The growing interest in red clover isoflavones can be attributed to research on the estrogenic effects of *Trifolium* species on grazing and experimental livestock. Given that red clover isoflavones show structural similarities to endogenous 17β-estradiol, they can exert their biological effects via activating the estrogen receptor (ER), showing higher affinity for ER-β than for ER-α. In addition, numerous nonhormonal effects of isoflavones have been reported, including tyrosine kinase inhibition, antioxidant activity, and effects on ion transport [39–41].

Over the past few years, semipurified isoflavone supplements from red clover have been studied for the use in menopause, improving bone health and maintaining the cardiovascular system, and in the treatment of breast, ovarian, and prostate cancer. A summary of the bioactive compounds that have shown effects on the symptoms and common comorbidities of menopause is shown in Table 2.

Table 2. Bioactive compounds isolated from red clover (*Trifolium pratense* L.) plant parts with positive effects on the common symptoms and/or comorbidities during menopause.

Menopausal Symptoms and/or Comorbidities	Aglycones of Red Clover Isoflavone Extract	Quantity (mg/g dw)	Red Clover Plant Parts	References
Hot flashes	biochanin A	7.29 ± 1.29	leaves	[1,8,11–16,28,42,43]
		2.81 ± 0.04	stems	
	formononetin	8.15 ± 3.14	leaves	
		8.87 ± 2.61	stems	
	genistein	0.62 ± 0.23	leaves, stems	
		0.27 ± 0.17	stems	
		0.46 ± 0.12	aboveground parts	
	daidzein	0.55 ± 0.35	leaves	
		0.07 ± 0.06	stems	
Blood lipoprotein composition	formononetin	8.15 ± 3.14	leaves	[1,8,11–16,26,28,44]
		6.38 ± 1.95	stems	
	biochanin A	8.12 ± 1.21	leaves	
		2.80 ± 0.65	stems	
	genistein	0.69 ± 0.03	leaves	
		0.31 ± 0.22	stems	
		0.34 ± 0.11	aboveground parts	
	daidzein	0.55 ± 0.35	leaves	
		0.07 ± 0.06	stems	
Atherosclerosis	biochanin A	9.05 ± 2.69	leaves	[1,8,11–16,28,45]
		2.66 ± 0.79	stems	
	genistein	0.59 ± 0.30	leaves	
		0.13 ± 0.05	stems	
		0.22 ± 0.04	aboveground parts	
	biochanin A	7.29 ± 1.29	leaves	
		3.08 ± 0.96	stems	

Table 2. *Cont.*

Menopausal Symptoms and/or Comorbidities	Aglycones of Red Clover Isoflavone Extract	Quantity (mg/g dw)	Red Clover Plant Parts	References
Anticancer effects	biochanin A	6.46 ± 1.61	leaves	[1,8,11–16,28,45]
		2.33 ± 1.74	stems	
	formononetin	8.15 ± 3.14	leaves	
		10.32 ± 3.63	stems	
Bone health	formononetin	8.15 ± 3.14	leaves	[1,8,11–16,26,28,46]
		17.51 ± 2.24	stems	
	biochanin A	5.77 ± 0.81	leaves	
		1.39 ± 0.34	stems	
Cognitive effects	formononetin	8.15 ± 3.14	leaves	[1,8,11–26,28]
		10.53 ± 3.50	stems	
	biochanin A	8.30 ± 2.15	leaves	
		2.30 ± 1.36	stems	

In plant isoflavones, glycosides are prominent. The hydrolytic conversion of glycosides to aglycone analogs is required to facilitate absorption. Fermented isoflavone aglycone preparations show increased bioavailability compared to that of similar glycosides. The use of enzymatic techniques and probiotics has been proven to enhance the uptake of these compounds, thereby improving the efficacy of isoflavone treatment [47].

Despite the well-known benefits of hormone therapy in menopause, due to the potential serious side effects and the risk of breast cancer, the use of this alternative therapy, even in the treatment of hot flashes, remains controversial [45,48]. Many women have discontinued hormone therapy after the publication of the Women's Health Initiative study results, seeking an effective and safe alternative to alleviate menopausal symptoms [49,50].

The reluctance to accept hormone therapy, which has been associated with concerns about its safety, has led to the popularization of many alternative and complementary treatment methods [42,51,52]. For several years, red clover has been one such alternative that women use to treat menopausal symptoms, including hot flashes [40].

4.1. Hot Flashes

The most common symptom of menopause are hot flashes, which can last several years after menopause. About 70% of women report experiencing them, with differences in different populations [53,54]. The frequency of hot flashes depends on climate conditions, race/ethnicity, but also diet and lifestyle, as well as women's personal attitude toward the end of reproductive life and aging [55–57]. Sometimes, the intensity of hot flashes can be so significant that treatment should be considered, since women report sleep and mood disturbances that eventually negatively impact their daily activities at home or at work, and the overall quality of life [58,59]. Hot flashes are thought to be the result of the brain's response to progressive estrogen deficiency and fluctuations in neurotransmitter activity, especially in the serotonergic and noradrenergic pathways, leading to instability of the hypothalamus thermoregulation mechanism. Ultimately, this results in increased blood flow to the skin and the enhanced activity of sweat glands, causing these symptoms [54,60].

Lambert et al. [11] launched one of the first studies addressing the impact of combined red clover isoflavones and probiotics on the vasomotor symptoms of menopause. They included the objective skin conductance (SC) capture of hot flashes and the determination and standardization of the concentration and molecular form of the isoflavone component of red clover extract (RCE); with a combined methodology, they considered participant characterization and eligibility [61]. They conducted a parallel, double-blind, randomized controlled trial of 62 women from the northern Denmark region in perimenopause, aged 40–65 years, with a BMI of 20–40, who reported ≥ 5 hot flashes/day and had follicle-stimulating hormone (FSH) levels ≥ 35 IU/L.

Each participant, twice a day for 12 weeks, received treatment with bioavailable RCE, providing 34 mg/d of isoflavones and probiotics, or a masked placebo formulation. The bioavailable RCE was obtained by adding a heterogeneous culture of lactic acid probiotic bacteria to the red clover extract to facilitate cold fermentation and improve bioavailability [62]. The standardization of the aglycone content after fermentation was confirmed by liquid chromatography–mass spectrometry (LC-MS) by DB Lab A/S, (Odense, Denmark). With the purpose of improving the characteristic taste and appearance of RCE, stevia and sugar-free berry/orange aroma were added.

The primary outcome was the change in the daily frequency of hot flashes from baseline to 12 weeks using 24 h SC (ambulatory skin conductance). Secondary outcomes included changes in SC-determined intensity (HFI), self-reported HFF (rHFF), severity of hot flashes (rHFS), blood pressure, and plasma lipids [63]. A significant reduction in 24 h HFF was observed when comparing the change from baseline to 12 weeks of RCE treatment with that of the placebo. rHFF was also significantly reduced in the RCE group compared to that in the placebo group. Other parameters were not significantly different. RCE was well tolerated. The results suggested that moderate doses of RCE were more effective and superior to the placebo in reducing physiological and self-reported vasomotor symptoms (hot flashes, flushing, and night sweats).

Kanadys et al. [1] conducted a meta-analysis on the effectiveness of red clover isoflavones in alleviating hot flashes and symptoms in perimenopausal and postmenopausal women. Twelve randomized, placebo-controlled clinical trials were selected for the analysis, with trial durations ranging from 12 weeks to 2 years. The clinical studies were conducted in Australia, Peru, the Netherlands, the United States, the United Kingdom, Ecuador, Brazil, Austria, Iran, and Denmark. In total, 1179 menopausal women participated in the studies, with sample sizes ranging from 37 to 252 (1043 participants included in the final analysis). Eight trials included postmenopausal women, three studies included women in peri- and postmenopausal periods [64–66], and women in perimenopause were included in one study [11]. The average dose of red clover isoflavones was 65.1 mg/d of aglycone equivalent (range 37.1–160 mg/d). Two studies included two therapeutic groups with different doses of isoflavones [64,67]. The compositions of the isoflavones and their doses varied among the studies. Eight studies measured the daily frequency of hot flashes ($\geq$3/day). Ten studies evaluated the presence and/or severity of various somatic and psychological symptoms using menopause symptom assessment scales. In most studies, a dose of 40–80 mg/d of RCIE was used, except in [5], where 37.1 mg/d of aglycone was applied. Out of eight RCTs with ten comparisons assessing the frequency of hot flashes, six [11,42,64,67,68] demonstrated a reduction in hot flashes, including four significant reductions [11,42,68] in the isoflavone group compared to a placebo group; in four comparisons, no differences were observed between the groups [64,66,67,69].

The meta-analysis of eight studies (ten comparisons) demonstrated a statistically significant reduction in the daily incidence of hot flashes in women receiving red clover compared to that in those receiving a placebo [11,42,65–67,70,71]. Due to 87.34% homogeneity, the analysis showed a significant difference in postmenopausal women with $\geq$5 hot flashes per day when the follow-up period was 12 weeks, with an isoflavone dose of $\geq$80 mg/day, and when formulations contained a higher proportion of biochanin A.

Booth [14] conducted an analysis of studies administering semipurified red clover preparations to women for alleviating menopausal hot flashes. The examined studies provided short-term results ($\leq$12 weeks), and the achieved positive effects of the preparations required 8 weeks to manifest. Six short-term trials ($\leq$4 months) investigated the effectiveness of red clover isoflavone supplements in reducing hot flashes, of which only three studies, all using Promensil, yielded positive results. A two-month noncontrolled study provided 23 menopausal women (aged 40–65 years) one Promensil tablet per day, reaching a 56% reduction in the frequency and a 43% reduction in the intensity of hot flashes. In the second double-blind, randomized, placebo-controlled study, 51 postmenopausal women (aged 45–65 years) were randomized to receive one Promensil tablet per day or placebo over 3 months. The two groups did not differ significantly in the frequency of hot flashes. Among one or four Promensil tablets per day over the 3 months, 37 postmenopausal women (aged 40–65 years) also did not differ in terms of the frequency of hot flashes. The authors, however, noted a significant placebo response.

A 4-month, randomized, double-blind study evaluated the effect of one tablet of Promensil per day in comparison to that of a placebo on the frequency and intensity of hot flashes in 30 Peruvian postmenopausal women (younger than 60 years; median age 52 ± 0.7 years) [70]. The frequency of hot flashes decreased by 49% and 11% in the treated and placebo groups, respectively, and the intensity of hot flashes decreased by 47% compared to 0% in the placebo group. Another double-blind, randomized, placebo-controlled study encompassing 30 Dutch postmenopausal women (aged 49–65 years) showed that two Promensil tablets per day in comparison reduced the frequency of hot flashes by 44% compared to placebo after 12 weeks [68]. The authors also reported a small change in the mean body mass index (BMI) in the treated group in comparison to the placebo group [70].

Tice et al. [64] conducted a double-blind, randomized, placebo-controlled study on 252 women aged 45 to 60 years, finding a similar reduction in hot flashes after 12 weeks in the two treatment arms (Promensil, 82 mg isoflavones/day, high genistein + biochanin A; Rimostil, 57 mg isoflavones/day, high daidzein + formononetin) in comparison to a placebo. Although the overall results did not reach statistical significance, a slightly better response was found in women with BMI > 25 kg/m^2 regardless of active treatment arm, suggesting metabolic effect of red clover isoflavones.

Shakeri et al. [15] conducted a randomized, triple-blind, placebo-controlled clinical trial involving 72 healthy postmenopausal women over 12 weeks. The intervention arm received two capsules containing 40 mg of dried red clover leaves per day. The placebo was two capsules containing 40 mg of starch per day. The outcome measures were menopausal symptoms determined using the Menopause Rating Scale (MRS) [71]. The overall MRS score decreased from 20.41 to 10.08 in the intervention group and from 20.77 to 17.20 in the control group [72], which was attributed to scores in the vegetative-somatic and psychological categories of menopausal symptoms. In comparison to the placebo, dried red clover leaves were more effective in reducing the severity of vasomotor and menopausal symptoms, including mood disorders, especially symptoms of anxiety and depression.

Akbaribazm et al. [8] conducted a randomized, double-blind, placebo-controlled study. They reviewed 80 related articles on the beneficial effects of red clover on biological processes involving the participation of 190 postmenopausal women. The research results showed that the ethanolic extract of the aerial parts of red clover (398 mg/day standardized to 120 mg of isoflavones) significantly reduced hot flashes and vasomotor symptoms in the studied women after 12 months [64].

4.2. Blood Lipoproteins

The potential effect of red clover isoflavones on lipid metabolism has been proven in animal models [73,74], while human findings have been mixed. Some studies showed a blood-lipid-lowering effect [75–77], while others found no change in lipid status [44,78].

Kanadys et al. [13] conducted a systematic review and meta-analysis with the aim of explaining the effect of a specific standardized isoflavone extract of red clover on the lipid profile of peri- and postmenopausal women. Ten studies, with interventions between 12 weeks and 12 months, fit the inclusion criteria, encompassing a total of 910 menopausal women, average age 53.9 (±4.1), (range 40–85). The studies were conducted in the United Kingdom, Ecuador, Australia, Denmark, Serbia, and the United States [79]. The mean dose of red clover isoflavone (RCI) extract was 61.5 mg per day as aglycone equivalents (range,

33.8–160 mg per day). The meta-analysis confirmed that red clover extract is effective in reducing the concentration of total cholesterol but not the levels of HDL-C, LDL-C, and triglycerides [13].

Booth [14] reviewed several studies investigating the effects of red clover isoflavone preparations on arthrosis risk, namely serum concentrations of HDL, LDL, and triglycerides. Studies included in the review tested the effects of Promensil, Rimostil, and three experimental formulations, P-07, P-07(b), and P-083, on serum lipids. All observed products differed significantly in both the total contents of isoflavones and their ratios (daidzein + formononetin vs genistein + biochanin A). These differences prevented the proper comparison since, for example, Rimostil, in comparison to Promensil, contains more formononetin and daidzein relative to genistein and biochanin A.

Four out of seven studies that assessed the effects of Promensil on plasma lipid levels had positive results [65,67,78,79], and the remaining found no effect [44,80,81] on HDL, LDL, or triglycerides. Two of the four studies with positive results reported an increase in HDL in postmenopausal women who consumed one or two tablets per day of Promensil for at least one month [67,78]. Reduced triglycerides in perimenopausal women consuming one tablet of Promensil during 12 months was reported in another study [65]. In a more recent study, lower triglycerides were found in women who consumed two Promensil tablets for only three months, but the effect was limited to women with baseline triglyceride levels > 178 mg/dL [80].

Four clinical studies examined the effect of Rimostil on plasma lipids [69,76,80,82]. One study, with negative findings [83], was conducted on patients with type 2 diabetes, a population known to have imbalanced lipid levels and a high risk of heart disease [84,85]. Positive findings, specifically increased HDL levels, were reported from an uncontrolled study in postmenopausal women who consumed one, two, or three Rimostil tablets daily for six months [69]. Another study, which used the same treatment protocol and duration as that previously mentioned but included a placebo, reported an increase in in HDL levels in postmenopausal women [76]. Interestingly, both studies [69,76] also reported reduced levels of apolipoprotein B in the treatment groups, another specific risk factor for atherosclerosis. Possibly, the reason lies in the different balances of isoflavones in Rimostil as compared to Promensil. Reduced triglyceride levels in women with baseline values > 178 mg/dL after using Promensil were found, as previously mentioned [79].

In perimenopausal women, experimental formulation P-07, containing 86 mg of red clover isoflavones per day consumed over 3 menstrual cycles, showed no impact on the total cholesterol, LDL, HDL, triglycerides, or lipoproteins [77]. The use of P-07 for 3 months in women before and during perimenopause had no effect on plasma lipids [86].

One study reported gender-specific effect of red clover isoflavones on LDL levels. This randomized, placebo-controlled, crossover, double-blind study encompassed men and postmenopausal women. The study used a red clover formulation enriched with biochanin A [P-07(b)] and a formulation enriched with formononetin (P-083) [44]. Only the P-07(b) product showed a 9.5% reduction in LDL in men only. Neither formulation affected the plasma lipid levels in postmenopausal women.

Clinical studies assessing the effects of the use of Promensil and Rimostil on serum lipids were reviewed by Akbaribazm et al. [8]. In a double-blind, randomized, placebo-controlled trial, supplementation with a 50 mg Rimostil tablet for 2 years reduced triglyceride and LDL levels and increased HDL levels in 189 menopausal women [87]. In another randomized, double-blind, placebo-controlled prospective study on 37 postmenopausal women with symptoms of estrogen deficiency, treatment with 40 mg/kg of Promensil in comparison to a placebo for 12 weeks had the same results: significantly lower triglyceride and LDL levels and increased HDL levels [67].

The administration of 40 mg of biochanin A to 19 women with premenstrual syndrome over 12 weeks in a randomized, double-blind, placebo-controlled trial had reduced triglyceride and LDL levels and increased HDL levels and reported significant reductions in premenstrual syndrome symptoms such as fatigue and swelling [88].

4.3. Atherosclerosis

Atherosclerosis is a general term for several conditions in which the artery wall becomes thinner and less elastic. The most common and important form is atherosclerosis, where the accumulation of fatty material occurs beneath the inner sheath (endothelium) of the arterial wall [5]. As a late symptom of menopause, as a consequence of a change in fat metabolism, lipoproteinemia alters the intensity of the atherosclerosis process. Consequently, high blood pressure is a common presentation. These factors contribute to increased risks of heart attack and stroke during menopause [89].

Arterial compliance, a measure of arterial stiffness, correlates with the presence of atherosclerotic plaques in major blood vessels. The first study [44] administered one Promensil tablet daily for 5 weeks. The dose was then doubled to two tablets per day (80 mg isoflavones/day) for an additional 5 weeks. Both treatment groups (both doses) showed an increase in arterial compliance. The second study was a randomized, double-blind crossover study that gave two tablets of two different products to normotensive men and postmenopausal women for 6 weeks per treatment. One product was significantly enriched with biochanin A [P-07(b); Novogen, Ltd.; noncommercial formulation], and the other was enriched with formononetin (P-083; Novogen, Ltd.; noncommercial formulation). Product P-083, enriched with formononetin, had a more favorable effect on arterial compliance than product P-07(b), which was enriched with biochanin A.

Vascular endothelial function has not been firmly linked to the development of atherosclerosis or hypertension but is assumed to play major role in its pathology. Based on the findings of two studies, supplementation with red clover isoflavones showed no effect on platelet adhesion or aggregation factors. After 6 weeks in the study by Teede et al. [90], the plasma levels of vascular cell adhesion molecule-1 (VCAM-1) were reduced in the group receiving 80 mg/day of red clover extract enriched with formononetin (P-083). The administration of up to 85.5 mg/day of isoflavones (Rimostil) to postmenopausal women for 6 months did not result in changes in coagulation factors V, VII, VIII, antithrombin III, or fibrinogen in the blood [69]. The serum taken from 25 healthy postmenopausal women with mild menopausal symptoms, receiving a daily dose of a combination of soy and red clover isoflavone product (Phytogyn, Gynea, Barcelona, Spain; 17 mg soy isoflavones, 38 mg red clover isoflavones) for 6 months, showed that the treatment stimulated the release of prostacyclin in the endothelial cells of human umbilical veins [91]. Since prostacyclin can inhibit platelet adhesion and aggregation in the endothelium, this could be one of the mechanisms by which red clover isoflavones improve vascular health. Red clover isoflavones also increased the activity of endothelial nitric oxide synthase (eNOS), eNOS expression, and nitrite levels in the endothelial cells of human umbilical veins after 48 h of exposure [92]. This is another potential (direct, genomic) vascular mechanism of action of red clover isoflavone supplements.

Akbaribazm et al. [8] conducted a systematic review of clinical studies to explain the effects of a specific standardized extract of red clover isoflavones on the cardiovascular system of postmenopausal women. Flavonoids have various effects, including anti-inflammatory, antiangiogenic, and antiallergic effects. The transcription factor NF-kB plays a role in inducing the expressions of inflammatory mediators such as cytokines, cell surface receptors, adhesion molecules, and acute-phase proteins. Treatment with genistein (0.3 mg/kg), a component of red clover, for 8 weeks inhibited the development of atherosclerosis by inhibiting the expressions of NF-kB, thrombin, TNF-α cytokines, and vascular cell adhesion molecule-1 (VCAM-1) in LDL receptor knockout mice [93]. These factors (i.e., NF-KB, TNF-a, and VCAM-1) contribute to the development of atherosclerosis by inducing monocyte accumulation and promoting monocyte adhesion to the vessel wall at sites prone to atherosclerotic lesion formation [94]. Furthermore, the Fas/Fas ligand system has been identified to be under the control of the estrogen receptor in monocytes. This suggests a link between estrogen and many disorders, including atherosclerosis, vascular inflammation (vasculitis), and rheumatoid arthritis [95].

The risk of cardiovascular diseases increases after menopause. In a 20-year follow-up of 2873 women younger than 55 years within the Framingham study, the annual incidence of cardiovascular diseases (coronary heart disease, stroke, and congestive heart failure) was reported as 0.6–5% per 1000 women in perimenopausal women aged <40 and 50–54 years [96]. A randomized, double-blind, placebo-controlled trial [97] found that the administration of drospirenone (2 mg) plus β-estradiol (1 mg) over a 13-month period significantly lowered blood pressure and endometrial bleeding in 1142 postmenopausal women. The most likely explanation for these observations is the protective effect of endogenous female sex steroids, especially estrogen, during the years before menopause [98]. In various studies, the effects of raw extracts and compounds isolated from red clover on heart and vascular diseases have been investigated. The use of commercial products containing the active ingredients from red clover, including biochanin A and formononetin, increased vascular vasoconstriction, SAC, and the speed of the pulmonary artery. The reason for these effects may be associated with the release of vasoconstrictor compounds, including NO, prostaglandins (PGs), and endothelial hyperpolarization factor. Calcium is essential for the contraction of the smooth muscle cells in the vascular wall; however, formononetin inhibits the intracellular influx of Ca into vascular smooth muscle cells [48].

4.4. Breast and Endometrium Carcinoma

Breast cancer stands as the leading cause of death among women from all cancerous diseases [99]. The introduction of the antiestrogen drug tamoxifen has significantly improved the health and overall survival of women with estrogen/progesterone diseases [100,101]. However, in perimenopausal women, antiestrogens induce early menopausal syndrome, which are often poorly tolerated [102], leading to therapy discontinuation in a considerably high percentage of cases [103,104].

Ferraris et al. [12] conducted a systematic review of clinical studies with the aim of determining whether a red clover preparation together with a dietary intervention applied in perimenopausal women with breast cancer (BC) improves menopausal symptoms due to antiestrogen treatment and therefore promotes tamoxifen compliance, prevents weight gain, and is safe.

Surgically treated perimenopausal women with estrogen receptor-positive (ER+) diseases taking tamoxifen were engaged in a prospective, double-blind, randomized trial [99]. The red clover group (n = 42) received one oral tablet daily of an 80 mg pharmaceutical extract of red clover for 24 months. The placebo group (n = 39) received one oral tablet daily without an active ingredient. All women were encouraged to adopt a Mediterranean-type diet and to remain active. Outcomes included menopause assessment (MRS), body mass index (BMI), waist and hip circumference, insulin resistance, and levels of cholesterol, triglycerides, and sex hormones. The safety indicators investigated were endometrial thickness, breast density, and the effects of patient serum on ER-positive breast cancer cell lines [105].

The red clover group exhibited significantly greater reductions in BMI and waist circumference. HDL cholesterol significantly increased in both groups. The levels of total cholesterol, LDL cholesterol, triglycerides, insulin resistance, and sex hormones did not significantly vary during the study period and did not differ between groups. Snack and red/processed meat consumption significantly decreased in both groups, while the consumption of unrefined grains, legumes, fish, nuts, and seed oils significantly increased. Physical activity also significantly increased. Endometrial thickness remained constant. Breast density significantly decreased in both groups. Proliferation and estrogen-regulated gene expression did not differ in the cell lines treated with the serum from each group.

The first study assessing red clover in tamoxifen-treated patients with BC, containing isoflavones, proved the treatment was clinically and in vitro safe and was associated with reduced BMI and waist circumference. However, dietary and lifestyle interventions likely improved menopausal symptoms.

Booth [14] analyzed five clinical studies, none designed to specifically assess the effect of red clover isoflavone supplementation in patients with breast cancer. In a study involving women with increased breast density, 177 participants (aged 49–65 years) with Wolfe P2/DY mammographic breast density patterns received one tablet of Promensil per day for one year. The study showed no statistically significant changes in estradiol, FSH, or luteinizing hormone (LH) levels [65]. Importantly, there were no significant differences in the breast density patterns between the treated and placebo groups. This is a positive outcome, as it is known that hormone replacement therapies (e.g., conjugated equine estrogens plus medroxyprogesterone acetate) increase mammographic density, a risk factor for breast cancer [106]. These results suggest that the consumption of red clover isoflavones by women at high risk of breast cancer may be safe, although studies have not been conducted on patients with confirmed breast cancer.

A randomized, placebo-controlled, crossover pilot study provided two Promensil tablets to 16 premenopausal women and 7 postmenopausal women for one month. It reported a nonsignificant reduction in insulin-like growth factor 1 (IGF-1) levels in perimenopausal but not postmenopausal women [78]. High serum IGF-1 levels are associated with an increased risk of breast cancer [107], making these results interesting for analysis through a larger study.

A study addressing cyclic mastalgia (breast pain) involved a placebo period of two menstrual cycles, and participants with an average pain reduction of 30% compared to baseline (i.e., a low placebo response) were then randomized and given one or two Promensil tablets (40 or 80 mg of red clover isoflavones) per day during three menstrual cycles [108]. Breast pain was significantly reduced by 44% in the 40 mg group and 31% in the 80 mg group compared to the placebo. A three-day increase in menstrual cycle length was observed in the 80 mg group compared to the placebo group. These findings imply a positive effect on breast pain in women experiencing normal hormonal fluctuations during their menstrual cycle.

A three-month study examined the endometrial effects in perimenopausal women of receiving 50 mg of red clover isoflavones daily (P-07) and found no change in the proliferative index Ki-67 in approved biopsy samples taken during the late follicular phase. There were also no changes in plasma estradiol, FSH, progesterone, or endometrial thickness [86]. Increasing the dose to 85.5 mg of isoflavones/day with Rimostil over 6 months in postmenopausal women did not result in increased endometrial thickness or breakthrough bleeding [87]. These results indicate the inability of red clover isoflavone supplements to stimulate endometrial hyperplasia in women, at least when taken over a short period.

Phytoestrogens influence the sensitivity of breast cancer cells to analogs of vitamin D3 [1]. Breast mammographic density (MBD) is considered an indicator of breast cancer progression or treatment failure [109]. This marker relates to the radiodensity of connective, epithelial, and radiolucent fatty breast tissue. Diet influences MBD, potentially affecting the level of endogenous estrogen. For example, it was observed that a low-fat, high-carbohydrate diet containing isoflavones reduced MBD [51]. Although antiestrogenic drugs like tamoxifen reduce MBD, hormone replacement therapy (HRT) increases this parameter [110]. Insulin-like growth factor-I (IGF-I) also increases the risk of developing breast cancer.

Akbaribazm et al. [8] analyzed a clinical study that was randomized, placebo-controlled, and double-blind, aiming to investigate the use of Promensil tablets daily (equivalent to 86 mg/day of total isoflavones) for 2 months in four Dutch hospitals. The study showed a reduction in IGF-I in 23 postmenopausal women who received 40 and 80 mg/day. Promensil tablets alleviated breast pain by 44% and 31% in 18 perimenopausal women over three menstrual cycles.

The presence of phytoestrogens in different parts of red clover can influence reproductive tissues and estrogen receptors. In a randomized, double-blind, placebo-controlled study, 30 postmenopausal women were treated with Promensil (equivalent to 80 mg of red clover isoflavones) [111]. There was no significant change in their plasma levels of estradiol, FSH, or LH hormones or endometrial thickness; yet, hot flashes were reduced in 44% of the participants [68]. ER-α is expressed in breast, uterus, and ovarian tissues; ER-β is expressed in bones and blood vessels [112]. Furthermore, treatment with genistein (375 and 750 µg/g) for 21 days increased uterine weight and reduced bone loss and osteoporosis. One systematic review [45] concluded that isoflavones such as daidzein and genistein enhance the proliferation of endometrial gland cells, increase the expression of ER-β, and reduce the expression of ER-α. In a normal state, endometrial cells express higher levels of ER-α than ER-β. In an in vitro study, endometrial gland cells isolated from premenopausal and nonpregnant women in the proliferative phase after incubation with genistein (1.15 µmol/mL) and daidzein (2.4 µmol/mL) showed significant expression reduction in ER-α and an increase in ER-β mRNA [113]. The secretion of cytokines such as TNF-a and IL-1a also decreased. Finally, phytoestrogens modulate the expression of ER-α/β at the mRNA and protein levels in endometrial gland cells [114].

4.5. Osteoporosis and Ostopenia

The modern way of life increasingly eliminates movement and a healthy diet, which contribute to the reduction in bone density, resulting in osteopenia and osteoporosis [115]. Osteopenia is defined as a bone mineral density (BMD) where the T-value is between −1.0 and −2.5, and above −2.5 is the state of osteoporosis [116]. Osteopenia occurs more often in postmenopausal women due to the loss of estrogen. Osteoporosis is a chronic disease of the bone system in women, which, with its progressive course and complication, accompanied by a high degree of disability and mortality, has significant medical, economic and social consequences for both the individual and society as a whole [117]. The decrease in bone mass first affects the trabecular bone (vertebrae, ribs), which leads to kyphosis and a consequent decrease in height. Only after five or more years do the changes affect the cortical bone (neck of the femur, distal radius), which increases the possibility of fractures. The risk of osteoporosis also depends on genetic predisposition and race, eating habits, constitution, and caffeine consumption, as well as on gynecological factors, such as parity, breastfeeding, regularity of menstruation in the reproductive period, previous gynecological surgery, other diseases (hyperthyroidism, hyperparathyroidism), and taking medications such as glucocorticoids [16]. Studies have shown that taking dried red clover, an extract, or a pharmaceutical product combined with calcium can significantly slow the loss of BMD in postmenopausal women with osteoporosis [118].

In a 12-month, double-blind study with parallel design, 78 postmenopausal osteopenic women were supplemented two times a day with calcium (1200 mg/day), magnesium (550 mg/day), calcitriol (25 mg/day), and a red clover extracts rich in isoflavone aglycones and probiotics (RCE, 60 mg isoflavone aglycones/day and probiotics) or a masked placebo. RCE in combination with supplementation was more effective than supplementation alone. Over 12 months of use, RCE prevented menopause-associated decreases in BMD, normalizing bone turnover, promoting a favorable estrogen metabolite profile (2-OH:16α-OH), and stimulating equal production in postmenopausal women with osteopenia [11].

Five studies examined the effects of red clover isoflavone extracts on BMD or markers of bone turnover. Three studies reported bone preservation measured using BMD. After a 1-month placebo run-in period, a 6-month study documented increases in the proximal forearm but not in the distal forearm BMD for 25, 50, and 75 mg daily doses of red clover isoflavones in menopausal women. Another 6-month study found 4.1% and 3.0% increases in the proximal radius and ulna BMD, respectively, in postmenopausal women taking 57 mg and 85 mg of red clover isoflavones per day [6]. One-year treatment with a daily dosage of one pharmaceutical tablet significantly reduced the loss of BMD of the lumbar spine in perimenopausal women [66]. No effect was found in postmenopausal women nor was there an effect on hip BMD in either group. Possibly, red clover isoflavones prefer cortical over trabecular bone, which has yet to be confirmed.

None of the studies found effects on the levels of urinary N-telopeptide (a putative marker of bone resorption) or serum osteocalcin (a putative marker of bone formation). However, the correlation of these markers with the actual change in BMD is still unclear. The first study included perimenopausal women who took 50 mg of red clover isoflavones per day (product P-07, noncommercial, Novogen, Ltd.) for 3 months [86]. The second study used Promensil and Rimostil for 3 months in menopausal women [79]. The previously mentioned 1-year study by Atkinson et al. [65] also found no effect on bone resorption, despite positive outcomes for BMD and bone mineral content, implying that physiological levels of bone markers are inaccurate markers of BMD and bone turnover rate.

4.6. Cognitive Effects

Menopause negatively affects cognitive functions [119]. By reducing the concentration of inhibin B, it leads to a lack of negative feedback on the secretion of pituitary FSH [2,120].

From a neurobiological standpoint, the monoamine neurotransmitters serotonin, norepinephrine, and dopamine appear to modulate hot flashes and mood disorders [121]. Damage to these regulatory pathways favors the onset of depression when this dysregulation occurs within the prefrontal cortex and limbic system, which are the SNC regions involved in mood control [122]. Moreover, the deregulation of hypothalamic thermoregulatory centers affects the appearance of vasomotor symptoms. Therefore, treating hot flashes with menopausal hormone therapy (MHT) may prevent or reduce depressive symptoms in vulnerable women, rather than treating only mood disorders without controlling vasomotor symptoms [123]. A total of 40% of women reported episodes of forgetfulness in perimenopause. Women in the menopausal transition have symptoms of depression up to four times more often, which are associated with hormonal fluctuations and the presence of vasomotor symptoms [124]. Several risk factors are associated with depression, including lower socioeconomic status, stress, a previous history of depression, and a higher body mass index. A longer reproductive period is associated with a lower risk of developing depression [125].

Only one study examined the effect of red clover isoflavones on the cognitive function in postmenopausal women [126]. Thirty women over the age of 60 years received two tablets of Rimostil or a placebo for 6 months. Better results on visual–spatial intelligence tests was found in women receiving Rimostil in comparison to the placebo (12% increase vs. 3% decrease, respectively).

5. Possible Side Effects

Red clover is generally recognized as safe by the Food and Drug Administration (FDA), and most studies have found it to be well tolerated. Nevertheless, possible side effects, drug interactions, and risks for certain populations have been identified [127].

The use of red clover should be avoided in patients taking hormonal drugs, because the chemical components of red clover and hormones compete for the same hormone receptor sites and can lead to inflammation of the eyes, mouth, and penis [17].

Users of thrombolytic agents and low-molecular-weight heparin should be aware of the possibility of increased bleeding, as the plant may contain coumarins, which have anticoagulant properties. The concurrent use of red clover and contraceptives may alter the effectiveness or enhance the contraceptive's side effects. Additionally, the use of red clover with progesterone may result in reduced effectiveness. Experiments on animals conducted in vitro showed that the concomitant use of red clover and the anticancer drug tamoxifen led to the reduced effectiveness of the drugs. The plant, herbal extract, or semipurified isoflavonoids from the extract should not be used during pregnancy until further studies confirm their safety for use [128].

Additionally, no safety data on red clover among children or pregnant or breastfeeding women exist, so it should be avoided in these groups (34 trusted sources).

Finally, red clover may slow blood clotting and should be avoided in case bleeding disorders are present [129].

6. Conclusions

This narrative review provides insights into the bioactive compounds of red clover (*T. pratense* L.), particularly focused on its commercially available isoflavone extracts that have shown efficacy in the treatment of various symptoms and comorbidities in menopausal women. This review, however, did not look at other red clover extracts that are under intense research, and this is its main limitation.

Despite its well-documented traditional medicinal uses, a limited number of clinical studies tested its health effects. The majority of research has focused on red clover isoflavones, which, due to their structural similarity to endogenous 17β, estradiol can activate estrogen receptors and alleviate common menopausal symptoms. However, the other bioactive compounds isolated from red clover leaves, aglycones biochanin A and formononetin, reduce hot flashes, improve blood lipid composition, reduce atherosclerosis, maintain bone mineral density, show anticancer activity, and have cognitive effects in menopausal women. Polyphenols like daidzein isolated from red clover leaves have been shown to affect hot flashes and blood lipid composition; genistein, isolated from the leaves, stems, and aerial parts of the plant, has an effect on hot flashes, blood lipid composition, and atherosclerosis.

Despite the limited clinical data, various bioactive components isolated from red clover plant show different activities with regard to menopausal symptoms, and as such should receive more research interest with regard to developing new pharmacological strategies to improve women's health during menopause.

Author Contributions: Conceptualization, M.Z., I.T. and I.B.; Methodology, M.Z., I.T. and I.B.; Resources, M.Z., I.T. and I.B.; Writing—Original Draft Preparation, M.Z.; Writing—Review and Editing, I.T. and I.B.; Visualization, M.Z.; Supervision, I.T. and I.B. All authors have read and agreed to the published version of the manuscript.

Funding: This research received no external funding.

Institutional Review Board Statement: Not applicable.

Informed Consent Statement: Not applicable.

Data Availability Statement: Not applicable.

Conflicts of Interest: The authors declare no conflicts of interest.

References

1. Kanadys, W.; Barańska, A.; Błaszczuk, A.; Polz-Dacewicz, M.; Drop, B.; Kanecki, K.; Malm, M. Evaluation of Clinical Meaningfulness of Red Clover (*Trifolium pratense* L.) Extract to Relieve Hot Flushes and Menopausal Symptoms in Peri- and Post-Menopausal Women: A Systematic Review and Meta-Analysis of Randomized Controlled Trials. *Nutrients* **2021**, *13*, 1258. [CrossRef] [PubMed]
2. Davis, S.R.; Lambrinoudaki, I.; Lumsden, M.; Mishra, G.D.; Pal, L.; Rees, M.; Santoro, N.; Simoncini, T. Menopause. *Nat. Rev. Dis. Primers* **2015**, *1*, 15004. [CrossRef] [PubMed]
3. Monteleone, P.; Giulia Mascagni, G.; Andrea Giannini, A.; Genazzani, A.R.; Simoncini, T. Symptoms of menopause—Global prevalence, physiology and implications. *Nat. Rev. Endocrinol.* **2018**, *14*, 199–215. [CrossRef] [PubMed]
4. Jankowska, K. Premature ovarian failure. *Menopause Rev.* **2017**, *16*, 51–56. [CrossRef]
5. Ivančević, Ž.; Rumbolt, Z.; Bergovec, M.; Silobrčić, V.; Sardelić, S. *Principi Interne Medicine*; Placebo d.o.o.: Split, Croatia, 2002.
6. Howard, S.R. Interpretation of reproductive hormones before, during and after the pubertal transition—Identifying health and disordered puberty. *Clin. Endocrinol.* **2021**, *95*, 702–715. [CrossRef]
7. Alemany, M. The Roles of Androgens in Humans: Biology, Metabolic Regulation and Health. *Int. J. Mol. Sci.* **2022**, *23*, 11952. [CrossRef]
8. Akbaribazm, M.; Khazaei, M.R.; Khazaei, M. *Trifolium pratense* L. (red clover) extract and doxorubicin synergistically inhibits proliferation of 4T1 breast cancer in tumor-bearing BALB/c mice through modulation of apoptosis and increase antioxidant and anti-inflammatory related pathways. *Food Sci. Nutr.* **2020**, *8*, 4276–4290. [CrossRef] [PubMed]
9. Wang, X.; Wang, L.; Di, J.; Zhang, X.; Zhao, G. Prevalence and risk factors for menopausal symptoms in middle-aged Chinese women: A community-based cross-sectional study. *Menopause J. N. Am. Menopause Soc.* **2021**, *28*, 1271–1278. [CrossRef]
10. Fistonic, I. Menopause in Croatia. Socio-demographic characteristics, women's attitudes and source of information, compliance with HRT. *Maturitas* **2004**, *47*, 91–98. [CrossRef]

11. Lambert, M.N.T.; Thorup, A.C.; Hansen, E.S.S.; Jeppesen, P.B. Combined red clover isoflavones and probiotics potently reduce menopausal vasomotor symptoms. *PLoS ONE* **2017**, *12*, e0176590. [CrossRef]
12. Ferraris, C.; Ballestra, B.; Listorti, C.; Cappelletti, V.; Reduzzi, C.; Scaperrotta, G.P.; Pulice, I.; Ferrari, E.G.A.; Folli, S.; Mariani, L.; et al. Red clover and lifestyle changes to contrast menopausal symptoms in premenopausal patients with hormone-sensitive breast cancer receiving tamoxifen. *Breast Cancer Res. Treat.* **2020**, *180*, 157–165. [CrossRef] [PubMed]
13. Kanadys, W.; Baranska, A.; Jedrych, M.; Religioni, U.; Janiszewska, M. Effects of red clover (*Trifolium pratense*) isoflavones on the lipid profile of perimenopausal and postmenopausal women—A systematic review and meta-analysis. *Maturitas* **2020**, *132*, 7–16. [CrossRef] [PubMed]
14. Booth, N.L. *Red Clover (Trifolium pratense) as a Botanical Dietary Supplement*; University of Illinois at Chicago, Health Sciences Center ProQuest Dissertations Publishing: Ann Arbor, MI, USA, 2005.
15. Shakeri, F.; Taavoni, F.; Goushegir, A.; Haghani, H. Effectiveness of red clover in alleviating menopausal symptoms: A 12-week randomized, controlled trial. *Climacteric* **2015**, *18*, 568–573. [CrossRef]
16. Trusunović, A. Diet and dietary supplements in the prevention of osteoopenia and osteoporosis in women in menopause. *Food Health Dis. Sci.-Prof. J. Nutr. Diet.* **2016**, *5*, 67–72.
17. Çölgeçen, H.; Koca, U.; Büyükkaral, H.N. Use of red clover (*Trifolium pretense* L.) seeds in human therapeutics. *Nuts Seeds Health Dis. Prev.* **2020**, *2*, 421–427.
18. Grlić, L. *Encyclopedia of Wild Edible Plants*; August Cesarec: Zagreb, Croatia, 1990; p. 2.
19. Inostroza, L.; Ortega-Klose, F.; Vásquez, C.; Wilckens, R. Changes in Root Architecture and Aboveground Traits of Red Clover Cultivars Driven by Breeding to Improve Persistence. *Agronomy* **2020**, *10*, 1896. [CrossRef]
20. Naidu, V.S.G.R. *Hand Book on Weed Identification Directorate of Weed Science Research*; Department of Agriculture and Cooperation, Ministry of Agriculture: Jabalpur, India, 2012; p. 354.
21. Ulloa-Aguirre, A.; Timossi, C.; Barrios-de-Tomasi, J.; Maldonado, A.; Nayudu, P. Impact of Carbohydrate Heterogeneity in Function of Follicle-Stimulating Hormone: Studies Derived from in Vitro and in Vivo Models. *Biol. Reprod.* **2003**, *69*, 379–389. [CrossRef] [PubMed]
22. Boelt, B. Legume seed production and research in Europe. *Forage Seed* **2002**, *9*, 33–34.
23. Prati, S.; Baravelli, V.; Fabbri, D.; Schwarzinger, C.; Brandolini, V.; Maietti, A.; Tedeschi, P.; Benvenuti, S.; Macchia, M.; Marotti, I.; et al. Composition and content of seed flavonoids in forage and grain legume crops. *Sep. Sci. Food Anal.* **2007**, *30*, 491–501. [CrossRef]
24. Oleszek, W.; Stochmal, A. Triterpene saponins and flavonoids in the seeds of Trifolium species. *Phytochemistry* **2002**, *61*, 165–170. [CrossRef]
25. Hayashi, S.; Imai, K.; Suga, K.; Kurihara, T.; Higashi, Y.; Nakachi, K. Two promoters in expression of estrogen receptor messenger RNA in human breast cancer. *Carcinogenesis* **1997**, *18*, 459–464. [CrossRef] [PubMed]
26. Saviranta, N.M.; Julkunen-Tiitto, R.; Oksanen, E.; Karjalainen, R.O. Red clover (*Trifolium pratense* L.) isoflavones: Root phenolic compounds affected by biotic and abiotic stress factors. *J. Sci. Food Agric.* **2010**, *90*, 418–423. [CrossRef] [PubMed]
27. Tsao, R.; Papadopoulos, Y.; Yang, R.; Young, J.C.; McRae, K. Isoflavone Profiles of Red Clovers and Their Distribution in Different Parts Harvested at Different Growing Stages. *J. Agric. Food Chem.* **2006**, *54*, 5797–5805. [CrossRef] [PubMed]
28. Lim, L.W.; Li, J.; Gong, Y.; Jin, A.; Yuan, J.-M.; Yong, E.L.; Koh, W.-P. Serum estrogen receptor bioactivity and breast cancer risk among postmenopausal women. *Endocr. Relat. Cancer* **2014**, *21*, 263–273. [CrossRef] [PubMed]
29. Tava, A.; Ramella, D.; Grecchi, M.; Aceto, P.; Paoletti, R.; Piano, E. Volatile constituents of *Trifolium pratense* and *T. repens* from N.E. Italian alpine pastures. *Nat. Prod. Commun.* **2009**, *4*, 835–838. [CrossRef] [PubMed]
30. Mikulić, M.; Atanackovic Krstonošić, M.; Kladar, N.; Vasiljević, S.; Katanski, S.; Mamlić, Z.; Rakić, D.; Cvejić, J. Phytochemical Composition of Different Red Clover Genotypes Based on Plant Part and Genetic Traits. *Foods* **2024**, *13*, 103. [CrossRef]
31. Murota, K.; Terao, J. Antioxidative flavonoid quercetin: Implication of its intestinal absorption and metabolism. *Arch Biochem Biophys* **2003**, *417*, 12–17. [CrossRef]
32. Grieve, M.A. *A Modern Herbal*; Dover Publications: New York, NY, USA, 1978; pp. 207–208.
33. de Bairacli, J. *Common Herbs for Natural Health*; Faber & Faber: London, UK, 1974.
34. Ossadcha-Janata, N. *Herbs Used in Ukrainian Folk Medicine*; Mimeographed Series #21; Research Program on the USSR and the NY Botanical Garden: New York, NY, USA, 1952.
35. Hamel, P.; Chiltoskey, M. *Cherokee Plants and Their Uses—A 400 Year History*; Herald Publishing Company: Sylva, NC, USA, 1975.
36. Herrick, J.W. Iroquois Medical Botany. Ph.D. Dissertation, The State University of New York at Albany, Albany, NY, USA, 1977.
37. Krag, K.J. Plants Used as Contraceptives by the North American Indians: An Ethnobotanical Study: Senior Honors. Ph.D. Thesis, Harvard University, Cambridge, MA, USA, 1976; 117p.
38. Felter, H.W.; Lloyd, J.U. *King's American Dispensatory*; Ohio Valley Company: Cincinnati, OH, USA, 1999.
39. Mintziori, G.; Lambrinoudaki, I.; Goulis, D.G.; Ceausu, I.; Depypere, H.; Erel, C.T.; Pérez-López, F.R.; Schenck-Gustafsson, K.; Simoncini, T.; Tremollieres, F.; et al. EMAS position statement: Non-hormonal management of menopausal vasomotor symptoms. *Maturitas* **2015**, *81*, 410–413. [CrossRef]
40. Booth, N.L.; Piersen, C.E.; Banuvar, S.; Geller, S.E.; Shulman, L.P.; Farnsworth, N.R. Clinical studies of red clover (*Trifolium pratense*) dietary supplements in menopause. *Menopause* **2006**, *13*, 251–264. [CrossRef] [PubMed]

41. Lemežienė, N.; Padarauskas, A.; Butkutė, B.; Ceseviciene, J. The concentration of isoflavones in red clover (*Trifolium pratense L.*) at flowering stage. *Zemdirbyste-Agriculture* **2015**, *102*, 443–448. [CrossRef]
42. Lipovac, M.; Chedraui, P.; Gruenhut, C.; Gocan, A.; Kurz, C.; Neuber, B.; Imhof, M. The effect of red clover isoflavone supplementation over vasomotor and menopausal symptoms in postmenopausal women. *Gynecol. Endocrinol.* **2012**, *28*, 203–207. [CrossRef]
43. Gartoulla, P.; Worsley, R.; Bell, R.J.; Davis, S.R. Moderate to severe vasomotor and sexual symptoms remain problematic for women aged 60 to 65 years. *Menopause* **2015**, *22*, 694–701. [CrossRef] [PubMed]
44. Nestel, P.J.; Pomeroy, S.; Kay, S. Isoflavones from red clover improve systemic arterial compliance but not plasma lipids in menopausal women. *J. Clin. Endocrinol. Metab.* **1999**, *84*, 895–898. [CrossRef] [PubMed]
45. Marjoribanks, J.; Farquhar, C.; Roberts, H.; Lethaby, A.; Lee, J. Long-term hormone therapy for perimenopausal and post-menopausal women. *Cochrane Database Syst. Rev.* **2017**, *17*, CD004143. [CrossRef] [PubMed]
46. Santell, R.C.; Chang, Y.C.; Nair, M.G.; Helferich, W.G. Dietary Genistein Exerts Estrogenic Effects upon the Uterus, Mammary Gland and the Hypothalamic/Pituitary Axis in Rats. *J. Nutr.* **1997**, *127*, 263–269. [PubMed]
47. Lipovac, M.; Pfitscher, A.; Hobiger, S.; Laschitz, T. Red clover isoflavone metabolite bioavailability is decreased after fructooligosaccharide supplementation. *Fitoterapia* **2015**, *105*, 93–101. [CrossRef]
48. Tao, L.; Marzecová, A.; Taft, M.; Asanowicz, D.; Wodniecka, Z. The efficiency of attentional networks in early and late bilinguals: The role of age of acquisition. *Front. Psycol.* **2011**, *2*, 123. [CrossRef] [PubMed]
49. French, S.; Smith, A.L.; Walker, E., II; Hurley, K.S. Peer Relationship Profi les and Motivation in Youth Sport. *J. Sport Exerc. Psychol.* **2006**, *28*, 362–382.
50. Schonberg, M.A.; Davis, R.B.; Wee, C.C. After the women's health initiative: Decision making and trust of women taking hormone therapy. *Women's Health Issues* **2005**, *15*, 187–195. [CrossRef]
51. Nedrow, A.; Miller, J.; Walker, M.; Nygren, P.; Huffman, L.H.; Nelson, H.D. Complementary and alternative therapies for the management of menopause-related symptoms: A systematic evidence review. *Arch. Intern. Med.* **2006**, *166*, 1453–1465. [CrossRef]
52. Wong, W.W.; Lewis, R.D.; Steinberg, F.M.; Murray, M.J.; Cramer, M.A.; Amato, P.; Young, R.L.; Barnes, S.; Ellis, K.J.; Shypailo, R.J.; et al. Soy isoflavone supplementation and bone mineral density in menopausal women: A 2-y multicenter clinical trial. *Am. J. Clin. Nutr.* **2009**, *90*, 1433–1439. [CrossRef] [PubMed]
53. Avis, N.E.; Crawford, S.L.; Greendale, G.; Bromberger, J.T.; Everson-Rose, S.A.; Gold, E.B.; Hess, R.; Joffe, H.; Kravitz, H.M.; Tepper, P.G. Duration of menopausal vasomotor symptoms over the menopause transition. *JAMA Intern. Med.* **2015**, *175*, 531–539. [CrossRef] [PubMed]
54. Freeman, E.W.; Sherif, K. Prevalence of hot flushes and night sweats around the world. *Climacteric* **2007**, *10*, 197–214. [CrossRef] [PubMed]
55. Gold, E.B.; Colvin, A.; Avis, N.; Bromberger, J.; Greendale, G.A.; Powell, L.; Sternfeld, B.; Matthews, K. Longitudinal analysis of the association between vasomotor symptoms and race/ethnicity across the menopausal transition. *Am. J. Public Health* **2006**, *96*, 1226–1235. [CrossRef]
56. Duffy, O.K.; Iversen, L.; Hannaford, P.C. Factors associated with reporting classic menopausal symptoms differ. *Climacteric* **2013**, *16*, 240–251. [CrossRef] [PubMed]
57. Herber-Gast, G.-C.M.; Mishra, G.D.; van der Schouw, Y.T.; Brown, W.J.; Dobson, A.J. Risk factors for night sweats and hot flushes in midlife. *Menopause* **2013**, *20*, 953–959. [CrossRef] [PubMed]
58. Chung, H.-F.; Pandeya, N.; Dobson, A.J.; Kuh, D.; Brunner, E.J.; Crawford, S.L.; Avis, N.E.; Gold, E.B.; Mitchell, E.S.; Woods, N.F. The role of sleep difficulties in the vasomotor menopausal symptoms and depressed mood relationships. InterLACE consortium. *Psychol. Med.* **2018**, *48*, 2550–2561. [CrossRef]
59. Geukes, M.; van Aalst, M.P.; Robroek, S.J.; Laven, J.S.E.; Oosterhof, H. The impact of menopause on work ability in women with severe menopausal symptoms. *Maturitas* **2016**, *90*, 3–8. [CrossRef]
60. Rossmanith WGRuebberdt, W. What causes hot flushes? The neuroendocrine origin of vasomotor symptoms in the menopause. *Gynecol. Endocrinol.* **2009**, *25*, 303–314. [CrossRef]
61. Franco, O.H.; Chowdhury, R.; Troup, J.; Voortman, T.; Kunutsor, S.; Kavousi, M.; Oliver-Williams, C.; Muka, T. Use of Plant-Based Therapies and Menopausal Symptoms. *JAMA* **2016**, *315*, 2554–2563. [CrossRef]
62. Harlow, S.D.; Gass, M.; Hall, J.E.; Lobo, R.; Maki, P.; Rebar, R.W.; Sherman, S.; Sluss, P.M.; De Villiers, T.J.; STRAW+ 10 Collaborative Group. Executive summary of the Stages of Reproductive Aging Workshop + 10: Addressing the unfinished agenda of staging reproductive aging. *Menopause* **2012**, *19*, 387–395. [CrossRef] [PubMed]
63. Freedman, R.R. Menopausal hot flashes: Mechanisms, endocrinology, treatment. *J. Steroid Biochem. Mol. Biol.* **2014**, *142*, 115–120. [CrossRef] [PubMed]
64. Tice, J.A.; Ettinger, B.; Ensrud, K.; Wallace, R.; Blackwell, T.; Cummings, S.R. Phytoestrogen supplements for the treatment of hot flashes: The isoflavone clover extract (ICE) study: A randomized controlled trial. *JAMA* **2003**, *290*, 207–214. [CrossRef]
65. Atkinson, C.; Compston, J.E.; Day, N.E.; Dowsett, M.; Bingham, S.A. The effects of phytoestrogen isoflavones on bone density in women: A double-blind, randomized, placebo-controlled trial. *Am. J. Clin. Nutr.* **2004**, *79*, 326–333. [CrossRef] [PubMed]
66. del Giorno, C.; da Fonseca, A.M.; Bagnoli, V.R.; de Assis, J.S.; Soares, J.M., Jr.; Baracat, E.C. Effects of *Trifolium pratense* on the climacteric and sexual symptoms in postmenopause women. *Rev. Assoc. Médica Bras.* **2010**, *56*, 558–562.

67. Knight, D.C.; Howes, J.B.; Eden, J.A. The effect of Promensil, an isoflavone extract, on menopausal symptoms. *Climacteric* **1999**, *2*, 79–84. [CrossRef]
68. van de Weijer, P.H.; Barentsen, R. Isoflavones from red clover (Promensil) significantly reduce menopausal hot flush symptoms compared with placebo. *Maturitas* **2002**, *42*, 187–193. [CrossRef] [PubMed]
69. Baber, R.J.; Templeman, C.; Morton, T.; Kelly, G.E.; West, L. Randomized placebo-controlled trial of an isoflavone supplement and menopausal symptoms in women. *Climacteric* **1999**, *2*, 85–92. [CrossRef] [PubMed]
70. Jeri, A.R. The use of an isoflavone supplement to relieve hot flashes. *Female Patient* **2002**, *27*, 35–37.
71. Schneider, H.; Heinemann, L.; Rosemeier, H.P.; Potthoff, P.; Behre, H.M. The Menopause Rating Scale (MRS): Reliability of scores of menopausal complaints. *Climacteric* **2000**, *3*, 59–64. [CrossRef]
72. Potthoff, P.; Heinemann, L.A.; Schneider, H.P.; Rosemeier, H.P.; Hauser, G.A. The Menopause Rating Scale (MRS II): Methodological standardization in the German population. *Zentralbl. Gynakol.* **2000**, *122*, 280–286. (In German) [PubMed]
73. Mu, H.; Bai, Y.H.; Wang, S.T.; Zhu, Z.M.; Zhang, Y.W. Research on antioxident effects and estrogenic effect of formononetin from *Trifolium pratense* (red clover). *Phytomedicine* **2009**, *16*, 314–319. [CrossRef] [PubMed]
74. Pakalapati, G.; Li, L.; Gretz, N.; Koch, E.; Wink, M. Influence of red clover (*Trifolium pratense*) isoflavones on gene and protein expression profiles in liver of ovariectomized rats. *Phytomedicine* **2009**, *16*, 845–885. [CrossRef]
75. Samman, S.; Wall, P.M.L.; Chan, G.S.M.; Smith, S.J.; Petocz, P. The effect of supplementation with isoflavones on plasma lipids and oxidisability of low density lipoprotein in premenopausal women. *Atherosclerosis* **1999**, *147*, 277–283. [CrossRef]
76. Clifton-Bligh, P.B.; Baber, R.J.; Fulcher, G.R.; Nery, M.L.; Moreton, T. The effect of isoflavones extracted from red clover (Rimostil) on lipid and bone metabolism. *Menopause* **2001**, *8*, 259–265. [CrossRef] [PubMed]
77. Blakesmith, S.J.; Lyons-Wall, P.M.; George, C.; Joannou, G.E.; Petocz, P.; Samman, S. Effects of supplementation with purified red clover (*Trifolium pratense*) isoflavones on plasma lipids and insulin resistance in healthy premenopausal women. *Br. J. Nutr.* **2003**, *89*, 467–474. [CrossRef] [PubMed]
78. Campbell, M.J.; Woodside, J.V.; Honour, J.W.; Morton, M.S.; Leathem, A.J.C. Effect of red clover-derived isoflavone supplementation on insulin-like growth factor, lipid and antioxidant status in healthy female volunteers: A pilot study. *Eur. J. Clin. Nutr.* **2004**, *58*, 173–179. [CrossRef]
79. Burger, H.G.; Hale, G.E.; Robertson, D.M.; Dennerstein, L. A review of hormonal changes during the menopausal transition: Focus on findings from the Melbourne Women's Midlife Health Project. *Hum. Reprod. Update* **2007**, *13*, 559–565. [CrossRef] [PubMed]
80. Schult, T.M.; Ensrud, K.E.; Blackwell, T.; Ettinger, B.; Wallace, R.; Tice, J.A. Effect of isoflavones on lipids and bone turnover markers in menopausal women. *Maturitas* **2004**, *48*, 209–218. [CrossRef]
81. Nachtigall, L.B.; La Grega, L.; Lee, W.W.; Fenichel, R.; Nachtigall, L. The effects of isoflavones derived from red clover on vasomotor symptoms and endometrial thickness. In Proceedings of the 9th International Menopause Society World Congress on the Menopause, Yokohama, Japan, 17–21 October 1999.
82. Howes, J.B.; Sullivan, D.; Lai, N. The effects of dietary supplementation with isoflavones from red clover on the lipoprotein profiles of postmenopausal women with mild to moderate hypercholesterolaemia. *Atherosclerosis* **2000**, *152*, 143–147. [CrossRef]
83. Howes, J.B.; Tran, D.; Brillante, D.; Howes, L.G. Effects of dietary supplementation with isoflavones from red clover on ambulatory blood pressure and endothelial function in postmenopausal type 2 diabetes. *Diabetes Obes. Metab.* **2003**, *5*, 325–332. [CrossRef] [PubMed]
84. Steiner, G. Treating lipid abnormalities in patients with type 2 diabetes mellitus. *Am. J. Cardiol.* **2001**, *88*, 37N–40N. [CrossRef] [PubMed]
85. Kaukua, J.; Turpeinen, A.; Uusitupa, M.; Niskanen, L. Clustering of cardiovascular risk factors in type 2 diabetes mellitus: Prognostic significance and tracking. *Diabetes Obes. Metab.* **2001**, *3*, 17–23. [CrossRef] [PubMed]
86. Hale, G.E.; Hughes, C.L.; Robboy, S.J.; Agarwal, S.K.; Bievre, M. A double-blind randomized study on the effects of red clover isoflavones on the endometrium. *Menopause* **2001**, *8*, 338–346. [CrossRef]
87. Clifton-Bligh, P.B.; Nery, M.L.; Clifton-Bligh, R.J.; Visvalingam, S.; Fulcher, G.R.; Byth, K.; Baber, R. Red clover isoflavones enriched with formononetin lower serum LDL cholesterol—A randomized, double-blind, placebo-controlled study. *Eur. J. Clin. Nutr.* **2015**, *69*, 134–142. [CrossRef] [PubMed]
88. Taher, Y.A.; ben Emhemed, H.M.; Tawati, A.M. The Menopausal Experience of Libyan Women. *JMJ* **2009**, *9*, 184–190.
89. Zhang, S.; Zhou, J.; Li, L.; Pan, X.; Lin, J.; Li, C.; Leung, W.T.; Wang, L. Effect of dehydroepiandrosterone on atherosclerosis in postmenopausal women. J-STAGE home. *BioSci. Trends* **2021**, *15*, 6. [CrossRef] [PubMed]
90. Teede, H.J.; McGrath, B.P.; DeSilva, L.; Cehun, M.; Fassoulakis, A.; Nestel, P.J. Isoflavones reduce arterial stiffness: A placebo-controlled study in men and postmenopausal women. *Arterioscler. Thromb. Vasc. Biol.* **2003**, *23*, 1066–1071. [CrossRef]
91. Garcia-Martinez, M.C.; Hermenegildo, C.; Tarin, J.J.; Cano, A. Phytoestrogens increase the capacity of serum to stimulate prostacyclin release in human endothelial cells. *Acta Obstet. Gynecol. Scand.* **2003**, *82*, 705–710. [CrossRef]
92. Simoncini, T.; Caruso, A.; Garibaldi, S.; Fu, X.D.; Giretti, M.S.; Baldacci, C.; Scorticati, C.; Fornari, L.; Mannella, P.; Genazzani, A.R. Activation of Nitric Oxide Synthesis in Human Endothelial Cells Using Nomegestrol Acetate. *Obstet. Gynecol.* **2006**, *108*, 969–978. [CrossRef]
93. Wang, Y.; Gho, W.M.; Chan, F.L.; Chen, S.; Leung, L.K. The red clover (*Trifolium pratense* L.) isoflavone biochanin A inhibits aromatase activity and expression. *Br. J. Nutr.* **2008**, *99*, 303–310. [CrossRef] [PubMed]

94. Ley, K.; Huo, Y. VCAM-1 is critical in atherosclerosis. *J. Clin. Investig.* **2001**, *107*, 1209–1210.85. [CrossRef] [PubMed]
95. Mori, A.; Murano, T.; Izumi, S.; Kika, G.; Haque, S.F.; Okuwaki, S.; Suzuki, T.; Matsubayashi, H.; Ikeda, M.; Goya, K.; et al. Impact of Menopause on Lipid and Bone Metabolism and Effect of Hormone Replacement Therapy. *Tokai J. Exp. Clin. Med.* **2003**, *28*, 109–119.
96. Kannel, W.B.; Hjortland, M.C.; Mcnamara, P.M.; Gordon, T. Menopause and Risk of Cardiovascular Disease. *Ann. Intern. Med.* **1976**, *85*, 447–452. [CrossRef]
97. Archer, D.F.; Thorneycroft, I.H.; Foegh, M.; Hanes, V.; Glant, M.D.; Bitterman, P.; Kempson, R.L. Long-term safety of drospirenone-estradiol for hormone therapy: A randomized, double-blind, multicenter trial. *Menopause* **2005**, *12*, 716–727. [CrossRef] [PubMed]
98. Nasr, A.; Breckwoldt, M. Estrogen replacement therapy and cardiovascular protection: Lipid mechanisms are the tip of an iceberg. *Gynecol. Endocrinol.* **1998**, *12*, 43–59. [CrossRef]
99. Bray, F.; Ferlay, J.; Soerjomataram, I.; Siegel, R.L.; Torre, L.A.; Jemal, A. Global cancer statistics 2018: GLOBOCAN estimates of incidence and mortality worldwide for 36 cancers in 185 countries. *Cancer J. Clin.* **2018**, *20*, 313. [CrossRef] [PubMed]
100. Fisher, B.; Redmond, C.; Dimitrov, N.V.; Bowman, D.; Legault-Poisson, S.; Wigkerham, D.L.; Wolmark, N.; Fisher, E.R.; Margolese, R.; Sutherland, C.; et al. A Randomized Clinical Trial Evaluating Sequential Methotrexate and Fluorouracil in the Treatment of Patients with Node-Negative Breast Cancer Who Have Estrogen-Receptor-Negative Tumors. *N. Engl. J. Med.* **1989**, *320*, 473–478. [CrossRef]
101. EBCTC Group; Zhou, Y.; Dai, R.P.; Gao, R.L.; Lü, S.Z.; Chen, Y.D. Clinical evaluation of intracoronary in-stent stenosis by electron-beam CT single flow mode study. *Zhonghua Xin Xue Guan Bing Za Zhi* **2005**, *33*, 687–690. [PubMed]
102. Paranjpe, R.; John, G.; Trivedi, M.; Abughosh, S. Identifying adherence barriers to oral endocrine therapy among breast cancer survivors. *Breast Cancer Res. Treat.* **2019**, *174*, 297–305. [CrossRef]
103. Hershman, D.L.; Tager, F.A.; McKinley, P.S.; Schnabel, F.R.; El-Tamer, M.; Cheung, Y.K.K.; Fang, Y.; Golden, C.R.; Frosch, M.E.; Habif, U.; et al. The cognitive effects of chemotherapy in post-menopausal breast cancer patients: A controlled longitudinal study. *Breast Cancer Res. Treat.* **2010**, *123*, 25–34.
104. Wassermann, J.; Rosenberg, S.M. Treatment Decisions and Adherence to Adjuvant Endocrine Therapy in Breast Cancer. *Curr. Breast Cancer Rep.* **2017**, *9*, 100–110. [CrossRef]
105. Sundaresan, A.; Radhiga, T.; Deivasigamani, B. Biological activity of biochanin A: A review. *Asian J. Pharm. Pharmacol.* **2018**, *4*, 1–5. [CrossRef]
106. Greendale, G.A.; Reboussin, B.A.; Sie, A. Effects of estrogen and estrogen-progestin on mammographic parenchymal density. Postmenopausal estrogen/progestin interventions (PEPI) investigators. *Ann. Intern. Med.* **1999**, *130*, 262–269. [CrossRef] [PubMed]
107. Hankinson, S.E.; Willett, W.C.; Colditz, G.A. Circulating concentrations of insulin-like growth factor-I and risk of breast cancer. *Lancet* **1998**, *351*, 1393–1396. [CrossRef]
108. Ingram, D.M.; Hickling, C.; West, L.; Mahe, L.J.; Dunbar, P.M. A double-blind randomized controlled trial of isoflavones in the treatment of cyclical mastalgia. *Breast* **2002**, *11*, 170–174. [CrossRef] [PubMed]
109. Maskarinec, S. The Effect of Phytoestrogens on Hot Flashes. *Nutr. Bytes* **2003**, *9*.
110. Mullooly, M.; Bejnordi, B.E.; Pfeiffer, R.M.; Fan, S.; Vacek, P.M.; Weaver, D.L.; Herschorn, S.; Brinton, L.A.; van Ginneken, B.; Karssemeijer, N.; et al. Using deep convolutional neural networks to identify and classify tumor-associated stroma in diagnostic breast biopsies. *Mod. Pathol.* **2018**, *31*, 1502–1512.
111. Johnson, E.B.; Muto, M.G.; Yanushpolsky, E.H.; Mutter, G.L. Phytoestrogen supplementation and endometrial cancer. *Obstet. Gynecol.* **2001**, *98*, 947–950.
112. Paterni, I.; Granchi, C.; Katzenellenbogen, J.A.; Minutolo, F. Estrogen receptors alpha (ERα) and beta (ERβ): Subtype-selective ligands and clinical potential. *Steroids* **2014**, *90*, 13–29. [CrossRef]
113. Szukiewicz, D.; Stangret, A.; Ruiz-Ruiz, C.; Olivares, E.G.; Soriţău, O.; Suşman, S.; Szewczyk, G. Estrogen- and Progesterone (P4)-Mediated Epigenetic Modifications of Endometrial Stromal Cells (EnSCs) and/or Mesenchymal Stem/Stromal Cells (MSCs) in the Etiopathogenesis of Endometriosis. *Stem Cell Rev. Rep.* **2021**, *17*, 1174–1193. [CrossRef] [PubMed]
114. Staar, S.; Richter, D.U.; Makovitzky, J.; Briese, V.; Bergemann, C. Stimulation of Endometrial Glandular Cells with Genistein and Daidzein and their Effects on ERα- and ERβ-mRNA and Protein Expresion. *Anticancer. Res.* **2005**, *25*, 1713–1718. [PubMed]
115. Papadopoulou, S.K.; Papadimitriou, K.; Voulgaridou, G.; Georgaki, E.; Tsotidou, E.; Zantidou, O.; Papandreou, D. Exercise and Nutrition Impact on Osteoporosis and Sarcopenia—The Incidence of Osteosarcopenia: A Narrative Review. *Nutrients* **2021**, *13*, 4499. [CrossRef] [PubMed]
116. WHO. Assessment of fracture risk and its application to screening for postmenopausal osteoporosis. Report of a WHO Study Group. *World Health Organ. Tech. Rep. Ser.* **1994**, *843*, 1–129.
117. Wiktorowicz, M.E.; Goeree, R.; Papaioannou, A. Economic Implications of Hip Fracture: Health Service Use, Institutional Care and Cost in Canada. *Osteoporos. Int.* **2001**, *12*, 271–278. [CrossRef] [PubMed]
118. Corletto, V.; Palma, S.D.; Lavarino, C.; Birindelli, S.; Pilotti, S. Unilateral aneuploid dedifferentiated acinic cell carcinoma associated with bilateral-low grade diploid acinic cell carcinoma of the parotid gland. *Virchows Arch.* **1999**, *434*, 361–365.
119. Davis, S.R.; Zeleke, B.M.; Fradkin, P.; Bell, R.J. Vasomotor symptoms and urogenital atrophy in older women: A systematic review. *Climacteric* **2015**, *18*, 112–120.

120. Harlow, S.D.; Gass, M.; Hall, J.E.; Lobo, R.; Maki, P.; Rebar, R.W.; Sherman, S.; Sluss, P.M.; de Villiers, T.J. Summary of workshops on stages of reproductive aging. Dealing with the unfinished stage plan of reproductive aging. *J. Clin. En-Docrinol. Metab.* **2012**, *97*, 1159–1168. [CrossRef]
121. Giannini, A.; Caretto, M.; Genazzani, A.R.; Simoncini, T. Neuroendocrine Changes during Menopausal Transition. *Endocrines* **2021**, *2*, 405–416. [CrossRef]
122. Farrag, A.F.; Khedr, E.M.; Abdel-Aleem, H.; Rageh, T.A. Effects of Surgical Menopause on Cognitive Functions. *Dement. Geriatr. Cogn. Disord.* **2002**, *13*, 193–198. [CrossRef]
123. Orentreich, N.; Brind, J.L.; Vogelman, J.H.; Andres, R.; Baldwin, H. Long-term longitudinal measurements of plasma dehydroepiandrosterone sulfate in normal men. *J. Clin. Endocrinol. Metab.* **1992**, *75*, 1002–1004. [PubMed]
124. Takahashi, R.; Miyazaki, H.; Ochiya, T. The Roles of MicroRNAs in Breast Cancer. *Cancers* **2015**, *7*, 598–616. [CrossRef] [PubMed]
125. Potter, B.; Schrager, S.; Dalby, J.; Torell, E.; Hampton, A. Menopause. *Prim. Care* **2018**, *45*, 625–641. [CrossRef] [PubMed]
126. Howes, J.B.; Bray, K.; Lorenz, L.; Smerdely, P.; Howes, L.G. The effects of dietary supplementation with isoflavones from red clover on cognitive function in postmenopausal women. *Climacteric* **2004**, *7*, 70–77. [CrossRef] [PubMed]
127. Red Clover. Natural Medicines Website. Available online: https://naturalmedicines.therapeuticresearch.com (accessed on 1 April 2020).
128. Adams, N.R. Organizational and activational effects of phytoestrogens on the reproductive tract of the ewe. *Proc. Soc. Exp. Biol. Med.* **1995**, *208*, 87–91. [CrossRef] [PubMed]
129. Karimpour-Reihan, S.; Firuzei, E.; Khosravi, M.; Abbaszade, M. Coagulation Disorder following Red Clover (*Trifolium pratense*) Misuse: A Case Report. *Adv. J. Emerg. Med.* **2018**, *2*, e20. [PubMed]

nutraceuticals

MDPI

Review

Nutraceutical Aspects of Selected Wild Edible Plants of the Italian Central Apennines

Francesca Fantasma †, Vadym Samukha †, Gabriella Saviano, Maria Giovanna Chini, Maria Iorizzi and Claudio Caprari *

Department of Biosciences and Territory, University of Molise, C.da Fonte Lappone Snc, 86090 Pesche, Isernia, Italy; fantasma@unimol.it (F.F.); v.samukha@studenti.unimol.it (V.S.); saviano@unimol.it (G.S.); mariagiovanna.chini@unimol.it (M.G.C.); iorizzi@unimol.it (M.I.)

* Correspondence: claudio.caprari@unimol.it; Tel.: +39-874-404152

† These authors contributed equally to this work.

Abstract: All over the world, wild edible plants are an essential source of chemical components that justify their use in folk medicine. The aim of this review is to document and summarize the knowledge of ten wild plants analyzed in a previous study for their ethnomedical significance. *Achillea millefolium, Borago officinalis, Foeniculum vulgare, Gentiana lutea, Juniperus communis, Laurus nobilis, Malva sylvestris, Satureja montana, Silybum marianum* and *Urtica dioica* were the subjects of our study. They are commonly found in the central Italian Apennines and the Mediterranean basin. Phytochemicals contained in wild plants, such as phenols, polyphenols, flavonoids, condensed tannins, carotenoids, etc., are receiving increasing attention, as they exert a wide range of biological activities with resulting benefits for human health. Based on the 353 studies we reviewed, we focused our study on the following: (a) the ethnobotanical practices and bioactive phytochemicals; (b) the composition of polyphenols and their role as antioxidants; (c) the methodologies commonly used to assess antioxidant activity; (d) the most advanced spectroscopic and spectrometric techniques used to visualize and characterize all components (metabolomic fingerprinting). The potential of pure compounds and extracts to be used as nutraceuticals has also been highlighted through a supposed mechanism of action.

Keywords: wild edible plants; ethnobotany; medicinal food; nutraceuticals; functional foods; Italian Apennines; Mediterranean basin

Citation: Fantasma, F.; Samukha, V.; Saviano, G.; Chini, M.G.; Iorizzi, M.; Caprari, C. Nutraceutical Aspects of Selected Wild Edible Plants of the Italian Central Apennines. *Nutraceuticals* **2024**, *4*, 190–231. https://doi.org/10.3390/nutraceuticals4020013

Academic Editors: Ivan Cruz-Chamorro and Guillermo Santos Sánchez

Received: 14 December 2023
Revised: 16 March 2024
Accepted: 2 April 2024
Published: 9 April 2024

1. Introduction

In recent decades, several epidemiological studies have shown a progressive growth in the incidence of chronic degenerative diseases in the population, mainly due to an incorrect diet. The main factors responsible for the pathogenesis of degenerative diseases are believed to be oxidative stress and inflammatory processes, which are involved in cardiovascular diseases, rheumatoid arthritis, and diabetes mellitus [1]. Thus, medicinal plants rich in antioxidants, such as polyphenols, flavonoids, and carotenoids, may contribute to the prevention of chronic diseases [2,3]. Natural products are important therapeutic agents and are becoming an attractive option as they have a lower incidence of adverse reactions and lower costs than synthetic pharmaceuticals [4].

The supplementation of natural products with antioxidant activity into the diet is therefore considered the main solution to reduce the occurrence of many health problems. For this reason, there is a growing interest in unexplored plants or wild plants characterized by bioactive molecules with potential health-beneficial effects [5]. Wild plants have been known for centuries in folk medicine for their therapeutic properties, and many scientific studies have determined the chemical composition of plant extracts and highlighted their side effects. In folk medicine, commonly used extraction procedures include conventional methods such as maceration, percolation, decoction, and infusion. Most wild plants known for their therapeutic effects are often used as food. They are especially common in salads as green vegetables, but are also cooked, fried, or boiled, used in omelet dishes, etc. [6].

Wild plants may play an essential role in a healthy diet as an alternative source of minerals, vitamins, and phytochemicals with antioxidant capacities, or they can be used as supplements to formulate functional foods. The use of wild plants in popular cuisine is a common practice in Italy [7], but their consumption varies in different regional districts. Generally, all plant organs are consumed, including the leaves, tender stems, bulbs, seeds, and roots [8]. In Italian tradition, as well as in other countries, wild plants are generally eaten in salads or as boiled vegetables in soups, herb omelets or drink preparations [8]. The Central Apennine mountains represent one of the world's biodiversity hotspots, with a rich flora characterized by the presence of numerous endemics [9].

In recent decades, social and economic changes have caused the depopulation of mountain villages in the Apennines; as a result, wild food plants are declining due to the lack of oral transmission of traditional knowledge from generation to generation. Modern lifestyles are quickly transforming traditions, and the consumption of wild foods is not as common as it was in the past [10]. Minerals and the primary metabolites of plants are essential for humans. In contrast, the secondary metabolites of plants are not vital to humans, but experimental research has shown that they promote health and longevity [10].

The ten wild edible plants reported in this analysis are the result of a selection based on these criteria:

(1) An analysis of about 90 wild plants reported for their ethnopharmacological properties by Fortini et al. [11];

(2) The selection of the ten most common wild edible plants identified in the flora of the south–central Apennines that are used as food or beverages without adverse health effects [8,11].

The aim of the present review is as follows:

(a) To provide an overview of wild food plants typical of the Mediterranean basin and traditionally used in the gastronomy of the central Apennine area (Italy), characterized by their high biological diversity and whose cultural heritage is well known;

(b) To summarize the current knowledge on the potential use of edible wild plant extracts in the prevention or treatment of some of the most widespread diseases in developed countries, such as cardiovascular diseases, cancer, neurodegenerative diseases, diabetes, obesity, and liver disease;

(c) To illustrate and discuss the chemical composition of edible wild plants, their nutritional properties, and their relationship with the biological effects reported in the literature for the development of nutraceuticals or functional foods;

(d) To identify the methodologies that are commonly used to assess antioxidant activity in vitro and in vivo, and to highlight widely used advanced spectroscopic and spectrometric methods that are able to identify all primary and secondary metabolites contained in an extract (metabolic fingerprinting).

2. Strategy of Searching Articles

A comprehensive phytochemical analysis of the ethnobotanical literature and a biological activity search on the food plants used in the Apennine and Mediterranean area were carried out using existing online scientific databases, such as Scopus, Web of Science, Wiley Online Library, and Science Direct, as well as Google Scholar keywords. The search was limited from 2000 to November 2023. The information summarized in Tables 1 and 2 was obtained from research articles (in vivo or in vitro studies). A total of 353 studies were selected and included in this review.

3. Role of Nutraceuticals

The term nutraceutical is commonly used in marketing, but there is no regulatory definition. Nutraceuticals are substances that are not recognized as nutrients but that have positive physiological effects on the human body and possess multiple therapeutic benefits. Epidemiological studies indicate that a diet rich in plant-based foods significantly reduces the risk of chronic-degenerative diseases, suggesting that some natural components found in plants may be effective agents for the prevention of diabetes, hypertension, heart disease, Alzheimer's disease, and arteriosclerosis [12,13].

Some nutrients, herbal products, probiotics, polyunsaturated fatty acids, and phytochemicals are considered nutraceuticals. Since ancient times, many plant extracts, now marketed as herbal products, have provided hundreds of remedies to treat acute and chronic diseases.

Herbal nutraceuticals are foods prepared from plants, and some examples include Yarrow (*Achillea millefolium*), which contains bioactive components useful for treating lack of appetite, gastric disorders, or diarrhea, or Chamomile (*Matricaria recutita*), which is widely used to treat insomnia, gastrointestinal disorders, inflammation, wounds, ulcers, muscle spasms, etc. [14].

The term 'probiotics' refers to live microorganisms that, when administered in sufficient quantities, provide health benefits by regulating the balance of human intestinal microorganisms and inhibiting the colonization of pathogenic bacteria in the gut. In addition, probiotics help the body build a healthy protective layer of the intestinal mucosa, enhancing the intestinal barrier effect and improving immunity [15].

Polyunsaturated fatty acids (PUFAs), mainly omega-3 and omega-6, have been shown to decrease the production of inflammatory eicosanoids, cytokines, and reactive oxygen species (ROS), and to possess immunomodulatory effects. They are, therefore, able to alleviate inflammatory pathologies and are effective in the prevention and treatment of coronary heart disease, hypertension, diabetes, arthritis, and other inflammations [16].

Phytochemicals are non-nutritive bioactive plant components that have attracted interest in human nutrition due to their potential effects as antioxidants, as well as their anti-inflammatory, immunomodulatory, and anticarcinogenic effects. They are naturally occurring secondary metabolites that impart color, taste, odor, and texture to plants [17]. Many vegetables, wild plants, legumes, whole grains, fruit, fruit and vegetable juices, tea/coffee, and spices have nutraceutical properties as they contain compounds with antioxidant activity, such as flavonoids, phenolic acids, anthocyanins, terpenoids, tannins, carotenoids, phenylethanoid glycosides, etc. [18]. Antioxidants used in the diet consist of different phytochemical molecules that are present in low concentrations [18], and their consistent inclusion in the diet has a protective effect against free-radical-related disease [19,20].

Polyphenols are found more abundantly in the edible parts of plants and are considered one of the main classes of plant compounds responsible for antioxidant activity, as they can scavenge free radicals such as reactive oxygen species; thus, they are of particular interest to the food and pharmaceutical industries [21]. Polyphenols are considered the principal agents responsible for several biological [22] and pharmacological functions, as they exhibit anti-inflammatory, antimicrobial, anti-allergic, antiviral, antithrombotic, and hepatoprotective activity, and they are involved as signaling molecules in some biochemical reactions [23], and in modulating a range of cancer signaling pathways [24].

3.1. Antioxidant Activity

In biological systems, oxidative stress is a complex process characterized by an imbalance between the production of free radicals (ROS) and the body's ability to eliminate these reactive species through endogenous and exogenous antioxidants. During metabolic processes, a variety of reactions take place in which the initiators are reactive oxygen species (ROS), such as hydrogen peroxide (H_2O_2), the superoxide radical anion (O_2^-), and many others. Endogenous antioxidants are enzymes, such as superoxide dismutase (SOD), catalase (CAT), and glutathione peroxidase, while examples of non-enzymatic compounds include bilirubin and albumin [25]. When an organism is exposed to a high concentration of ROS, the endogenous antioxidant system fails, and to compensate for this antioxidant deficit, the body can use exogenous antioxidants provided through food, dietary supplements, or pharmaceuticals. The main characteristic of an antioxidant compound is that it can prevent or break the chain of oxidative propagation by stabilizing the radical that is generated, thus reducing oxidative damage in the human body. Phenolic compounds, through their distinctive chemical structure, reduce or inhibit free radicals by reducing oxidative stress and, thus, inflammatory processes [26].

Data from the literature suggest that plant polyphenols (PPs), especially phenolic acids and flavonoids, can inhibit the inflammation process by regulating the production of pro-inflammatory molecules, such as cytokines like tumor necrosis factor (TNF-α), nitric oxide (NO), and leucocyte adhesion, which are produced during inflammatory reactions. PPs have been shown to play a crucial role in the immune-inflammatory response [27,28]. Hence, inhibition of the production of such pro-inflammatory molecules is expected to have therapeutic value against inflammatory diseases. By blocking reactive oxygen species, flavonoids can mitigate photo-oxidative damage in plants [29,30].

The mechanisms of action of these compounds in the human organism have not been fully elucidated [5]. Studies have indicated that the mechanism underlying the radical-scavenging activity of polyphenols is related to the high reactivity of the phenolic OH-groups through hydrogen atom donation. Radicals can be inactivated through the following equation [26]:

$$\mathrm{PPs(OH)} + \mathrm{R}^{\bullet} \rightarrow \mathrm{PPs(O^{\bullet})} + \mathrm{RH}$$

where $R^{\bullet}$ is a free radical and $O^{\bullet}$ is a reactive oxygen species.

In flavonoids, structure–activity relationship (SAR) studies have shown that to achieve the best antioxidant activity, the following functions are required in the chemical structure: an ortho-hydroxy substitution in the B ring, a C2-C3 double bond, and a carbonyl function at C4 in the C ring (Figure 1) [31,32]. The free hydroxyl groups on the B ring donate hydrogen atoms to a radical to obtain neutral derivatives with stable molecular structures, interrupting the chain reaction. At the same time, a relatively stable flavonoid radical is produced. Flavonoids with the C2-C3 double bond in conjugation with a C4 carbonyl group are planar; this structural feature allows for a charge delocalization from the A ring to the B ring throughout the aromatic system. In flavonoids with the ortho-dihydroxy group (catechol), the formation of flavonoid phenoxy radicals can be stabilized via the mesomeric equilibrium with the ortho-semiquinone structures [33]. Moreover, some flavonoids can chelate transition metal ions (pro-oxidants), which are responsible for the production of reactive oxygen species and inhibit the lipoxygenase reaction [5].

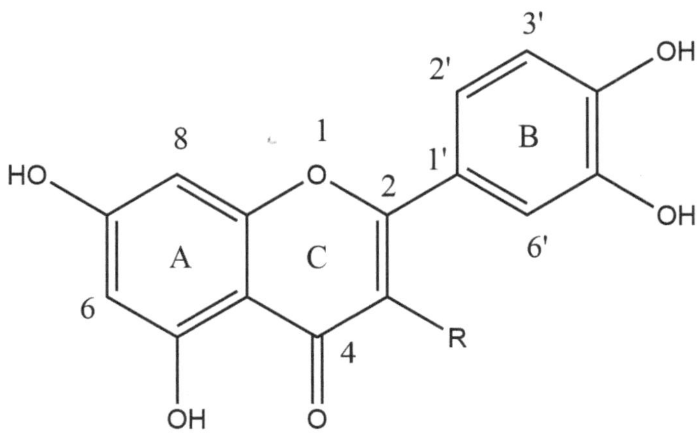

Figure 1. Basic structure of flavonoids. R=OH flavonols; R=H flavones.

Phytoestrogens belong to the class of flavonoids but, due to their structural similarity to estrogens, are able to interact with the estrogen receptor [32,34]. Although polyphenols and flavonoids are generally associated with health-promoting properties, it has been reported that, when consumed in high doses, flavonoids can act as pro-oxidants and mutagens, and are thus cytotoxic [35].

The method based on the Folin–Ciocalteu reagent is commonly used to determine and quantify total phenols in many matrices. This method evaluates the ability of phenols to react with oxidizing agents, but is not very selective as it reacts with any phenol [36].

3.2. *Methods of Estimation of Total Antioxidant Activity (TAC)*

The antioxidant activity of an extract can be evaluated in vitro or in vivo by means of simple experiments. Several in vitro methods are proposed and described in the literature to determine the effectiveness of antioxidant compounds in different matrices (plant extracts, blood, etc.) using lipophilic, hydrophilic, and amphiphilic media (emulsions). Because of their mechanism, in vitro methods can be divided into two main groups: (a) hydrogen atom transfer (HAT) reactions and (b) transfer reactions of a single electron (SET) [37].

Reducing compounds donate electrons or hydrogen atoms to compounds which have higher reduction/oxidation (redox) potentials. The latter group of compounds includes free radicals and other oxidants occurring in living systems. These methods are widely used due to their high speed and sensitivity. When assessing the antioxidant capacity/activity of a sample, more than one method should be used [37]. Diverse tests have been developed to evaluate the potential antioxidant activity of plant extracts or pure secondary metabolites in vitro. The most popular assay includes 2,20-azinobis (3-ethylbenzothiazoline-6-sulphonic acid (ABTS) [38,39]. The antioxidant activity (colorimetric method) measured by ABTS• reduction is usually referred to as that of Trolox (6-hydroxy-2,5,7,8-tetramethylchroman-2-carboxylic acid), a standard antioxidant. This allows for the results to be expressed in Trolox equivalents (TE).

2,2-diphenyl-1-picrylhydrazyl (DPPH) radical scavenging activity, a colorimetric method, is one of the most stable free radicals and is frequently used in the evaluation of radical scavengers in natural foods. The DPPH assay method is straightforward and can be used to perform a quick manual analysis of antioxidant contents [40]. The ferric reducing antioxidant power assay (FRAP), a colorimetric method, is based on the reduction of the 2,4,6-tripyridyl-s-triazine (TPTZ)–Fe^{3+} to the deep blue TPTZ-Fe^{2+} complex [41,42]. The cupric-reducing antioxidant capacity (CUPRAC) method uses copper 2+-neocuproine (2,9-dimethyl-1,10-phenanthroline), which can be reduced by antioxidants [43]. The oxygen radical absorbance capacity (ORAC) assay is based on the inhibition of the oxidation of a fluorescent substrate (and fluorescence loss) by peroxyl radicals. Commonly used peroxyl radical generators in this assay are represented by azo-compounds that decompose at elevated temperatures [44,45]. Chemiluminescence (CL) assays are based on the reaction of ROS/RNS with detection reagents to generate species in an excited state that emit light upon de-excitation to the ground state [46]. The main chemiluminescence reagents that are used include luminol, lucigenin, and peroxyoxalate [37].

Less popular methods include the potassium ferricyanide reducing power (PFRAP) assay [47], the total reactive antioxidant potential (TRAP) test [48], and the β-carotene-linoleic acid bleaching (BCLB) assay [49].

For all in vivo methods, the samples that are to be tested are usually administered to mice, rats, etc., following methods that ensure strict compliance with recommended doses and administration times. Upon completion, the animal is typically sacrificed, and the blood and/or tissues are used for analysis [50]. The main methods used with plasma are the Ferric-Reducing Ability of Plasma (FRAP) [51] and the measurement of γ-Glutamyl transpeptidase activity (GGT), which is important for glutathione homeostasis [52]. The estimate of reduced glutathione (GSH) is used as an index of cell protection from free radicals, peroxides, and other toxic compounds [53]. Estimating glutathione peroxidase (GSHPx) activity is important; this activity may indicate a disturbance of the prooxidant/antioxidant balance [54]. Glutathione-S-transferase (GST) is believed to play a physiological role in initiating the detoxification of potential alkylating agents, including pharmacologically active compounds. Under these circumstances, GST is important to carcinogenesis and chemoresistance [55]. Erythrocyte lysate can be used as a substrate for the evaluation of antioxidant enzymes such as SOD and CAT activity [50]. The glutathione reductase (GR) assay is important for maintaining the supply of reduced glutathione [56]. Finally, lipid peroxidation (LPO) is the most important test, which considers the quantity and physiological importance of biological membranes and lipoproteins. LPO is commonly used to estimate the oxidative state of the cell [57].

3.3. *Metabolomics Analysis*

The phytochemicals contained in plant food sources or non-food plants are an unlimited reservoir of nutraceutical compounds with a broad range of biological activities [58,59]. The plant metabolome consists of primary metabolites, secondary metabolites, vitamins, organic acids, alkaloids, etc.; therefore, the metabolome of plants is the main target in the search for new nutraceuticals at present [60].

Metabolomic analysis allows for the simultaneous detection of all primary and secondary metabolites in a biological system and provides qualitative and quantitative information on its components. Metabolomics is, therefore, a powerful tool for defining the phytochemical profile in an extract and allows several phenomena to be monitored. It can help to understand plant responses to stress, assess changes in natural products in different tissues/organs or during growth or ripening, etc. [61]. Metabolomics can be undertaken using two different approaches: non-targeted and targeted methods.

Currently, the two main analytical techniques used for these purposes are nuclear magnetic resonance (NMR) spectroscopy in both 1D and 2D experiments and mass spectrometry (MS), often coupled with separation techniques such as liquid or gas chromatography (LC or GC) [62–64].

NMR methods provide information on a wide range of compounds present in the plant extract in a single experiment, offering advantages in terms of the simplicity of sample preparation, the time required for analysis, high reproducibility, and the acquisition of a large amount of data in a relatively short time [65]. NMR spectroscopy is a non-destructive technique because the sample can be recovered and used in a further experiment. However, the main drawback of NMR spectroscopy is its relative lack of sensitivity, coupled with the overlapping of signals in the ^{1}H NMR spectrum of biological samples, which limits the identification of metabolites. The acquisition of 2D NMR experiment series (TOCSY, HSQC, HMBC) in metabolomics workflows can reduce the signal overlap and provide crucial information to elucidate the structure of metabolites.

Over the years, efforts have been made to improve sensitivity and resolution in NMR experiments with ultra-high-field magnets [66].

The richness of this information often results in high spectral complexity, so multivariate data analysis is required to study the spectra and extract meaningful information.

NMR is an evolving field, and many new techniques are emerging in NMR-based metabolomics analysis. Among these, high-resolution magic-angle sample spinning (HRMAS) has been increasingly applied in recent years [67] due to its potential in solid-state sample analyses without previous extraction. Other new NMR applications include hyperpolarization methods, ultrafast 2D NMR methods, pure-shift NMR techniques, and hybrid NMR approaches [67].

MS-based techniques are the most widely implemented strategies for metabolomics purposes, especially UPLC-MS with electrospray ionization (ESI), thanks to the greater sensitivity that this technique offers. In recent years, further developments have taken place using high-resolution (HR)-MS techniques, with the possibility of accurately determining the mass of a compound. However, these techniques are less robust than NMR techniques. A difference can be found between targeted and non-targeted methods. The results of the former are generally comparable across studies, whereas non-targeted methods require careful quality-control procedures to assess robustness and repeatability over time [68]. Moreover, the high sensitivity offered by MS, especially HRMS techniques, has disadvantages, such as ion suppression, meaning that other strategies are required to increase the number of metabolites.

Multivariate methods are routinely used to visualize biological data, identify possible clusters, and build predictive models based on the amount of data obtained from previous spectral datasets [69]. Multivariate data analysis (MVA) can be divided into two main categories: unsupervised analyses to explore data without any class membership and supervised analyses to discriminate among known groups of interest. Techniques such as Principal Component Analysis (PCA) and Orthogonal Partial Least Squares Discriminant Analysis (OPLS-DA) provide an essential platform for the rapid interpretation of information-rich spectral datasets to infer biological conclusions. Through the proper application of preprocessing transformations, the optimal choice of analysis algorithms, and the judicious application of validation metrics, MVA can lend a powerful hand in the biological understanding and exploration of complex metabolic systems.

4. Botanical Information, Ethnobotanical Practices, and Bioactive Phytochemicals in Wild Edible Plants

The characteristic botanical information (systematics, etymology, distribution, habitat, etc.) was obtained for each species analyzed in this study. Table 1 provides references showing where the complete information can be obtained.

Table 1. Botanical information, including scientific name and important morphological characteristics, systematics, etymology, distribution, and habitat.

Scientific Name and Important Morphological Characters	Systematics	Etymology	Distribution and Habitat	Ref.
Achillea millefolium L. Perennial herbaceous plant, roots in the rhizome; hairy stem; simple or branched; leafy; ascending; can reach up to 80 cm in height. The hairy leaves have contours, both lanceolate and linear. The flowers are white or pink, whitish achenes. The fruits are achenes.	Domain Eukaryota, Kingdom Plantae, Division Magnolio-phyta, Class Magnoliopsida, Subclass Asteridae, Order Asterales, Family Asteraceae, Subfamily Asteroideae, Tribe Anthemideae, Subtribe Achilleinae, Genus *Achillea*, Species *A. millefolium*.	Tradition (Pliny) states that Achilles healed some wounds of his comrades in arms during the siege of Troy, with the plant, hence the name of the genre. The name (milfoil) is due to its deeply indented leaves; in fact, the epithet refers to the numerous foliar laciniae that characterize this plant.	The *A. millefolium* is native to Europe; it grows in temperate regions all over the world up to 2500 m. It prefers sunny places, meadows, and the edges of paths and railways; it also adapts well to dry, stony, and acidic soils.	[70,71]
Borago officinalis L. is an herbaceous plant; it can reach up to 80 cm. It has elliptic oval leaves and petioles, with rough hair and a dark green color, collected in a basal rosette 10–15 cm long, which are then smaller on the stem. The flowers have five petals arranged in a blue-purple star. The fruits are achenes that contain small seeds.	Domain Eukaryota, Kingdom Plantae, Division Magnolio-phyta, Class Magnoliopsida, Order Solanales, Family Boraginaceae, Genus *Borago* and specie *B. officinalis*.	The etymology of its name is uncertain. Some suppose that it derives from the Arabic "abou" and "rash". Others assume that it comes from the Latin "wad" or from the Celtic "barrach", meaning "brave man". The Italian name Borage comes from the Latin Borago.	This herb is well adapted to the Mediterranean basin and widespread throughout Italy, where it grows spontaneously up to 1800 m above sea level. It prefers a rich soil, without stagnant water.	[72,73]
Foeniculum vulgare Mill is complex and difficult to summarize. It derives from the distinction between the varieties of wild fennel and "sweet" fennel (horticultural production).	Eukaryota Domain, Kingdom Plantae, Division Magnolio-phyta, Class Magnoliopsida, Subclass Rosidae, Order Apiales, Family Apiaceae, *Foeniculum* Genus *and F. vulgare* Specie.	The names comes from foenum, meaning hay, due to the subtlety of the leaves and its intense aromatic odor. Vulgare means that the plant is quite widespread (vulgar = common).	Fennel is a typical Mediterranean plant. It is mainly found in southern regions and islands, from sea level up to about 1000 m altitude. It prefers sunny, unspoilt, dry, and pebbled places.	[74,75]

Table 1. *Cont.*

Scientific Name and Important Morphological Characters	Systematics	Etymology	Distribution and Habitat	Ref.
Juniperus communis L. is an evergreen shrub or tree with a twisted trunk, of 1–10 m tall, with linear, needle-like, pungent leaves, gathered in verticils of 3. Male flowers are yellowish; female flowers are small greenish cones, which produce berries (called cuddles).	Domain Eukaryota, Kingdom Plantae, Subkingdom Tracheo bionta, Superdivision Sperma tophyta, Division Pinophyta, Class Pinopsida, Order Pinales, Family Cupressaceae, Genus *Juniperus*, Specie *J. Communis.*	The term Juniperus derives from iúnix, heifer, and pário, meaning giving birth. This is due to its presumed properties favoring childbirth. Communis epithet obviously means common, banal.	It is widespread, from marine regions to mountainous areas, and is found in dry pastures, as well as on moors or scrubland. It is a very long-lived species in the temperate regions of the northern hemisphere. It is resistant to low temperatures, and tolerates aridity and strong wind.	[76,77]
Malva sylvestris L. is an annual herbaceous plant that is biennial or perennial. The stem can grow to 60–80 cm. The leaves of palminervia have 5–7 lobes and an irregularly serrated margin. The flowers are grouped axils of leaves. The fruit is a circular poliachenio.	Eukaryota Domain, Kingdom Plantae, Division Magnolio-phyta, Class Magnoliopsida, Order Malvales, Family Malvaceae Genus *Malva* and Specie *M. sylvestris.*	The genus name, the consonant with the greek "Malatto" and "malákhe", means emollient, benevolent, with reference to the soothing properties of these plants.	The plant is native to Europe and temperate Asia; it can be found in fields and uncultivated places.	[78,79]
Gentiana lutea L. is an herbaceous, perennial species with very slow growth. It can reach up to 150 cm, with a single stem that is hollow inside, and green leaves. The flowers are yellow, sometimes punctuated with darker color, star-shaped, and gathered in bundles to the axil of the upper leaves. The fruit is an oblong oval capsule, which opens at maturity in two parts, containing brown oval seeds.	Domain Eukaryota, Kingdom Plantae, Phylum Magnolio-phyta, Class Magnoliopsida, Order Gentianales, Family Gentianaceae, Genus *Gentiana*, Specie *G. lutea*	According to Pliny, Gentiana derives from Gentius (in Greek, , Gentios) Genzio, last king of the Illyrians (II century BC), discoverer of the antimalaric properties of the roots of *G. lutea*. The lutea name derives from lúteus (yellow); that is, the floral color.	Gentian is a plant that grows in meadows and low-humidity pastures, as well as in calcareous soils, rich in organic substances, with heavy sunlight. In Italy, it grows in the central-southern Apennines, at an altitude that varies between 1000 and 2200 m above sea level.	[80,81]
Laurus nobilis L. The laurel often appears in shrubs when pruned. In natural conditions, it becomes a tall tree reaching up to 10 m. It is an evergreen plant. The leaves are ovate, dark green, leathery, glossy on the top, and dull underneath. The fruits of the laurel are black and shiny berries with only one seed.	Domain Eukaryota, Kingdom Plantae, Subkingdom Tracheo bionta, Superdivision Sperma tophyta, Division Magnolio phyta, Class Magnoliopsida, Subclass Magnoliidae, Order Laurales, Family Lauraceae, Genus *Laurus*, Specie *L. nobilis* L.	The name of this plant comes from the Latin "laus", meaning praise, to highlight the curative properties of the plant, which have been "praised" from ancient times. "Nobilis" stands for illustrious, important, famous. For others, the vulgar name would be derived from the Celtic root "laur", meaning green.	*L. nobilis* is a common species along the northern coastal areas of the Mediterranean basin. In Italy, it grows spontaneously in the central and southern areas along the coast, while in the northern regions it is cultivated.	[82,83]
Satureja montana L. is an herbaceous species which grows to 50 cm. The stems are woody at the base, tetragonal, erect, and have short back hairs when pubescent. They are usually widely branched from the bottom to form a small bush. The leaves are bright green, opposite, and subsessile. The fruit is formed by 4 oval achenes dotted with small grains.	Domain Eukaryota, Kingdom Plantae, Subkingdom-Tracheo bionta, Superdivision Sperma tophyta, Division Magnolio phyta, Class Magnoliopsida, Subclass Asteridae, Order Lamiales, Family Lamiaceae, Tribe Mentheae, Genus *Satureja*, Specie *S. montana.*	The term Satureja is of uncertain etymology. The specie name "mountain" comes from mons montis, mountain, meaning "of the mountains", because it grows 1000–1400 m above sea level.	Winter savory is a perennial semi-evergreen species native to the mountainous regions of central–southern and western Europe. Its habitat is that of calcareous, rocky, arid lands, at the edge of mountain roads, at up to 1400 m altitude.	[84]

Table 1. *Cont.*

Scientific Name and Important Morphological Characters	Systematics	Etymology	Distribution and Habitat	Ref.
Silybum marianum (L) Gaertn is an herbaceous species with vigorous bearing that can reach up to 150 cm. The plant is completely glabrous and spiny. The scape is robust, streaked, and branched, with erect branches. The plant has hermaphroditic flowers, with a tubular red-purple corolla; these are united in large globular end heads, covered with strong bracts.	Domain Eukaryota, Kingdom Plantae, Subkingdom Tracheo bionta, Superdivision Sperma tophyta, Division Magnolio phyta, Class Magnoliopsida, Subclass Asteridae, Order Asterales, Family Asteraceae, Subfamily Cichorioideae, Tribe Cardueae, Subtribe Carduinae, Genus *Silybum*, Specie *S. marianum*.	The term Silybum comes from the Greek sîlybon/síllybon, the name which Dioscorides called some edible thistles, which was taken over by Pliny to denote sillybus, a type of thistle. The name "marianum" derives from the Virgin Mary.	Milk thistle is a wild species, widespread in all the Mediterranean regions from sea level to submountain areas. Its habitat is in ruins, along roads, and in uncultivated areas, and it is common in desert and sub-desert areas ranging from the Mediterranean basin to Central Asia.	[85,86]
Urtica dioica L. Nettle is a perennial, deciduous herbaceous plant, 30–250 cm tall. It has an erect, densely hairy, striated, and grooved stem. The leaves are large, ovate, and opposite; lanceolate, jagged, and pointed; dark green on the upper side, and lighter and hairier on the lower side. The female flowers are collected in long hanging spikes, while the male flowers are grouped in erect spikes.	Domain Eukaryota, Kingdom Plantae, Division Magnolio phyta, Class Magnoliopsida, Subclass Rosidae, Order Urticales, Family Urticaceae, Genus *Urtica*, Specie *U.dioica*.	The name "nettle" probably derives from the Latin "urere" (Urtica), meaning burn, indicating the effect of the irritating substances contained in stinging hairs.	*U. dioica* is widespread in Europe, most of Asia, North Africa, and North America. In Italy, it is found in all regions: uncultivated land, woods, urbanized areas, roadside, and places in the half-shade of nitrate-rich soil, ranging from the plains to 2300 m above sea level.	[87,88]

Table 2 lists the main ethnobotanical uses, phytochemical constituents, and biological activity of the selected wild edible plants (see also Figure 2).

Table 2. Phytochemical components, ethnobotanical uses, and biological activity of selected wild edible plants.

Scientific Name	Ethnobotanical Uses	Phytochemical Components	Biological Activity	Ref.
A. millefolium L.	Tea for gastrointestinal disorders. Essential oils (from flowers) against influenza. Infusions, decoctions, or fresh juices against hemorrhage, hemorrhoidal, menstrual problems, and dysmenorrhea, toothache, headache, diuretic, wounds, and burns (hemostatic).	Rutin; luteolin 7-O-glucoside; apigenin 4′-O-glucoside; apigenin 7-O-glucoside; luteolin 4′-O-glucoside	Anti-inflammatory activity, treatment of gastrointestinal and hepato-biliary disorder and skin inflammation. The in vitro anti-inflammatory activity was established through the inhibition of matrix metalloproteinases (MMP-2 and -9), which are involved in psoriasis and atopic dermatitis and in inflammatory bowel diseases such as ulcerative colitis and Crohn's disease.	[89–93]
B. officinalis L.	Diuretic; promotes perspiration; emollient; lenitive; mild laxative; diuretic. Decoction of leaves against rheumatism and as a diuretic. Leaf poultice against tooth abscess. Digestive; depurative.	Flavonoids; phenolic acids; rosmarinic acid; syringic; sinapic; chlorogenic acids. β−sitosterol, oleuropein; lithospermic acid (leaves); tocopherols; sterols; squalene.	Anti-inflammatory properties (HaCaT and BJ cell lines) and anti-ageing properties. Weak anti-inflammatory activity in murine RAW 264.7 macrophage cell. Cytotoxic effects of extracts by MTT assay against human liver (HPG2), prostate (LNCaP) and colon (HT-29) cancer.	[11,94–104]
F. vulgare Mill.	Antispasmodic and carminative effects. Promotes intestinal peristalsis. Diuretic action. Cures respiratory diseases as an expectorant.	Cirsiliol, 4-O-caffeoylquinic acid (4-CQA); vanillic acid; O-coumaric acid; rosmarinic acid; kaempferol; resveratrol; rutin; myricetin; catechin; quercetin.	Antioxidant; antimicrobial; anti-inflammatory. Protection against cardiovascular diseases, neurological disorders, and diabetes, and hepatoprotective effects.	[105–120]

Table 2. *Cont.*

Scientific Name	Ethnobotanical Uses	Phytochemical Components	Biological Activity	Ref.
G. lutea L.	In folk medicine, it is known for its digestive and appetite-stimulating effects. Other uses include antipyretic, hepatoprotective, hypoglycemic, antianemic, and cardiotonic activity; for sores and minor wound healing; for stomach ulcers, as an immune stimulant.	Isovitexin, isosaponarin, isoorientin, isoorientin-2′-O-glucoside, and isoorientin-4′-O-glucoside	Anti-inflammatory properties, with the rate of enzyme inhibition increasing with time.	[121–127]
J. communis L.	Urinary antiseptic for acute and chronic cystitis; diuretic; emmenagogue; sudorific; digestive; anti-inflammatory. Used as a stimulant and disinfectant against constipation, chronic Bright's disease, migraine, dropsy, rheumatic swellings, and infantile tuberculosis.	Quercetin, kaempferol, myricetin, isorhamnetin, and patuletin derivatives in their composition. Quinic acids, 5-O-caffeoylquinic, catechin, epicatechin, luteolin, apigenin, naringenin, amentoflavone, and their derivatives.	Antidiabetic, anti-inflammatory, antihypercholesterolemic, antihyperlipidemic, and hepatoprotective effects. Anticancer properties alleviate cardiovascular disorders. Anticataleptic activity alleviates neuropathologies and improves the mental state of individuals.	[128–134]
L. nobilis L.	In cooking recipes, it is used to provide an aroma and a spicy flavour to meat, fish, broths, and vegetables. It is a component of two typical Italian vegetable infusions: one used as a digestive, called "canarino", and one for the treatment of respiratory aliments, called Ricotto or Ricuotto. It is used in treatments for gastro-intestinal disorders, carminative, diarrhea, hemorrhoids, stomach aches, and kidney diseases.	Isoquercitrin, luteolin, rutin, apigenin derivatives, catechin, cinnamtannin B1, epicatechin hexoside, (+)-catechin, (−)-epicatechin, epigallocatechin, and methyl eugenol. Gallic; vanillic; rosmarinic acid; ferulic acid; coumaric acid. Costunolide; santamarine; reynosine.	Anti-inflammatory: reduction in lung inflammation caused by LPS and in skin lesions and inflammation caused by Propionobacterium acnes.	[135–142]
M. sylvestris L.	Used to treat various ailments, such as colds, antiseptic, colic, constipation, cough, cystitis, high fever, migraines, puerperal mastitis, stomachic, wounds, and abscesses. Leaves, flowers, fruits, roots, shoots, and seeds are applied in infusions, decoctions, poultices, liniments, lotions, baths, and gargles.	Gossypetin 3-sulphate-8-O-β-Dglucoside; hypolaetin 3′-sulphate; isoscutellarein 8-O-ß-D-glucuronpyranoside; hypolaetin 8-O-ß-D-glucuronopyranoside; 3-O-β-D-glucopyranosyl-8-O-β-D-glucuronopyranoside; hypolaetin 4′-methyl ether 8-O-β-D-glucuronopyranoside.	Antibacterial, anti-inflammatory, antioxidant, and anti-inflammatory activity on carragenin-induced edema in rats. Antiproliferative activity on cancer cell lines. Reduction in nephrotoxicity induced by gentamicin.	[11,143–155]
S. montana L.	Effective against colds, asthma, antitussive and expectorant, cough, bronchitis, and inflammation of the respiratory tract.	Rosmarinic acid, caffeic acid and its glycoside derivatives. Quercetin, catechin or luteolin derivatives.	High antimicrobial potential, together with antioxidant and anxiolytic capacity. Hepatoprotective effects, protection against cardiovascular ailments, and uses in cancer treatment.	[156–161]
S. marianum (L) Gaertn	Antihypertensive; stimulates milk production in rats and insect (flies) repellents. Used in the treatment of liver dysfunctions and gallbladder disorders, laxatives, and breast cancer treatment.	Flavonoids. Flavonolignan complex composed of isosilychristin, isosilybin A and isosilybin B, silybin A and silybin B, silychristin, silydianin, and taxifolin mariamide A and mariamide B (seeds).	Antidiabetic agent (α-glucosidase and PTP1B inhibitory activities), used in the treatment of chronic hepatitis, cirrhosis, and hepatic toxic lesions, with choleretic and cholagogue effects. Applied in Italy to treat liver and gastrointestinal disorders and as a laxative, with anti-inflammatory, antioxidant, cardiovascular protective, anti-cancer, and neuroprotective effects.	[11,162–168]

Table 2. *Cont.*

Scientific Name	Ethnobotanical Uses	Phytochemical Components	Biological Activity	Ref.
U. dioica L.	In folk medicine, it has been used to treat rheumatism, arthritis, gout, eczema, anemia, urinary tract infections, kidney stones, hay fever, and the early stages of an enlarged prostate.	3-O-caffeoylquinic acid; 4-O-caffeoylquinic acid; 5-O-caffeoylquinic acid; caffeoylmalic acid; *p*-coumaroylmalic acid; quercetin O-rutinoside.	Antiviral, antimicrobial, antioxidant, anti-inflammatory, antiaging, and cytotoxic/anticancer effects, as well as benign prostatic, hyperplasia, antidiabetic, antiendometriosis, and nephroprotective effects.	[169–171]

4.1. Ethnobotanical and Ethnomedicinal Relevance

A. millefolium L. (Asteraceae family) is commonly known as yarrow or milfoil. It is the best-known and most widespread species, and has been one of the most used plants in both folk and conventional medicine for over 3000 years [172]. *A. millefolium* is native to Europe and western Asia, widespread in most temperate regions, and represented by about 85 species, which are mostly found in Europe, Asia, and North America [173]. In Bosnia–Herzegovina, the flowers are used as vegetables or in the preparation of liqueurs [129]. The leaves can be eaten cooked or raw. As they have a slightly bitter taste and a strong licorice-like scent, they can be added to mixed salads [6,11]. The plant is used in traditional medicine against wounds, burns, and internal and external bleeding due to its hemostatic properties. Almost all the pharmacological activities are attributed to the flowering tops and leaves [174]. Several cultures use this plant for different treatments. In Italy, it is mainly used against gastrointestinal problems but is also used also for urinary problems such as diuretics, menstrual problems, toothache, and sedatives [89]. These properties are attributed to the essential oils, sesquiterpenes, and phenolic compounds [92,93]. The recipe states that three cup of *Achillea* tea a day, prepared with 1.5 g crude drug, equal to a 900 mg extract, would cause an anti-inflammatory effect [90].

The *Borago* genus belongs to the Boraginaceae family. It comprises only five species, *B. officinalis*, *B. trabutii*, *B. longifolia*, *B. pygmaea*, and *B. morisiana*, all of them native to the Mediterranean basin. In folk medicine, most medicinal plants are used as aqueous extracts, which provide raw materials for different medicinal purposes. *B. officinalis* seeds and the aerial parts are traditionally used to treat respiratory, cardiovascular, and gastrointestinal diseases [103]; borage juice and tea are used to treat flu, colds, injuries, and ulcers. Borage is generally cultivated for its culinary and medicinal uses; currently, the preferentially consumed parts are the seeds, which are marketed for their oil. Therefore, borage is considered an oilseed crop [104]. As a vegetable, borage is included in many recipes from different countries: in Germany, it is used for the preparation of green sauce; in the Italian region, Liguria is an ingredient in the famous "Preboggion"; in Crete, in France, in Great Britain, and in Spain, it is boiled and sautéed with garlic. *Borago* flowers are used as snacks, salads, vinegar aromatizers, fritters, and soups in Italy, Libya, and Spain [175].

Figure 2. Some examples of plant wild species traditionally used in the central Apennines as food and/or medicine: (**a**) *A. millefolium* [70]; (**b**) *B. officinalis* [72]; (**c**) *F. vulgare* [74]; (**d**,**e**) *G. lutea* L. subsp. *Lutea*, reprinted with permission from Ref. [176]; copyright 2008–2024, Giuliano Mereu, (**f**,**g**) *L. nobilis*, reprinted with permission from Ref. [177]; copyright 2008, Marinella Zepigi (**h**) *J. communis*, reprinted with permission from Ref. [177]; copyright 2008, Marinella Zepigi (**i**) *M. sylvestris*, reprinted with permission from Ref. [177]; copyright 2008, Marinella Zepigi (**j**) *S. montana*, reprinted with permission from Ref. [178]; copyright 2024, Enzo de Santis (**k**) *S. marianum*; [179] (**l**) *U. dioica* [180].

Fennel (*Foeniculum vulgare* Mill.) is a plant belonging to the Apiaceae family and represents one of the most used plants in traditional medicine. It is a plant that grows spontaneously as a native aromatic plant in the large area around the Mediterranean basin, particularly in Israel, Egypt, and Tunisia, and on both coasts of the Adriatic Sea: Montenegro, Croatia, and Italy [108,112–116,181,182]. Alcoholic beverages are often enriched with aromatic components, such as herbs (leaves, roots, seeds, and flowers), fruits (whole fruit, peel, and hazel) and natural sweetening agents [117,183]. The preparation of these drinks dates to ancient Mediterranean history, and aromatic plants and essential oils are still used today [118]. In popular Italian cuisine [159], fennel liqueur, reaching 30% alcohol content, called "Finocchietto", is produced by macerating the fruits of *F. vulgare* in alcohol [119]. The characteristic flavor of fennel essential oil is related to the presence of anethole, fenchone, and estragole, which are the main chemical components. Anethole has a sweet, anise-like note, while estragole has a bitter flavor. The composition and concentration of the individual components depend on the geographical origin of the plant. Generally, southern European plants produce sweeter-tasting extracts, while plants from central and northern Europe have more bitter flavors [112]. Fennel can also be used as an herbal tea for stomach aches due to its antispasmodic and carminative effects, aids intestinal peristalsis, has a diuretic effect, acts as an expectorant, and is also recommended for respiratory diseases [120].

Gentiana lutea L. is also known as yellow gentian, bitter root, and bitter herb. Belonging to the Gentianaceae family, it grows wild in hilly areas of Europe as far as Japan and is present in the traditional medicine of many countries [184], especially due to its antiseptic and anti-inflammatory properties [185]. The roots are the most widely used part of the plant as they are rich in bitter-tasting molecules such as amarogentin and gentiopicroside, and they have been known for their medicinal properties since antiquity. *G. lutea* L. radix is often used in the preparation of bitter liqueurs to stimulate the appetite and improve digestion. "Amaro di genziana" is a typical liqueur produced in the Abruzzo region of Italy. The quality of the gentian root is assessed by evaluating the main bitter principle, gentiopicroside [186].

Gentiana is considered a pleiotropic drug as it has multiple properties, such as antimicrobial, antioxidant, anti-inflammatory, anti-atherosclerotic, antihypertensive, hepatoprotective, and antidepressant activity [125–127]. *Gentiana* is a plant included in the "Regional Red Lists of the Plants" of Italy and is a protected plant (L.R. 11.9.1979 no. 45).

The *Juniperus* genus is a member of the Cupressaceae family [187] and has about 68 species and 36 variants of the same species. The *J. communis*, also called Zimbro, is the only species of *Juniperus* that has been documented as existing in both hemispheres, and has been found in the arctic regions of both Asia and North America. The Alps, Scandinavia, Poland, northwest European lowlands, and the mountainous regions of the Mediterranean in Europe are home to more varieties [188,189]. The considerable variety in the morphological traits and chemical composition of secondary metabolites can be traced to the vast geographical dispersion of the species [187]. Berries have a fragrant, spicy aroma and a slightly bittersweet flavor. Mature, dark berries are used in cuisine to season sauces and stuffing, and in pickling meats, and are also used to flavor spirits like gin or grappa. *Juniperus* has been traditionally used in many countries as a diuretic, antiseptic, and digestive [131].

Laurus nobilis L. is a member of the family Lauraceae, which comprises 32 genera and about 2000–2500 species [141]. It is cultivated in temperate areas of the world, mainly in south Europe and the Mediterranean basin [190,191]. It is also known as laurel, bay laurel, or sweet bay, and is the laurel tree featured Greek and Roman mythology [142], where it was considered a symbol of peace and a sign of victory in both military and sports competitions. Laurel is used in cooking as a flavoring and provides a spicy taste for meat, fish, broths, and vegetables. It is a component of a typical Italian plant infusion used as a digestive, named "canarino" [135]. Based on the classifications of diseases and remedies in ethnomedicine and ethnopharmacology suggested by Staub et al. [136], the major uses of *L. nobilis* include treatments for gastro-intestinal complaints, including indigestion, constipation, and flatulence; it is also used for diarrhea, hemorrhoids, and stomach aches. The leaves are traditionally used to reduce blood glucose levels and for fungal and bacterial infections [136]. This species is also reported to treat kidney diseases and coughs, colds, flu, and sore throats. Laurel leaves are one of the main ingredients of a preparation used for the treatment of respiratory ailments that are often called "Ricotto" or "Ricuotto". They still used in this way today and can be found in the traditional phytotherapy of central and southern Italian regions [135]. Essential oils or fumigations with bay leaves are also used as repellents and insecticides against home insects and crop pests [138].

Malva sylvestris L. is a flowering plant belonging to the Malvaceae family. It is native to north Africa, southwest Asia and southern Europe (Mediterranean area), although it is widespread worldwide as a weed [192]. Its edible uses are found in popular gastronomy; the young leaves are eaten raw in salads or, together with the sprouts, are used in soups and as boiled vegetables [143]. Traditionally, the plant has been used to treat various ailments, such as coughs, colds, diarrhea, dysentery, hypertension, and skin diseases [11]. The Greeks and Romans noted its emollient and laxative properties, and several ethnobotanical surveys conducted in Europe highlight the potential of this neglected local resource, whose use is now on the brink of disappearance [143]. Roots, shoots, leaves, flowers, fruits, and seeds are applied in infusions, decoctions, poultices, liniments, lotions, baths, and gargles [98,149–155]. Also known as mallow, it is considered to have spasmolytic, lenitive, and choleretic effects. It is also used as a bronchondilator, an expectorant, in acne and skin care, and as an antiseptic, emollient, and demulcent [153–155].

Satureja montana L., known as winter savory, is a plant belonging to the Lamiaceae family. This family includes approximately 236 genera and more than 6000 species, some of them important medicinal plants [193]. The *Satureja* genus is mainly distributed in the Mediterranean area. Reports have been registered in Italy [194–196], Spain [197], France [198], Montenegro [199], Slovenia [200], Croatia [201], Serbia [202], Bosnia, and Herzegovina [203]. Winter savory is often used in Mediterranean recipes, and, recently, the use of its essential oil as a natural antibacterial agent in food packaging has been reported. [204,205]. The enrichment of olive oil with winter savory essential oil (EO) has led to low values of lipid oxidation and a higher concentration of antioxidants (total phenols and pigments) [206]. Winter savory dried leaf is used as an herbal tea [207]; this is an aqueous preparation that is extemporaneously prepared for oral administration for therapeutic purposes. In some regions of Italy, it is used for nervous gastric pains, bloating, and vomiting [161].

Silybum marianum L. Gaertn. (milk thistle) is a medicinal plant widespread in southern Europe, northern Africa, and parts of southern Russia, and found in North and South America and southern Australia. The aerial parts of the plant are edible and are cooked like the artichoke [208]. In Italy, its use has been reported in the treatment of liver and gastrointestinal disorders and as a laxative [168]. The leaves, unripe fruits, roots, and the bark are used in the treatment of gastroenteritis, diarrhea, and dysentery, while the leaves are applied on sores and for hemorrhoidal pains [11]. The leaves also act as a choleretic and cholagogue [74] and the Greeks have suggested that *S. marianum* could be used to treat gallstones and allergic coughs, and for "blood purification" [164].

Urtica dioica L. (stinging nettle) is an herbaceous perennial flowering plant growing in temperate and tropical wasteland areas around the world. It grows 1–2 m high and produces pointed leaves and white to yellowish flowers. Nettle has a well-known reputation for giving a savage sting when the skin touches the hairs and bristles on the leaves and stems [169]. *U. dioica* is certainly one a primitive vegetables that has been consumed since time immemorial. For a long time, in folkloric medicine, it has been used to treat rheumatism, arthritis, gout, eczema, anemia, urinary tract infections, kidney stones, hay fever, and the early stages of an enlarged prostate [169,209].

4.2. Bioactive Phytochemicals

Many secondary metabolites were isolated from each plant; Figures 3 and 4 show some of the most significant natural products from the polar, apolar, and essential oil (EO) extracts.

The MeOH extract of Italian *A. millefolium* (Vercelli, Italy) was shown to contain chlorogenic acid, 1,3-dicaffeoylquinic acid, 1,4-dicaffeoylquinic acid, 3,4-dicaffeoylquinic acid, 3,5-dicaffeoylquinic acid, and phenolic compounds like rutin, luteolin 7-O-glucoside, apigenin 4′-O-glucoside, apigenin 7-O-glucoside, and luteolin 4′-O-glucoside [91]. Major volatile compounds found in the Italian *A. millefolium* include the following: α-pinene, 17.2%; sabinene, 3.9%; β-pinene, 2.1%; (E)-methyl isoeugenol, 8.8%; β-bisabolene, 16.6%. *Achillea* oils were shown to have antifungal activity [210] and cytotoxicity activity against cancer cells [211]. Other compounds, like alkaloids, choline, achillinin A, azulene, chamazulene, salicylic acid, artemetin, lignans, tannins, and flavonoids, were found in this plant [172,174].

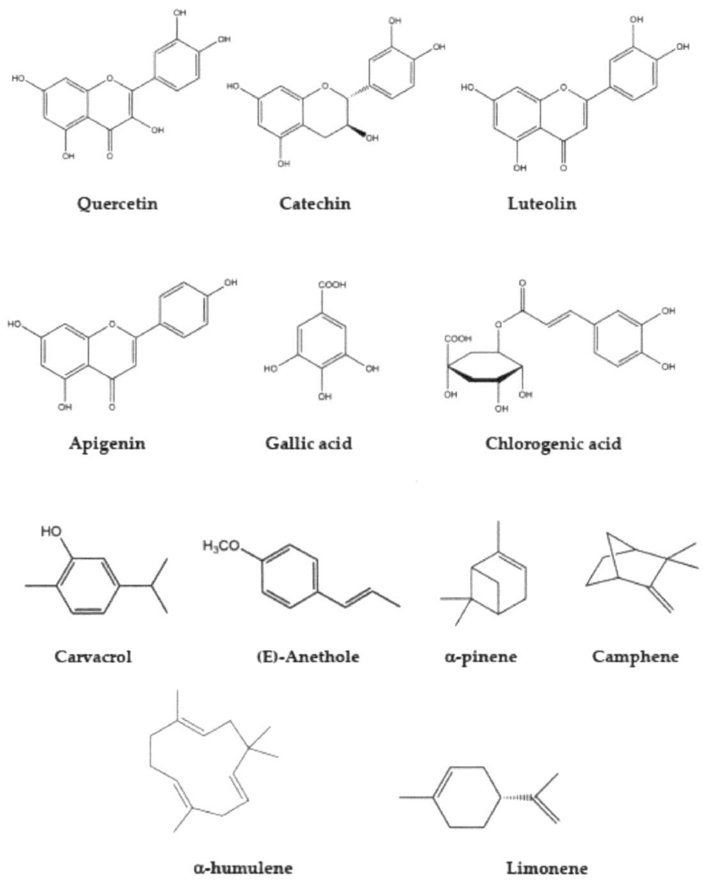

Figure 3. Common plant phenolic components and some volatile essential oil components (carvacrol; (*E*)-anethole; α-pinene; camphene; α-humulene; limonene).

An analysis of the aerial parts (methanolic extracts) of *B. officinalis* from Algeria revealed the presence of many flavonoids, rosmarinic acid, gallic acid, and chlorogenic acid, while, in the essential oil, spathulenol was the main component [212]. A lignane derivative, officinalioside, was isolated from the polar extract of borage's aerial parts from Egypt, along with megastigmane derivatives (actinidioionoside, (6S,9R)-roseoside, crotalionoside C), and kaempferol 3-O-β-D-galactopyranoside [213]. The RP-HPLC analyses of the methanolic, ethanolic, and aqueous extracts of Borage flowers from Iran confirmed the presence of phenolics (gallic acid; pyrogallol; salicylic acid; caffeic acid), flavonoids (myricetin; rutin), and isoflavonoid (daidzein). The EO from borage flowers was also prepared, and the major individual fatty acids were α-linolenic, stearidonic, palmitic, linoleic, and γ-linolenic acids [214]. As a member of the Boraginaceae family, *B. officinalis* is also known for its pyrrolizidine alkaloid content and its toxic properties [215]; oleuropein and litospermic acid were also identified in the aqueous extract of the plant [101]. The seed oil of *B. officinalis* is considered one of the richest natural sources of γ-linolenic acid (GLA, 18:3 n-6), ranging from 20 to 23% of the total fatty acid composition. GLA displays interesting medicinal properties, such as anti-inflammatory and anti-cancer properties, and can be used as an emollient of the skin and mucous membranes [104]. The main polyunsaturated fatty acids that were identified were linoleic acid (18:2 n-6), α-linolenic acid (18:3 n-3), γ-linolenic acid

(18:3 n-6, GLA), and stearidonic acid (SDA, 18:4, n-3), which account for approximately 70% of the polyunsaturated fatty acids [216].

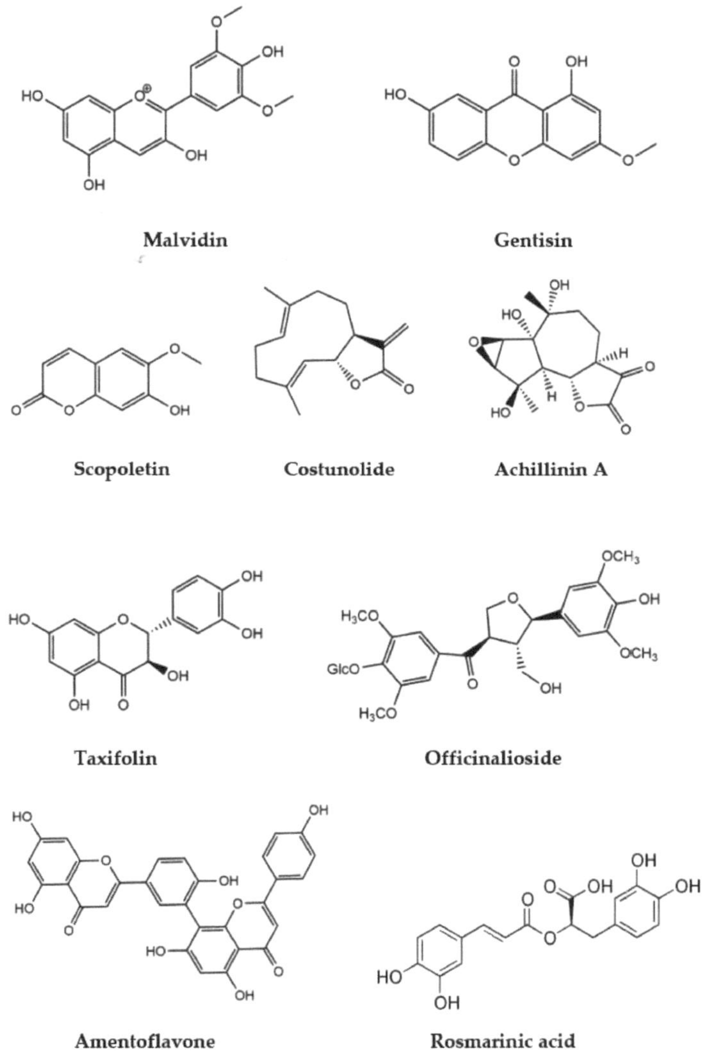

Figure 4. Some natural products from wild plants: malvidin from *M. sylvestris*; gentisin from *G. lutea*; scopoletin from *U. dioica*; costunolide from *L. nobilis*; achillinin A from *A. millefolium*; taxifolin from *S. marianum*; officinalioside from *B. officinalis*; amentoflavone from *J. communis*; rosmarinic acid from *F. vulgare*.

In general, the main ingredients of *F. vulgare* EOs are anethole (40–70%), fenchone (1–20%), and estragole (2–9%). *Trans*-anethole is also a common main component of fennel populations, especially cultivated populations. Alpha-pinene, camphene, methyl chavicole, and limonene are also presented in essential oils [112,115,217,218]. Differences in the quality of the essential oils' composition have been observed. These differences may be caused by different chemotypes, phenological stages, drying conditions, distillation modes, and geographic and climatic factors [115,181]. The main composition of fennel essential oil from mid–southern Italy includes α-pinene (33.75%), β-pinene (5.13%), myrcene (5.25%), 3-carene (6.12%), γ-terpinene-like (9.45%), estragole (25.06%), and (*E*)-

anethole (5.30%) [113]. Flavonoids and some important hydroxycinnamic acids are the most abundant polyphenols in fennel waste. Methanolic fennel seed extracts from Saudi Arabia were shown to contain vanillic acid, o-coumaric acid, and rosmarinic acid. Among the flavonoids, kaempferol, resveratrol, and rutin were found in higher concentrations, followed by myricetin, catechin, and quercetin [111]. In an aqueous extract of *F. vulgare* waste, 24 phenolic compounds were found, and 4-O-caffeoylquinic acid (4CQA) had the highest concentration of 1949 mg/g and 5824 mg/g of total polyphenols [110]. These exert evident antioxidant activity and other important biological properties, such as anti-inflammatory and anti-tumor activities, as well as the ability to modulate cell signaling and gene expression in different experimental models [5], making them good candidates for nutraceutical applications [219]. Complex trimers of stilbenes diglucosides were isolated from polar extracts of the fruit of *F. vulgare*, which were tested for their in vitro antioxidant activity [142].

Several secondary metabolites have been identified from *G. lutea*, such as iridoids, secoiridoids, xanthones, and flavones, which are distributed in different concentrations between the aerial parts and rhizomes. Gentiopicroside and stemoside were found to be abundant in the roots; iso-vitexin predominated in the leaves, and the amount of isogentisin was found to be ten times higher in the flowers than in the leaves. [220–223]. Iridoids and secoiridoids are a broad group of cyclopentane [c] pyran monoterpenoids found in the *Gentiana* genus, particularly in the leaves of *G. lutea*. Loganic acid, sweroside, amarogentin, gentiopicroside, swertiamarin, and their derivatives belong to these classes of compounds. They have been shown to have a large variety of pharmacological properties, including hepatoprotective, antitumor, and anti-inflammatory effects. Isovitexin, isosaponarin, isoorientin, and its glycosides are the main flavonoids isolated from *Gentiana* and known for their antioxidant and anti-inflammatory activities [224]. Xanthones such as isogentisin, gentisin, mangiferin, and gentiol are compounds of great interest due to their antibacterial, antifungal, hepatoprotective, and antioxidant activity. They have mainly been isolated in roots as mono- or polymethyl ethers or as glycosides. [225–228].

J. communis species contain a complex mixture of secondary metabolites that are responsible for both organoleptic characteristics, such as aroma and color, and beneficial health effects. These metabolites can be divided into several major categories, such as carotenoids and chlorophylls, phenolic compounds, and Volatile Organic Compounds (VOC's) [229]. Generally, most phenols reported in the *Juniperus* plant include caffeoylquinic acids with their corresponding derivatives, amentoflavone, catechin, epicatechin, quercetin, and their derivatives (see Table 2) [132]. Flavonols and flavones have been shown to act as radical scavengers and are associated with anti-inflammatory, antimicrobial, anti-proliferative, and pro-apoptotic properties. Furthermore, flavanones acting in synergy with flavones can inhibit the development of estrogen-dependent colon tumors. [133]. Several anthocyanins have also been isolated in juniper berries and, generally, occur in the form of cyanidin, delphinidin, peonidin, and pelargonidin glycosides. Anthocyanins also act as radical scavengers and exhibit anti-inflammatory activity, interacting with related pathways, increasing antioxidant defenses, and diminishing proinflammatory biomarkers, thus preventing the occurrence of many oxidative-stress-related disorders [130]. The chemical composition of the essential oil of *J. communis* differs according to the part of the plant that is extracted (berries, leaves, flowers) and the berries' stage of ripening [134].

L. nobilis leaves contain flavonoids such as isoquercitrin, luteolin, and rutin, and apigenin derivatives and flavonols such as catechin, cinnamtannin B1, epicatechin hexoside, (+)-catechin, (−)-epicatechin, epigallocatechin, and methyl eugenol. Many phenolic acids have been detected: rosmarinic acid, caffeic acid, 3,4-dihydroxybenzoic acid, 2-hydroxycinnamic acid, and others (see Table 2). Several cyclic terpenoids were found in *L. nobilis*, such as gazaniolide, spirafolide, reynosin, costunolide, santamarine, and lauroxepine [139]. The chemical composition of the EO from laurel leaves has been analyzed in different studies, and 1,8-cineole was found to be the major component [139]. Other compounds were present in appreciable amounts, such as camphene (0.05–13.4%), linalool (0.37–47.21%), methyl eugenol (3.3–7.8%), D- limonene (21.6–32.4%), sabinene (0.34–14.05%), neoiso-isopulegol (2.5%), eugenol (0.22–2.47%), γ-terpinene (0.23–3.48%), α-pinene (1.39–8.92%), β-pinene (3.0–6.22%), and terpinen-4-ol (1.21–5.2%). α-terpinyl acetate (5.9–15.33%) and α-humulene (0.51–8.58%) were major constituents [141,230–233]. Anthocyanins were found in berries from *L. nobilis* with cyanidin 3-O-glucoside (41%) and cyanidin 3-O-rutinoside (53%) [139]. The main compounds of berries' EO were 1,8-cineole, α-phellandrene, β-pinene, α-pinene, α-terpinyl acetate, sabinene, camphene, germacrene D, and β-caryophyllene [139,141]. Several megastigmane and phenolic components were also isolated from polar extracts of *L. nobilis* leaves that exhibited anti-inflammatory activity [142,234].

Many phenolic compounds were detected in extracts from various parts of *M. sylvestris*. Total phenolic compounds were identified in leaves (386.5 mg g^{-1}), flowering stems (317.0 mg g^{-1}), flowers (258.7 mg g^{-1}), and immature fruits (56.8 mg g^{-1}) [143]. The primary components of the leaves were identified as gossypetin 3-sulphate-8-O-β-D-glucoside (gossypin) and hypolaetin 3′-sulphate, followed by isoscutellarein 8-O-β-D-glucuronpyranoside, hypolaetin 8-O-β-D-glucuronopyranoside, 3-O-β-D-glucopyranosyl-8-O-β-D-glucuronopyranoside, and hypolaetin 4′-methyl ether 8-O-β-D-glucuronopyranoside [145,235,236]. Anthocyanins like malvin (malvidin 3,5-glucoside), found only in flavylium cationic form, were mostly identified in flowers [237–240]. Oenin (malvidin 3-O-glucoside), malvidin, delphinidin, delphinidin-3-O-glucoside, kaempferol derivatives, quercetin, apigenin, myricetin, and genistein were detected in the flowers, and the total anthocyanin content was found to be 0.42–7.3% of the dry weight [239,241,242]. The leaves of *M. sylvestris* contain γ-sitosterol, stigmasterol, and campesterol [243]. Vernolic acid, linoleic acid, palmitoleic acid, sterculic acid, myristic acid, lauric acid, malvalic acid, oleic acid, and palmitic acid are the primary fatty acids found in seed oil [244,245]. Sesquiterpenes, nor-terpenes, monoterpenes, and diterpenes were also found in *M. sylvestris* [246]. Linalool, linalool-1-oic acid, and linalool-2-oic acid were detected in aqueous extracts from fresh leaves, along with several megastigmane derivatives [247].

S. montana (winter savory) is used in cooking due to its distinctive aroma, which is related to the presence of essential oils. The EO yield from the fresh plant can vary between 0.12 and 0.7% [197,203,248,249], with the volatile fraction mainly being characterized by thymol and carvacrol (oxygenated monoterpenes), indicators of antimicrobial activity [250,251], associated with open-chain and/or monocyclic monoterpenes that exhibit an allelopathic effect [252]. Carvacrol has been classified as Generally Recognized As Safe (GRAS) and approved for use in food [253], while linalool and p-cymene (non-oxygenated monoterpenes) have been shown to have analgesic effects [204]. Polar extracts contain variable amounts of secondary metabolites, such as phenolic acids, phenylpropanoids, fatty acids, tannins, and tocopherols [254,255]. The main constituents of the phenolic acid components were caffeic acid (78.17 μg g^{-1}) and gallic acid (15.36 μg g^{-1}). Quercetin, p-coumaric acid, chlorogenic acid, and ferulic acid were represented at concentrations of 2.36, 1.59, 1.36, and 0.50 μg g^{-1}, respectively [255].

From *S. marianum*, a typical extract, namely silymarin, is used. This is a mixture of different flavonolignans, and, at present, the term "silymarin" is indicative of an extract of *S. marianum* that is rich in these compounds. It is composed of silicristin, isosilybin A and B, dehydroxylysilybin and silybin, and flavonoids such as taxifolin, with silybinin being the most active. Silybinin consists of two diastereoisomers: silybin A and silybin B [256]. Flavonolignans have been isolated from the seeds and fruits and are the biologically active constituents of the plant; to date, 23 compounds have been identified from this species [163]. Several other minor flavonolignans have also been found: silychristin, isosilychristin, and silydianin, with several flavonoids, such as taxifolin [167], and 3′-O-methyltaxifolin and dihydrokaempferol from plant seeds [257]. Polyphenolic compounds such as hydroxycinnamic acids (caffeic, chlorogenic, ferulic, and cynarinic acids) and flavonoids (apigenin; catechin; luteolin; luteolin-7-O-glucoside; quercetin) have also been identified [257]. The oil fraction is known to be rich in fatty acids, palmitic (C16:0), oleic

(C18:1) and linoleic (C18:2) organic acids, sterols (cholesterol, campesterol, and stigmasterol) and tocopherol (vitamin E), triacylglycerols, and phospholipids.

The active chemical part of *U. dioica* includes several compounds from the lipophilic and hydrophilic extracts of different parts of the plant [170]. In particular, the commonly known phytochemical components from *U. dioica* are flavonoids, tannins, volatile compounds, and sterols [171,258,259]. Hexahydrofarnesyl acetone, carvacrol, carvone, naphthalene, copaene-8-ol, anethol, geranyl acetone, β-ionone, α-ionone, and phytol are characterized as the main components of *U. dioica* essential oil [260,261]. Rhizomes of *U. dioica* contain other biologically active compounds, such as scopoletin, sterols, fatty acids, polysaccharides, and isolectins [262]. Extracts of *U. dioica* have been studied for their various potential therapeutic applications: antitumor, antimicrobial, analgesic, and anti-inflammatory activity has been evidenced. Extracts showed antioxidant properties, and experimental tests proved that the constituents of *U. dioica* may have neuroprotective effects [263].

4.3. Evaluation of In Vitro Antioxidant Activity of Selected Edible Wild Plants

The antioxidant activity of *A. millefolium* was tested by different assays. DPPH scavenging test of MeOH extract showed an IC_{50} = 1.18 mg mL^{-1}; total antioxidant capacity (TAC) based on Cu (II) reduction and lipid peroxidation measurements (TBARS/LDL) were also calculated [91]. In the same study, Total Polyphenol Content (TPC) for the Italian plant was determined by Folin–Ciocalteau assay, with a result of 281.7 mg g^{-1} reported as mEq of Gallic acid [91]. Dias et al., in 2013, reported the results of DPPH, TBARS, and inhibition of β-carotene bleaching tests [264]. Mohammed et al., in 2023, described the antioxidant activity of the essential oils using TAA, DPPH, FRAP, and Metal Chelating Assay (MCA) [211], and the Total Flavonoid Content (TFC) of the Iranian plant was measured using the standard curves of Rutin and reported as mg per g dry weight [265].

B. officinalis seed extracts from Poland were assessed for their total polyphenol content using the Folin–Ciocalteau method, followed by an evaluation of antioxidant potential using the FRAP assay and the free radical method with the DPPH reagent. The flavonoid content in borage seeds was much lower than that observed in borage flower and leaf oil; however, the antioxidant activity of the seed meal infusion was high [266].

The antioxidant activity of essential oil extracted from different parts of *F. vulgare* was evaluated, showing that the leaves have a better EC_{50} (12.37 mg mL^{-1} at 60 min incubation) than seeds and umbels [267]. A similar study conducted on Tunisian EOs from fennel, characterized by their richness in estragole, revealed an important antioxidant activity [182], while the antioxidant activity of Tajikistan fennel EO was moderate. EC_{50} values were between 30 and 210 mg L^{-1} [268]. Finally, the Italian wild cultivar of fennel has a better antioxidant activity in essential oil when compared to cultivated fennel [269]. The total phenolic content (TPC) of the methanolic fennel seed extract (FS) was 70.42 mg gallic acid equivalent (GAE) g^{-1}; Total Flavonoids (TFC) and Total Flavonols (TFL) were 4.83 and 4.93 mg quercetin equivalent (QE) g^{-1}, respectively. Antioxidant activity was 9.36 μmol of Trolox equivalent (TE) g^{-1} [111].

The most studied part of *G. lutea* is the roots due to the presence of characteristic secoiridoid glucosides. *G. lutea* root's antioxidant activity was evaluated through various assays, including total phenolic content (TPC), DPPH, ferric-reducing antioxidant power (FRAP), and oxygen radical absorbance capacity (ORAC). Both methanol and ethanol produced the extracts with the highest activity, but methanol is toxic and not suitable for human use. Ethanol is safe and environmentally friendly and should be the first choice when producing extracts from natural resources. pH also plays a significant role in antioxidant activity, with higher activity observed under acidic conditions while increased pro-oxidant action was observed under alkaline conditions [125]. Gentiopicroside and stemoside were not directly involved in antioxidant activity, but mass spectrometry data indicated that antiradical scavenging activity is probably associated with xanthones' glycosides [226].

Furthermore, in vivo studies have evaluated the antioxidant activity of *G. lutea* root extract against ketoconazole-induced testicular damage in rat models [270].

In vitro tests showed that the *J. communis* ethanolic extract of berries showed a half-maximal inhibitory concentration (IC_{50}) of 1.42 μg mL^{-1}. Berries' methanolic extracts and essential oils also exhibited the capacity to scavenge DPPH•, ferric species, and β-carotene species. Ethanolic extracts of *J. communis* berries also showed the ability to scavenge peroxyl radicals and reduce power potential [271]. The remarkable antioxidant capacities displayed by *J. communis* extracts are indeed linked to their phenol and terpenoid content, in particular, quercetin, which contains the flavan-3-ol system, with hydroxyl groups in the 'key positions' of its structure (see Section 3.1), making it a potent radical scavenger. Regarding the antioxidant activity of terpenes, it has already been reported that α-pinene, p-cymene, limonene, and linalool possess notable capacities to block lipid peroxidation and to avoid deoxyribose degradation [272].

L. nobilis L. is a source of monoterpenes and other antioxidant compounds, such as tocopherol. The ultrasound-assisted extract (UAE) of dried laurel leaves from Brazil presented total phenolic compounds (TPC) of 47 mg GAE g^{-1} per extract, and the hydrodistillation extract (HD) shows a TPC of 22 mg GAE g^{-1} extract and EC_{50} (35 $\pm$ 1 μg mL^{-1}). Although phenolic compounds are the primary natural antioxidants, they are not the only class of substances that contribute to the antioxidant performance of natural products, which explains the good EC_{50} results obtained for the HD extract, since this extract does not possess α-tocopherol [230]. A potent suppression of lipid peroxidation was observed in aqueous and ethanolic extracts, with 96.8% and 98.6% inhibition rates, respectively, when using a linoleic acid emulsion at a concentration of 60 μg mL^{-1} [139].

M. sylvestris contains phenolic compounds in its leaves and flowers, which may be responsible for the plant's antioxidant activity [273]. Several tests have also determined the antioxidant properties of the plant. In the DPPH test, the aqueous extract at concentrations of 20 g mL^{-1} and 100 g mL^{-1} showed scavenging activity by decreasing the DPPH radical by 24% and 30%, respectively. The 0.1 mg mL^{-1} aqueous extract demonstrated 87% antioxidant activity when tested using the ß-carotene-linoleic acid assay. Antioxidant activity was also found in the EO of *M. sylvestris* (77% antioxidant activity) [151,274]. Overall, the antioxidant activity in the seed extracts was moderate to poor [275–277].

Due to the presence of polyphenolic compounds, ethyl acetate fractions of *S. montana* from MeOH extract and the total EO demonstrated radical scavenging activity via DPPH and ABTS assays. A spray-dried hydroalcoholic extract of winter savory, in combination with 10% maltodextrin as a carrier and drying agent, also showed the same activity [278]. EOs show less antioxidant activity than aqueous or ethanol extracts; this is due to the presence of carvacrol and sesquiterpenes that do not give rise to antioxidant activity [279].

Several studies on antioxidant activity were carried out on methanol, ethanol, and aqueous extracts and EO. However, many papers agree that the remarkable antioxidant properties of *S. marianum* are significantly related to its flavonolignan content [165].

The aqueous extract of *U. dioica* leaves exhibited antioxidant activity, achieved by the DPPH radical scavenging (IC_{50} = 16.93 ug mL^{-1}), reducing power (EC_{50} = 30.07 ug mL^{-1}), and polarographic (HPMC = 243.2% mL^{-1}) assays [259]. Based on these results, *U dioica* extracts have been proposed as an antioxidant and as a source of anti-ageing phytochemicals for cosmetic applications [280]. A comparative study by Carvalho et al. demonstrated that the antioxidant properties of *U. dioica* are greater than those observed for the aerial parts of other nettle species [171].

5. Therapeutic Potential of Selected Edible Wild Plants

5.1. Anti-Inflammatory Activity

A. millefolium has shown anti-inflammatory activity. It is used in the treatment of gastrointestinal and hepato-biliary disorders and to treat skin inflammation. This activity is probably due to the presence of sesquiterpenes, known for their anti-inflammatory activity through the inhibition of arachidonic acid metabolism. Antiphlogistic activities have been observed in Yarrow fractions enriched with flavonoids and di-caffeoylquinic acids. The anti-inflammatory activity of this extract is ascribable to the inhibition of human neutrophil elastase, which is known to be associated with the inflammatory process. The in vitro anti-inflammatory activity was also established through the inhibition of matrix metalloproteinases (MMP-2 and -9). These proteases are involved in psoriasis, atopic dermatitis, and inflammatory bowel diseases such as ulcerative colitis and Crohn's disease. Azulenes, which make up about half of the chemical compounds in yarrow, are a potent anti-inflammatory agent [90].

The seed oil of *B. officinalis* has shown powerful anti-inflammatory and analgesic effects. Seed oil was tested against carrageenan-induced inflammation and compared with indomethacin, a known anti-inflammatory drug. The analgesic effect of *Borago* seed oil was tested in mice using two assays: the tail immersion test, which is used to determine the central analgesic effect, and the writhing test, which is used to determine the peripheral analgesic effect in mice [94]. Borage extracts have exhibited anti-inflammatory properties, as observed in the methanolic extracts, which result in a potent inhibition of collagenase and elastase activity. This characteristic underlines the anti-ageing properties, meaning that borage extracts are a source of valuable bioactive compounds with protective effects on skin cells [97].

The oral administration (200 mg kg^{-1}) of *F. vulgare* fruit methanolic extract exhibited inhibitory effects against acute and subacute inflammatory diseases and type IV allergic reactions and showed a central analgesic effect. At the same time, the administration of methanolic extracts of *F. vulgare* led to an increase in plasma SOD and catalase while MDA decreased. These data seem to demonstrate that the use of methanolic extracts from fennel seeds is effective in relieving inflammation [281]. Extracts and pure compounds from *F. vulgare* fruits showed antioxidant activity in vitro (DPPH, TBARS); the antioxidant activity was higher in the pure compounds than the crude extracts but was weaker than the reference compound, i.e., quercetin [282].

G. lutea L. has been used to prevent or treat inflammatory disorders for centuries. In vitro studies have shown that *G. lutea* root extracts have anti-inflammatory properties, with the rate of enzyme inhibition increasing with time. In vivo studies have also demonstrated that apolar and alcoholic extracts of *G. lutea* rhizomes have anti-inflammatory activity in different animal models [127].

The anti-inflammatory effects of *J. communis* have already been evaluated by in vitro and in vivo studies. The anti-inflammatory activity is closely associated with the presence of phenolic compounds and terpenes, such as amentoflavone, α-pinene, 1-octanol, and linalool. These compounds have been shown to inhibit inflammatory cytokine and prostaglandin expression [125]. Recently, loganic acid and gentiopicroside were tested in silico using an innovative technique named Inverse Virtual Screening (IVS) to highlight putative partners among a panel of proteins involved in inflammation and cancer events [283].

L. nobilis extracts have shown the ability to reduce edema caused by chemicals in the ears and paws and lung inflammation caused by LPS. Bay leaf extracts also induced a reduction in skin injuries and inflammation caused by *Propionobacterium acnes* [140]. Megastigmane and phenolic components able to inhibit nitric oxide production were isolated from polar extracts of *L. nobilis* leaves [142], and it has been proven that lauroside B induces apoptosis in human melanoma cell lines by inhibiting NF-κB activation [234].

Numerous studies have investigated the anti-inflammatory properties of *M. sylvestris* [284]. Their results claim that malvidin 3-glucoside, isolated from *M. sylvestris* leaves, is mainly responsible for the anti-inflammatory action on the skin. In carrageenan-induced oedema in rats, the anti-inflammatory properties of creams with various concentrations of mallow extract were evaluated. A 5% malva cream effectively reduced carrageenan-induced edema when compared to placebo therapy. This effect was greater than that of a cream containing 2% indomethacin, used as a positive control and a potent non-selective inhibitor of cyclo-oxygenase-2 (COX-2) [148]. The beneficial component rutin was isolated in a chemical investigation of *M. sylvestris* extract. This flavonoid is widely used in plant-based beverages, cuisine, and folk medicine [285]. Rutin has been shown to be an anti-inflammatory therapeutic candidate, with a novel mechanism for selective COX-2 inhibition [286].

Methanolic extracts from *S. montana* were evaluated for their anti-inflammatory activity using COX-1, COX-2, 5-LOX, and MPO inhibition assays. The alcoholic extract showed both a powerful anti-inflammatory activity and a strong antioxidant activity [287].

Silymarin extract from *S. marianum* showed an important anti-inflammatory effect in carrageenan-induced rat paw oedema, inhibiting the release of elastase proteinases from neutrophils as a response to normal and chronic inflammation [165]. The common molecular targets of *S. marianum* are the multiple signaling pathways associated with oxidative stress and inflammation. In addition, flavonolignans are potential PPARγ and ABCA1 agonists, PTP1B inhibitors, and metal chelators [288].

5.2. Anti-Microbial Activity

A. millefolium extracts show antimicrobial activity. They are used as an infusion for respiratory tract infection [172], against flu as an antiseptic [174], to treat gastrointestinal infections, and as an anti-acne [90]. Many papers reported the antifungal activity of the essential oils on various fungal strains, with an MIC ranging from 0.32 to 1.25 μL mL^{-1} against dermatophyte strains [210]. Vitalini et al. 2011 reported luteolin 7-O-glucoside and apigenin 7-O-glucoside as the most active compounds against the *Plasmodium falciparum* chloroquine-resistant strain (IC_{50} of MeOH extract = 44.6 μg mL^{-1}) [91].

In *B. officinalis* flower extracts, high (methanolic extract), moderate (ethanolic extract), and weak (aqueous extract) antimicrobial activity was reported [214]. Flavonoid-rich extracts and EO *Borago* aerial parts were tested on bacteria isolated from respiratory infections of clinical patients. Multiresistant hospital isolates were found to be sensitive to the flavonoid extracts and to the essential oil; interestingly, *Escherichia coli* (resistant), *Streptococcus pneumoniae* (sensitive to amoxicillin), and *Klebsiella pneumoniae* (sensitive to Imipeneme) were sensitive to flavonoids [289].

Fennel essential oils showed extensive antibacterial activity against Gram-positive bacteria and fungi such as *Aspergillus niger*. Gram-negative bacteria, particularly *E. coli*, are less sensitive to fennel essential oils, as well as *Listeria innocua* CECT910 and *Pseudomonas fluorescens* [290–292]. An antimicrobial role against *Giarda duodenalis* has been described by the *trans*-2,4-undecadienal (IC_{50} 72.1 μg mL^{-1}). However, it was not active as the positive control metronidazole (IC_{50} 0.5 μg mL^{-1}) against the parasite [293]. The activity of the essential oil of *F. vulgare* also varied considerably between Gram-negative and Gram-positive bacteria. This low antimicrobial activity also seems to be related to strain susceptibility and the essential oil's composition, which is poor in components such as carvacrol and thymol, which are usually associated with higher antimicrobial activity [294]. Other authors specifically describe controversial antimicrobial properties obtained using different extraction methods [295].

Interestingly, the experiment was designed to evaluate the effectiveness of fennel essential oil in controlling *Fusarium solani* infections on *Vicia faba* L. The growth of *F. solani* was inhibited both in vitro and in vivo, allowing for a reduction in disease incidence by 50%. The essential oil acted on both the fungus and the plant as an enhancer of defense reactions [296].

In vitro studies have shown that *Pseudomonas aeruginosa*, *Bacillus subtilis*, *Proteus mirabilis*, *Staphylococcus epidermidis*, and *Candida albicans* were the most sensitive to Gentiana leaf extract, with MIC values between 0.12 and 0.31 mg mL^{-1} [220,297,298]. The flower extract was not very active against the tested microorganisms, with the most sensitive being *Salmonella enteritidis* (MIC 0.15 mg mL^{-1}). Both leaf and flower extracts showed an antitubercular effect against *Mycobacterium bovis*. Tests performed on isolated pure compounds showed a broader spectrum of activity; gentiopicrin was active against *E. coli* (0.12 mg mL^{-1}), while it had a moderate effect against *Staphylococcus aureus* and *Salmonella typhimurium* (0.15 mg mL^{-1}). Xanthone isogentisin was particularly active against *M. bovis*, while moderate activity was observed against the gram-negative *E. coli* and *P. aeruginosa* (0.15 mg mL^{-1}) and the gram-positive *Micrococcus luteus* (0.15 mg mL^{-1}).

A different biological activity was observed in *J. communis* extracts. In phenol-rich extracts, antiparasitic action dominates, whereas essential oils exhibit antimicrobial effects. [129]. Generally, extracts with a more balanced composition in their components showed greater antibiotic effects against multiresistant hospital isolates belonging to the species *S. aureus*, *Serratia marcescens*, *Enterobacter cloacae*, *K. pneumoniae*, *P. aeruginosa*, *Acinetobacter baumanii*, and *Listeria monocytogenes*, as well as *C. albicans*. The use of essential oils obtained from *J. communis* biomass, without differentiating each of its parts, showed remarkable inhibitory activity against *E. coli* at concentrations between 1.25 and 2.5 mg mL^{-1} [299]. Notable inhibitory activities were observed against other Gram-negative bacteria, such as *P. mirabilis*, *K. pneumoniae*, *P. aeruginosa*, and *Morganella morganii*; however, only slight activity against *L. monocytogenes* and methicillin-resistant *S. aureus* was observed [130]. The different effects against various pathogenic fungi obtained using the EOs of *Juniperus* are believed to be related to the composition of the EOs, particularly the ratio of sesquiterpene hydrocarbons and oxygenated aromatic hydrocarbons [300].

Since ancient times, *L. nobilis* has been an important ingredient in traditional medicine for the treatment of different infectious diseases [301]. *L. nobilis* EO, seed oil, and a methanolic extract of seed oil showed antibacterial activity in vitro. However, the methanolic extract of the seed oil has higher antibacterial activity than the EO. *L. nobilis* was detected to have EO activity against *S. aureus*, *B. subtilis*, and *Staphylococcus intermedius*. One of the main constituents of bay leaf is 1,8 cineole, which may be responsible for its antibacterial activity. The antifungal activity of EO from leaves was examined on seven strains of plant pathogenic fungi in vitro at different concentrations. The highest antifungal activity was obtained against the fungus *Botrytis cinerea* at a concentration of 250 mg mL^{-1} [302]. Bay leaf EO has shown efficacy against a large panel of Gram-negative and Gram-positive bacteria and three fungi [303].

M. sylvestris exhibited moderate activity against selected microorganisms associated with typical antibiotics [304]. De Souza et al. showed the antimicrobial activity of *M. sylvestris* aerial part extracts against *C. albicans*, *S. aureus*, *M. luteus*, *B. subtilis*, *S. epidermidis*, *E. coli*, and *Saccharomyces cerevisiae*. Their study reported that ethanol extracts of *M. sylvestris* were active against *P. aeruginosa*, *B. subtilis*, and *E. coli*, whereas methanol extracts only showed activity against *S. cerevisiae* [305]. The antimicrobial activity of ethanolic extracts of the leaves and flowers against *Helicobacter pylori* strains ranged from moderate to low [306]. Other studies showed that the seed oil inhibited the growth of all tested microorganisms except the Gram-negative bacteria [307]. The only preparation of *M. sylvestris* that demonstrated substantial antimicrobial activity against fungi was an aqueous extract of the leaves. The aqueous extract prevented the growth of colonies of the *Fusarium culmorum*, *Aspergillus candidus*, *A. niger*, and *Penicillium* species [308].

S. montana EO (SEO) was demonstrated to have antimicrobial activity in several application fields, ranging from veterinary medicine [309,310] to plant pathology, with phytogenic bacteria such as *Xanthomonas euvesicatoria* [311]. In general, gram-positive bacteria proved to be more sensitive to EO treatment, while gram-negative bacteria were less sensitive—even more resistant yeasts and fungi [312,313]. The impact of SEO is certainly effective and important, but the disposal problem must be taken into consideration, as this can cause problems and modify the microbial communities of soil and water [314]. Encouraging results were obtained from a combined therapy of SEO and antibiotics. For example, associations between SEO and erythromycin and gentamicin have been described. This association improves the effectiveness of monotherapeutic treatments, reduces adverse effects by reducing the dose of the drug, and combats antibiotic-resistant bacteria [253,315]. Antiphytoviral activity was also described [316].

S. marianum extracts were tested against several pathogenic strains, such as *P. aeruginosa*, *E. coli*, *Salmonella typhi*, *S. epidermidis*, and *K. pneumoniae* [317]. In a recent study, Rakelly de Oliveira et al. [318] demonstrated an interesting antibacterial effect of silymarin and its major compound, silibinin. Indeed, silymarin inhibited *E. coli* at MIC = 512 μg mL^{-1}, while silibinin inhibited 64 μg/mL (MIC = 64 μg mL^{-1}), *P. aeruginosa* (MIC = 1024 μg mL^{-1}), and *S. aureus* (MIC = 1025 μg mL^{-1}). An important effect against *C. albicans* was also observed by Yun e Lee et al. Their results revealed that a possible mechanism of action of silymarin as an antifungal agent may involve an increase in the membrane permeability of *C. albicans* [319].

The antibacterial activity of ethanol and aqueous extracts of *U. dioica* has been demonstrated against both Gram-positive and Gram-negative bacteria and yeasts, including *Proteus mirabilis*, *P. aeruginosa*, *Enterobacter aerogenes*, *E. coli*, *Citrobacter koseri*, *S. pneumonia*, *S. aureus*, *M. luteus*, *S. epidermidis*, and *C. albicans*. The extracts were also active against *Mycobacterium tuberculosis* in cases of multiple drug resistance [170,320]. Notably, the aqueous (microwave-assisted, ultrasound-assisted, and subcritical water extraction) and ethanol extracts of *U. dioica* leaves also confirmed the antibacterial activity, with a minimal inhibitory concentration (MIC) of 9.76 ug mL^{-1} and 0.0625–0.500 mg mL^{-1} against methicillin-resistant (MRSA) and methicillin-sensitive (MSSA) *S. aureus* strains [259]; these observed effects are assumed to be due to the high content of hydroxycinnamic acids (chlorogenic, caffeic, and rosmarinic acids) and flavonoids (quercetin) [321].

5.3. Protection against Cardiovascular Diseases

The hydroalcoholic extract of *B. officinalis* leaves, rich in polyphenols and sterols, produced a concentration-dependent relaxation of spontaneous and K^+-induced contractions (80 mM) in isolated rabbit jejunum preparations, suggestive of a Ca^{++} antagonistic effect. In rabbit aorta preparations, *Borago* showed a vasodilator effect against phenylephrine- and K^+-induced contractions. When tested in guinea pig atria, *B. officinalis* inhibited the force and speed of atrial contractions. These results suggest the spasmolytic effects of *Borago* extracts [103].

Recent experiments in rats suggest that the inhalation of *F. vulgare* essential oil by experimental animals could lead to a reduction in blood pressure [322].

G. lutea root extracts may also have promising activity in the prevention and treatment of cardiovascular disease, particularly thromboembolic disorders, attributable to their bitter constituents, such as amarogentin and isovitexin [125].

L. nobilis decoction is utilized to reduce blood pressure and treat cardio-vascular illnesses [140]. The powdered leaves of *L. nobilis* have positive effects on lipid and blood sugar dysregulation. After treatment, there was a reduction in plasma glucose levels, a decrease in overall cholesterol levels, a significant decrease in low-density lipoprotein (LDL) levels, an increase in high-density lipoprotein (HDL) levels, and a decrease in triglyceride levels [323]. Extracts of *L. nobilis* leaves showed a vascular protective effect and angioprotective activity on rat liver capillaries, and they prevented the progression of necroinflammation. These results could be explained by the presence of flavonoids, terpenes, and terpenoids with antioxidant and antimicrobial properties [139].

The main components of the EO of *S. montana*, such as carvacrol and thymol, have been found to be responsible for reducing serum cholesterol levels. Carvacrol and other monoterpene hydrocarbons, flavonoids such as apigenin, and phenolic acids such as labiatic acid could contribute to the antiplatelet properties [324]. *Satureja* flavonoids also have antioxidant and anti-hyperlipidemic properties [325].

The various traditional uses of *S. marianum* have motivated several experimental investigations into the pharmacological properties of the plant. Antihypertensive and cardioprotective activities have been documented, which seem to be linked to the presence of taxifolin [165].

U. dioica (leaves extracts) and isolated flavonoids were active against thrombin-induced platelet aggregation (IC50 values of 0.25 $\pm$ 0.05 and 0.40 $\pm$ 0.04 mg/mL) [326].

5.4. Role of Wild Plants in Cancer Prevention and Treatment

Due to the considerable number of different secondary metabolites, the tested species exhibit toxic activity against the growth of tumor cell lines, and several experiments were conducted both in vivo and in vitro. *A. millefolium* showed activity on human cancer cell lines (MCF-7, NCI-H460, HCT-15, HeLa, and HepG2) with low toxicity to primary non-cancerous liver cells (PLP2) [264].

B. officinalis EO contains a high concentration of γ-linolenic acid, with several anti-cancer activities, inhibiting the p38 MAPK-dependent activator protein and the mitochondria-mediated apoptosis pathway [327].

The ethanolic extract of *F. vulgare* seeds significantly reduced the growth of lung cancer cells both in vitro and in vivo. The alcoholic extract reduced viability and triggered apoptosis in lung cancer cell lines NCI-H446 and NCI-H661 by targeting the Bcl-2 protein, which may suggest that it has potential as a therapeutic drug for lung cancer [328]. The role of anethole, found in fennel extracts, in anti-cancer activity was demonstrated in albino mice [329].

In vitro studies have investigated the cytotoxic effect of the *G. lutea* leaf extract on various cell lines, including human cervix adenocarcinoma (HeLa), breast cancer (MCF7), prostate cancer (PC3), and colon cancer (LS174). The Gentian methanolic leaf extract demonstrated a moderate cytotoxic effect against HeLa cells, with an IC50 value of $41.1 \pm 1.5\ \mu g\ mL^{-1}$, compared to cisplatin, used as a control [125].

Methanolic extracts of *J. communis* leaves have been found to block the growth and development of C6 rat-brain tumor and HeLa human-cervix carcinoma cells, PC3 human-prostate cancer cells, HCT 116 human-colon cancer cells, and MCF7 breast cancer cells. Essential oil and extracts from *J. communis* berries have also been found to suppress A549 human lung adenocarcinoma epithelial cells' growth and development, as well as suppress the development of SH-SY5Y human neuroblastoma cells [330,331].

L. nobilis seed extract was suitable for eliminating multidrug-resistant P-glycoprotein-expressing tumor cells [139]; fresh EO exhibited growth-inhibitory effects on the breast cell line, lung cell line, and brain cancer cell line. The cervix cell line exhibited the lowest sensitivity to essential oil (IC_{50} value of $1.8\ \mu g\ mL^{-1}$) [141].

The study of Alesiani et al. [332] demonstrated the cytotoxic activity of *M. sylvestris* leaf extracts on murine using an MTT assay and human cancer cell lines.

Studies have been conducted on the antiproliferative activity of *S. montana* extracts on "mice's model of induced Ehrlich ascites carcinoma (EAC)". The results show that the extracts had a positive role in inducing oxidative stress in malignant cells [279]. Carvacrol is confirmed to have an antitumor effect on liver cancer [333], as well as apoptosis [334], metastatic breast cancer cells (MDA-MB 231) [335], and on human colon adenocarcinoma (HT-29) and human breast adenocarcinoma (MVF-7) [160].

Silibinin, silymarin, and silybin A and B from *S. marianum* possess anticancer activity on several tumor cell lines [165].

U. dioica aqueous extracts have responded positively in studies on prostate and breast cancer [336,337].

5.5. Neurological Disorders and Wild Plants

The various metabolites may play an important role in neurological disorders. *A. millefolium* is used as a sedative and analgesic against pain (headache, toothache, menstrual pain, and dysmenorrhea) [338,339].

The analgesic effect of *B. officinalis* seed oil was tested in mice using two assays: a tail immersion test to determine the central analgesic effect, and a writhing test, used to establish the peripheral analgesic effect in mice [94].

Inran et al. investigated the role of fennel seed extracts in promoting functional recovery following a mechanical insult to the sciatic nerve of mice, concluding that *F. vulgare* may be a potential therapeutic candidate to accelerate functional recovery after peripheral nerve injury [340,341]. In both studies, the authors considered fennel extracts and *trans*-anethole to be suitable candidates for the prevention and treatment of stress-induced neurological disorders [342].

G. lutea extract and its compounds exert effects on the central nervous system (CNS). Iridoids such as geniposide have been found to exert beneficial effects on neuronal cell cultures due to their ability to activate protein kinase, leading to neuronal cell differentiation [343]. The same plant extracts also enhanced neurite outgrowth [344]. *G. lutea* extract significantly enhanced the viability of cells treated with vinblastine and prevented Bcl-2 phosphorylation induced by the antimitotic drug vinblastine. These results suggested that *G. lutea* may be a potential vegetable resource for preventing and treating Parkinson's and Alzheimer's disease thanks to its MAO-B inhibition activities [345].

A neuroprotective effect was observed in the n-hexane fraction of *L. nobilis*. Indeed, in an in vivo study using rodents with Parkinson's disease, the fraction exhibited a marked inhibition of 6-hydroxydopamine (6-OHDA)-induced cell loss of tyrosine hydroxylase (TH)-positive cells in the substantia nigra [139]. Additionally, bay leaf extracts showed promising results in reducing neuronophagia, localized gliosis, and neural necrosis in the rat brain, helping to resolve the lead-induced imbalance in brain acetylcholinesterase (AcChE) activity [140,346].

The dried methanolic extract from *S. montana* leaves was demonstrated to have significant anxiolytic activity in rats. Carvacrol and rosmarinic acid, used as controls, only showed a moderate anxiolytic effect in some tests [347]. *Satureja* genera essential oil may act as a neuroprotective agent in the early stage of Alzheimer's disease [348], as well as *S. marianum* extracts [165].

5.6. Diabetes and Hepatoprotective Effects of Edible Wild Plants

Recently, *A. millefolium* ethanolic extract was tested for its in vivo antidiabetic effects. The hydroalcoholic extract possesses an anti-diabetic effect in vivo through a multitarget activity involving α-glucosidase inhibition, insulin secretion, and potential insulin-sensitizing actions [165].

B. officinalis was shown to have relevant hypoglycemic activity in rat models [349].

Essential oil and aqueous extracts of *F. vulgare* were administered to rats with streptozotocin-induced diabetes, with hyperglycemia corrected from (162.5 + 3.19 mg dL^{-1}) to (81.97 + 1.97 mg dL^{-1}), and reducing the pathological abnormalities in diabetic-induced rats [350] A similar study showed that fennel seed extract and its active ingredient *trans*-anethole can protect the liver from diabetes-induced liver damage in rats, probably through its hypoglycemic and antioxidant effects [351]. Experimental treatments with the 80% methanolic extract of the wild and cultivated fennel showed hepatoprotective effects at a concentration of 12.5 μg mL^{-1} and hepatotoxic effects at a concentration of 1000 μg mL^{-1} [352]. In a further study, oxidative stress and the complications of hepatotoxicity caused by CCl_4 injection in rats were reversed by the administration of fennel seed extracts at 300 or 600 mg kg^{-1}. The improvement in liver function was monitored by following the attenuation values of the enzymes ALT, AST, and ALP [111].

Gentiana roots have been widely used in folk medicine, so scientific studies have focused on their choleretic and hepatoprotective properties, which make them a good remedy for stomach and liver inflammations. In pylorus-ligated mice treated with methanolic extract of gentian root in the duodenum, there was a decrease in gastric juice secretion and total acid production, with a noticeable dose-dependent effect at doses of 500 and 1000 mg kg^{-1}. The hepatoprotective activity of gentian root may be due to gentiopicroside, which has been reported in previous studies to resolve cholestasis [353,354]. As the incidence of liver problems has increased in recent years, the use of gentian root extracts may provide an alternative to synthetic drugs due to their hepatoprotective activity [354,355].

Ethyl acetate fractions of *J. communis* leaf extracts have been shown to be hepatoprotective agents, promoting favorable portal triads and central-vein arrangements [356].

Extracts of *L. nobilis* have been used in folk medicine due to their anti-diabetic effect. Diabetic rats treated with *L. nobilis* extracts showed a significant decrease in glucose concentration compared to untreated diabetic rats. A beneficial effect on pancreatic islet regeneration was also observed, and levels of liver enzymes, total protein, creatine kinase, calcium, urea, and ferritin returned to near-normal levels. The EO of *L. nobilis* also suggested the inhibition of alpha-glucosidase, which is an indication of its in vitro antidiabetic activity [139,140]. An extract of *L. nobilis* leaves proved to be a powerful free radical scavenger in vivo, preventing carbon tetrachloride-induced hepatotoxic effects in rats.

The methanolic extract of *M. sylvestris* protected liver tissue from the harmful effects of paracetamol in a dose-dependent manner by lowering the blood levels of liver enzyme markers. In animals treated with mallow, the dramatic lowering of the blood levels of liver enzyme markers was complemented by the regeneration of liver tissue, demonstrating the hepatoprotective properties. The traditional use of mallow in liver problems has been scientifically validated through the hepatoprotective activity of *M. sylvestris* [147,357].

The EO of savory products, through experiments on rats, induces a hepatoprotective effect and a decrease in inflammatory processes in the organs of the gastrointestinal tract [358].

The silymarin extract from *S. marianum* was found to promote hepatocyte regeneration and inhibit liver fibrosis by significantly increasing the survival time of rats with paracetamol-induced liver injury [359]. In addition to the hepatoprotective action, scientific evidence suggests that silibin exerts its activity by interacting with various tissues through the modulation of inflammation and apoptosis, which, together with its antioxidant power, are the key points that have led to its use in various diseases [360]. Oral administration of silymarin extract was able to generate a significant decrease in ALP, ALT, and AST in the liver tissue of rats with lead-induced liver toxicity [361].

U. dioica showed a hepatoprotective effect by increasing the activity of some liver enzymes (paraoxonase, arylesterase, and catalase). Treatment with *U. dioica* reduced oxidative stress, with a decrease in ceruloplasmin levels. Also, treatment with *U. dioica* extracts generated an antioxidant effect, preventing the formation of some oxidant agents such as LOOH and showing a protective effect on the liver in rats damaged by hepatic ischaemia-reperfusion [355]. *U. dioica* can prevent liver fibrosis and cirrhosis, suggesting that this plant probably protects the liver through immunomodulatory and antioxidant activities [362].

5.7. Other Biological Activity

The results of in vitro tests on human keratinocytes (HaCaT) and fibroblasts (BJ) showed that methanol and methanol/water extracts of *B. officinalis* can reduce the intracellular level of reactive oxygen species in skin cells. It has been proposed that oral or topical borage oil may be effective for the treatment of atopic dermatitis. Atopic dermatitis is believed to be associated with an abnormality in the metabolism of essential fatty acids (EFAs), particularly the altered production of gamma-linolenic acid (GLA), so nutritional supplementation with omega-6 essential fatty acids (ω-6 EFAs) is of potential interest for the treatment of atopic dermatitis. Borage oil is of interest because it contains two to three times more GLA than evening primrose oil (*Oenothera biennis* L.). Borage oil is well tolerated in the short term, but no long-term tolerability data are available [363].

In human studies, *G. lutea* extracts were effective in reducing increased intestinal permeability, a problem that causes a more significant absorption of endotoxins due to the loss of integrity of the epithelial cells of the intestine tenuous. In Complementary and Integrative Medicine (CIM), *G. lutea* reduced the time needed to resolve the alterations in intestinal permeability to 4–5 months compared to the expected 6 months [364]. The Committee on Herbal Medicinal Products approved the use of *G. lutea* for mild stomach and gut complaints. Still, they limited their approval to their traditional use and not long-term use, as there is poor evidence from clinical trials [125]. Gentian root and its hydroalcoholic extracts have shown potential in treating skin disorders, including atopic dermatitis and psoriasis. Further research is needed to understand the mechanisms behind these effects and their potential applications in various skin conditions.

Hydroethanolic extracts (90% ethanol, *v*/*v*) of *J. communis* berries displayed antiprogestational and antifertility activity at doses ranging from 50 to 450 mg/kg on female rats, without estrogenic or antiestrogenic effects. Moreover, the oral administration of hydroethanolic extracts (50% ethanol, v/v) from *J. communis* berries at doses of 300 and 500 mg/kg in albino female rats from day 1 to day 7 of pregnancy exhibited dose-dependent anti-implantation activity. Furthermore, these extracts, at the same concentrations, generated abortions when administered on days 14, 15, and 16 of pregnancy. No teratogenic effects were detected. [365].

M. sylvestris (aqueous or hydroalcoholic extract) increases skin hydration and prevents or alleviates skin dryness when used as a cream, lotion, serum, patch, emulsion, hydrogel, mask, etc. [2,143]. The extracts of mixed *Mentha piperita* and *M. sylvestris* have a substantial skin-whitening effect. Cosmetics made from the leaves and flowers of *M. sylvestris* and other plant extracts inhibit melanogenesis and tyrosinase activity, improving skin color and reducing pigmentation [2].

Silymarin from *S. marianum* protects the kidneys against renal ischemia/reperfusion injury in Wistar rats. The protective effect is associated with its antioxidant properties, as it possibly acts as a free-radical-scavenger and lipid peroxidation inhibitor. Thus, new therapeutic strategies, such as antioxidant supplementation with flavonoid silymarin, could be explored for protection against damage caused by ischemia and reperfusion [366]. Studies have shown that *S. marianum* extracts have an immunomodulatory effect in vitro and are able to increase lymphocyte proliferation. These properties were strongly associated with the increase in IF-γ, IL-4, and IL-10 [367]. In addition, extracts of *S. marianum* fruits have shown hyperprolactinemic activity, which exerts a stimulating effect on milk production in the mother [162]. An antiulcerogenic effect of *S. marianum* extract has been reported in rats. This activity involves a reduction in acid production and increase in mucin secretion, and the release of prostaglandin E_2 with a decrease in leukotrienes [368].

6. Conclusions

This review has gathered all the basic information on the phytochemical, nutritional, and pharmacological profile of the active ingredients known to date, as published in various books and journals on the wild plants under study from 2000 to November 2023. Most of the ailments treated with wild plants are common: digestive disorders, colds, coughs, circulatory problems, diarrhea, etc. However, there are some examples of treatments for more specific diseases, such as hypertension, hypercholesterolemia, hyperglycaemia, and others. Their use in folk medicine is supported by scientific investigations and, together with the knowledge of their side effects, makes these plants potential sources for nutraceutical purposes. Wild plants are often identified as functional foods because of their higher content of vitamins, antioxidants, trace elements and fibers compared to cultivated crops. The richness of natural antioxidants, mainly phenolic compounds with nutraceutical properties, is crucial in preventing acute and chronic diseases induced by improper nutrition, so wild plants lend themselves to the formulation of dietary supplements with benefits for human health and longevity, allowing for an improved quality of life.

Author Contributions: Conceptualization, M.I. and C.C.; methodology, G.S. and M.G.C.; software, M.G.C., F.F. and V.S.; validation, M.I. and C.C.; formal analysis, M.I.; investigation, F.F, V.S., G.S., M.G.C. and C.C.; resources, M.G.C.; data curation, C.C.; writing—original draft preparation, C.C.; writing—review and editing, M.I.; visualization, F.F and V.S.; supervision, C.C.; project administration, M.I.; funding acquisition, M.G.C. All authors have read and agreed to the published version of the manuscript.

Funding: University of Molise (Start-Up 2022), grant number "PROGET_20232024_BIOACTIVE_START_UP_CHINI—Isolation, characterization, activity and evaluation of natural and synthetic bioactive compounds" and MUR (PRIN 2022 PNRR), grant number "P2022MWY3P—Old but Gold! Identification of molecular platforms for age-associated diseases to promote healthy and active aging".

Acknowledgments: The authors would like to acknowledge Giuliano Mereu, Enzo De Santis, and Marinella Zepigi for providing the opportunity to publish the photos in Figure 2.

Conflicts of Interest: The authors declare no conflicts of interest.

References

1. Islam, M.A.; Alam, F.; Solayman, M.; Khalil, M.I.; Kamal, M.A.; Gan, S.H. Dietary Phytochemicals: Natural Swords Combating Inflammation and Oxidation-Mediated Degenerative Diseases. *Oxidative Med. Cell. Longev.* **2016**, *2016*, 1–25. [CrossRef] [PubMed]
2. Gasparetto, J.C.; Martins, C.A.F.; Hayashi, S.S.; Otuky, M.F.; Pontarolo, R. Ethnobotanical and scientific aspects of *Malva sylvestris* L.: A millennial herbal medicine. *J. Pharm. Pharmacol.* **2012**, *64*, 172–189. [CrossRef] [PubMed]
3. Cutrim, C.S.; Cortez, M.A.S. A review on polyphenols: Classification, beneficial effects and their application in dairy products. *Int. J. Dairy Technol.* **2018**, *71*, 564–578. [CrossRef]
4. Cassidy, L.; Fernandez, F.; Johnson, J.B.; Naiker, M.; Owoola, A.G.; Broszczak, D.A. Oxidative stress in alzheimer's disease: A review on emergent natural polyphenolic therapeutics. *Complement. Ther. Med.* **2020**, *49*, 102294. [CrossRef] [PubMed]
5. Rasouli, H.; Farzaei, M.H.; Khodarahmi, R. Polyphenols and their benefits: A review. *Int. J. Food Prop.* **2017**, *20*, 1700–1741. [CrossRef]
6. Paura, B.; Di Marzio, P.; Salerno, G.; Brugiapaglia, E.; Bufano, A. Design a Database of Italian Vascular Alimurgic Flora (AlimurgITA): Preliminary Results. *Plants* **2021**, *10*, 743. [CrossRef] [PubMed]
7. Aliotta, G. Edible wild plants of Italy. *Inf. Bot. Ital.* **1987**, *19*, 17–30.
8. Guarrera, P.M.; Savo, V. Wild food plants used in traditional vegetable mixtures in Italy. *J. Ethnopharmacol.* **2016**, *185*, 202–234. [CrossRef]
9. Conti, F.; Bartolucci, F. *The Vascular Flora of the National Park of Abruzzo, Lazio and Molise (Central Italy)*; Springer: Cham, Switzerland, 2015.
10. Łuczaj, Ł.; Pieroni, A.; Tardío, J.; Pardo-de-Santayana, M.; Sõukand, R.; Svanberg, I.; Kalle, R. Wild food plant use in 21st century Europe: The disappearance of old traditions and the search for new cuisines involving wild edibles. *Acta Soc. Bot. Pol.* **2012**, *81*, 359–370. [CrossRef]
11. Fortini, P.; Di Marzio, P.; Guarrera, P.M.; Iorizzi, M. Ethnobotanical study on the medicinal plants in the Mainarde Mountains (central-southern Apennine, Italy). *J. Ethnopharmacol.* **2016**, *184*, 208–218. [CrossRef]
12. Chiuve, S.E.; Fung, T.T.; Rimm, E.B.; Hu, F.B.; McCullough, M.L.; Wang, M.; Stampfer, M.J.; Willett, W.C. Alternative Dietary Indices Both Strongly Predict Risk of Chronic Disease. *J. Nutr.* **2012**, *142*, 1009–1018. [CrossRef] [PubMed]
13. Chaturvedi, S.; Sharma, P.; Garg, V.K.; Bansal, M. Role of nutraceuticals in health promotion. *Int. J. Pharm. Tech. Res.* **2011**, *3*, 442–448.
14. Bommakanti, V.; Puthenparambil Ajikumar, A.; Sivi, C.M.; Prakash, G.; Mundanat, A.S.; Ahmad, F.; Haque, S.; Prieto, M.A.; Rana, S.S. An Overview of Herbal Nutraceuticals, Their Extraction, Formulation, Therapeutic Effects and Potential Toxicity. *Separations* **2023**, *10*, 177. [CrossRef]
15. Dahiya, D.; Nigam, P.S. Nutraceuticals Prepared with Specific Strains of Probiotics for Supplementing Gut Microbiota in Hosts Allergic to Certain Foods or Their Additives. *Nutrients* **2023**, *15*, 2979. [CrossRef] [PubMed]
16. Shahidi, F. Nutraceuticals, functional foods and dietary supplements in health and disease. *J. Food Drug Anal.* **2020**, *20*, 78. [CrossRef]
17. Heleno, S.A.; Martins, A.; Queiroz, M.J.R.P.; Ferreira, I.C.F.R. Bioactivity of phenolic acids: Metabolites versus parent compounds: A review. *Food Chem.* **2015**, *173*, 501–513. [CrossRef] [PubMed]
18. Bhatt, I.D.; Rawat, S.; Rawal, R.S. Antioxidants in Medicinal Plants. In *Biotechnology for Medicinal Plants*; Springer: Berlin/Heidelberg, Germany, 2013; pp. 295–326.
19. Sdona, E.; Ekström, S.; Andersson, N.; Hallberg, J.; Rautiainen, S.; Håkansson, N.; Wolk, A.; Kull, I.; Melén, E.; Bergström, A. Fruit, vegetable and dietary antioxidant intake in school age, respiratory health up to young adulthood. *Clin. Exp. Allergy* **2021**, *52*, 104–114. [CrossRef] [PubMed]
20. Masullo, M.; Montoro, P.; Mari, A.; Pizza, C.; Piacente, S. Medicinal plants in the treatment of women's disorders: Analytical strategies to assure quality, safety and efficacy. *J. Pharm. Biomed. Anal.* **2015**, *113*, 189–211. [CrossRef]
21. Khan, J.; Deb, P.K.; Priya, S.; Medina, K.D.; Devi, R.; Walode, S.G.; Rudrapal, M. Dietary Flavonoids: Cardioprotective Potential with Antioxidant Effects and Their Pharmacokinetic, Toxicological and Therapeutic Concerns. *Molecules* **2021**, *26*, 4021. [CrossRef]
22. Kumar, N.; Goel, N. Phenolic acids: Natural versatile molecules with promising therapeutic applications. *Biotechnol. Rep.* **2019**, *24*, e00370. [CrossRef] [PubMed]
23. Sytar, O.; Hemmerich, I.; Zivcak, M.; Rauh, C.; Brestic, M. Comparative analysis of bioactive phenolic compounds composition from 26 medicinal plants. *Saudi J. Biol. Sci.* **2018**, *25*, 631–641. [CrossRef] [PubMed]
24. Cháirez-Ramírez, M.H.; de la Cruz-López, K.G.; García-Carrancá, A. Polyphenols as Antitumor Agents Targeting Key Players in Cancer-Driving Signaling Pathways. *Front. Pharmacol.* **2021**, *12*, 710304. [CrossRef] [PubMed]
25. Nimse, S.B.; Pal, D. Free radicals, natural antioxidants, and their reaction mechanisms. *RSC Adv.* **2015**, *5*, 27986–28006. [CrossRef]
26. Kumar, S.; Pandey, A.K. Chemistry and Biological Activities of Flavonoids: An Overview. *Sci. World J.* **2013**, *2013*, 162750. [CrossRef] [PubMed]
27. Luca, S.V.; Macovei, I.; Bujor, A.; Miron, A.; Skalicka-Woźniak, K.; Aprotosoaie, A.C.; Trifan, A. Bioactivity of dietary polyphenols: The role of metabolites. *Crit. Rev. Food Sci. Nutr.* **2019**, *60*, 626–659. [CrossRef] [PubMed]
28. Zhang, L.; Ravipati, A.S.; Koyyalamudi, S.R.; Jeong, S.C.; Reddy, N.; Smith, P.T.; Bartlett, J.; Shanmugam, K.; Münch, G.; Wu, M.J. Antioxidant and Anti-inflammatory Activities of Selected Medicinal Plants Containing Phenolic and Flavonoid Compounds. *J. Agric. Food Chem.* **2011**, *59*, 12361–12367. [CrossRef] [PubMed]
29. Zhuang, W.-B.; Li, Y.-H.; Shu, X.-C.; Pu, Y.-T.; Wang, X.-J.; Wang, T.; Wang, Z. The Classification, Molecular Structure and Biological Biosynthesis of Flavonoids, and Their Roles in Biotic and Abiotic Stresses. *Molecules* **2023**, *28*, 3599. [CrossRef] [PubMed]
30. Khalid, M.; Saeed ur, R.; Bilal, M.; Huang, D.-f. Role of flavonoids in plant interactions with the environment and against human pathogens—A review. *J. Integr. Agric.* **2019**, *18*, 211–230. [CrossRef]

31. Lago, J.; Toledo-Arruda, A.; Mernak, M.; Barrosa, K.; Martins, M.; Tibério, I.; Prado, C. Structure-Activity Association of Flavonoids in Lung Diseases. *Molecules* **2014**, *19*, 3570–3595. [CrossRef] [PubMed]
32. Khare, S.; Dewangan, R.P.; Kumar, A. Structure-Activity Relationship of Flavonoids: Recent Updates. In *The Chemistry inside Spices & Herbs: Research and Development*; Bentham Science Publishers: Sharjah, United Arab Emirates, 2022; pp. 237–259.
33. Jiang, L.; Yanase, E.; Mori, T.; Kurata, K.; Toyama, M.; Tsuchiya, A.; Yamauchi, K.; Mitsunaga, T.; Iwahashi, H.; Takahashi, J. Relationship between flavonoid structure and reactive oxygen species generation upon ultraviolet and X-ray irradiation. *J. Photochem. Photobiol. A Chem.* **2019**, *384*, 112044. [CrossRef]
34. Schloms, L.; Swart, A. Rooibos Flavonoids Inhibit the Activity of Key Adrenal Steroidogenic Enzymes, Modulating Steroid Hormone Levels in H295R Cells. *Molecules* **2014**, *19*, 3681–3695. [CrossRef] [PubMed]
35. Ahmad, A.; Singh, P.; Mishra, S.K.; Noel, S.; Sharma, S.; Rath, S.K. Acute Exposure of Apigenin Induces Hepatotoxicity in Swiss Mice. *PLoS ONE* **2012**, *7*, e31964. [CrossRef]
36. Peterson, G.L. Review of the folin phenol protein quantitation method of Lowry, Rosebrough, Farr and Randall. *Anal. Biochem.* **1979**, *100*, 201–220. [CrossRef] [PubMed]
37. Sadowska-Bartosz, I.; Bartosz, G. Evaluation of The Antioxidant Capacity of Food Products: Methods, Applications and Limitations. *Processes* **2022**, *10*, 2031. [CrossRef]
38. Re, R.; Pellegrini, N.; Proteggente, A.; Pannala, A.; Yang, M.; Rice-Evans, C. Antioxidant activity applying an improved ABTS radical cation decolorization assay. *Free Radic. Biol. Med.* **1999**, *26*, 1231–1237. [CrossRef] [PubMed]
39. Kut, K.; Cieniek, B.; Stefaniuk, I.; Bartosz, G.; Sadowska-Bartosz, I. A Modification of the ABTS$^{\bullet}$ Decolorization Method and an Insight into Its Mechanism. *Processes* **2022**, *10*, 1288. [CrossRef]
40. de Menezes, B.B.; Frescura, L.M.; Duarte, R.; Villetti, M.A.; da Rosa, M.B. A critical examination of the DPPH method: Mistakes and inconsistencies in stoichiometry and IC50 determination by UV–Vis spectroscopy. *Anal. Chim. Acta* **2021**, *1157*, 338398. [CrossRef] [PubMed]
41. Benzie, I.F.F.; Strain, J.J. The Ferric Reducing Ability of Plasma (FRAP) as a Measure of "Antioxidant Power": The FRAP Assay. *Anal. Biochem.* **1996**, *239*, 70–76. [CrossRef]
42. Benzie, I.F.F.; Choi, S.-W. Antioxidants in Food. In Advances in Food and Nutrition Research; Academic Press is an imprint of Elsevier: Waltham, MA, USA; Volume 71, pp. 1–53.
43. Özyürek, M.; Güçlü, K.; Tütem, E.; Başkan, K.S.; Erçağ, E.; Esin Çelik, S.; Baki, S.; Yıldız, L.; Karaman, Ş.; Apak, R. A comprehensive review of CUPRAC methodology. *Anal. Methods* **2011**, *3*, 2439–2453. [CrossRef]
44. Cao, G.; Alessio, H.M.; Cutler, R.G. Oxygen-radical absorbance capacity assay for antioxidants. *Free Radic. Biol. Med.* **1993**, *14*, 303–311. [CrossRef]
45. Ou, B.; Hampsch-Woodill, M.; Prior, R.L. Development and Validation of an Improved Oxygen Radical Absorbance Capacity Assay Using Fluorescein as the Fluorescent Probe. *J. Agric. Food Chem.* **2001**, *49*, 4619–4626. [CrossRef] [PubMed]
46. Fedorova, G.F.; Menshov, V.A.; Trofimov, A.V.; Vasil'ev, R.F. Facile chemiluminescence assay for antioxidative properties of vegetable lipids: Fundamentals and illustrative examples. *Analyst* **2009**, *134*, 2128–2134. [CrossRef] [PubMed]
47. Işıl Berker, K.; Güçlü, K.; Tor, İ.; Demirata, B.; Apak, R. Total Antioxidant Capacity Assay Using Optimized Ferricyanide/Prussian Blue Method. *Food Anal. Methods* **2009**, *3*, 154–168. [CrossRef]
48. Somogyi, A.; Rosta, K.; Pusztai, P.; Tulassay, Z.; Nagy, G. Antioxidant measurements. *Physiol. Meas.* **2007**, *28*, R41–R55. [CrossRef] [PubMed]
49. Marco, G.J. A rapid method for evaluation of antioxidants. *J. Am. Oil Chem. Soc.* **1968**, *45*, 594–598. [CrossRef]
50. Alam, M.N.; Bristi, N.J.; Rafiquzzaman, M. Review on in vivo and in vitro methods evaluation of antioxidant activity. *Saudi Pharm. J.* **2013**, *21*, 143–152. [CrossRef] [PubMed]
51. Hsieh, C.; Rajashekaraiah, V. Ferric reducing ability of plasma: A potential oxidative stress marker in stored plasma. *Acta Haematol. Pol.* **2021**, *52*, 61–67. [CrossRef]
52. Zhang, H.; Forman, H.J. Redox Regulation of γ-Glutamyl Transpeptidase. *Am. J. Respir. Cell Mol. Biol.* **2009**, *41*, 509–515. [CrossRef] [PubMed]
53. Kalinovic, S.; Stamm, P.; Oelze, M.; Daub, S.; Kröller-Schön, S.; Kvandova, M.; Steven, S.; Münzel, T.; Daiber, A. Comparison of three methods for in vivo quantification of glutathione in tissues of hypertensive rats. *Free Radic. Res.* **2021**, *55*, 1048–1061. [CrossRef] [PubMed]
54. Birben, E.; Sahiner, U.M.; Sackesen, C.; Erzurum, S.; Kalayci, O. Oxidative Stress and Antioxidant Defense. *World Allergy Organ. J.* **2012**, *5*, 9–19. [CrossRef] [PubMed]
55. Singh, R.R.; Reindl, K.M. Glutathione S-Transferases in Cancer. *Antioxidants* **2021**, *10*, 701. [CrossRef] [PubMed]
56. Couto, N.; Wood, J.; Barber, J. The role of glutathione reductase and related enzymes on cellular redox homoeostasis network. *Free Radic. Biol. Med.* **2016**, *95*, 27–42. [CrossRef] [PubMed]
57. Ito, F.; Sono, Y.; Ito, T. Measurement and Clinical Significance of Lipid Peroxidation as a Biomarker of Oxidative Stress: Oxidative Stress in Diabetes, Atherosclerosis, and Chronic Inflammation. *Antioxidants* **2019**, *8*, 72. [CrossRef] [PubMed]
58. Verpoorte, R.; Choi, Y.H.; Kim, H.K. NMR-based metabolomics at work in phytochemistry. *Phytochem. Rev.* **2007**, *6*, 3–14. [CrossRef]
59. Das, L.; Bhaumik, E.; Raychaudhuri, U.; Chakraborty, R. Role of nutraceuticals in human health. *J. Food Sci. Technol.* **2011**, *49*, 173–183. [CrossRef] [PubMed]

60. Valentino, G.; Graziani, V.; D'Abrosca, B.; Pacifico, S.; Fiorentino, A.; Scognamiglio, M. NMR-Based Plant Metabolomics in Nutraceutical Research: An Overview. *Molecules* **2020**, *25*, 1444. [CrossRef] [PubMed]
61. Nagana Gowda, G.A.; Raftery, D. Recent Advances in NMR-Based Metabolomics. *Anal. Chem.* **2016**, *89*, 490–510. [CrossRef] [PubMed]
62. Guo, S.; Qiu, S.; Cai, Y.; Wang, Z.; Yang, Q.; Tang, S.; Xie, Y.; Zhang, A. Mass spectrometry-based metabolomics for discovering active ingredients and exploring action mechanism of herbal medicine. *Front. Chem.* **2023**, *11*, 1142287. [CrossRef] [PubMed]
63. Mulder, F.A.A.; Tenori, L.; Licari, C.; Luchinat, C. Practical considerations for rapid and quantitative NMR-based metabolomics. *J. Magn. Reson.* **2023**, *352*, 107462. [CrossRef] [PubMed]
64. Saviano, G.; Paris, D.; Iorizzi, M. Editorial: Exploring metabolomic diversity in plant species by NMR-based and mass-based spectrometry. *Front. Plant Sci.* **2023**, *14*, 1248781. [CrossRef] [PubMed]
65. Takis, P.G.; Ghini, V.; Tenori, L.; Turano, P.; Luchinat, C. Uniqueness of the NMR approach to metabolomics. *TrAC Trends Anal. Chem.* **2019**, *120*, 115300. [CrossRef]
66. Wishart, D.S. NMR metabolomics: A look ahead. *J. Magn. Reson.* **2019**, *306*, 155–161. [CrossRef] [PubMed]
67. Emwas, A.-H.; Roy, R.; McKay, R.T.; Tenori, L.; Saccenti, E.; Gowda, G.A.N.; Raftery, D.; Alahmari, F.; Jaremko, L.; Jaremko, M.; et al. NMR Spectroscopy for Metabolomics Research. *Metabolites* **2019**, *9*, 123. [CrossRef] [PubMed]
68. Wilson, I.D.; Theodoridis, G.; Virgiliou, C. A perspective on the standards describing mass spectrometry-based metabolic phenotyping (metabolomics/metabonomics) studies in publications. *J. Chromatogr. B* **2021**, *1164*, 122515. [CrossRef] [PubMed]
69. Saccenti, E.; Hoefsloot, H.C.J.; Smilde, A.K.; Westerhuis, J.A.; Hendriks, M.M.W.B. Reflections on univariate and multivariate analysis of metabolomics data. *Metabolomics* **2013**, *10*, 361–374. [CrossRef]
70. Vélez-Gavilán, J. *Achillea Millefolium (Yarrow)*; CABI Compendium: Wallingford, UK, 2022. [CrossRef]
71. Available online: https://antropocene.it/en/2022/10/25/achillea-millefolium-en/ (accessed on 1 March 2024).
72. *Borago Officinalis (Borage)*; PlantwisePlus Knowledge Bank; CABI Compendium: Wallingford, UK, 2022. [CrossRef]
73. Available online: https://antropocene.it/en/2022/10/25/borago-officinalis-en/ (accessed on 1 March 2024).
74. Datiles, M.J.; Popay, I. *Foeniculum Vulgare (Fennel)*; CABI Compendium: Wallingford, UK, 2022. [CrossRef]
75. Available online: https://antropocene.it/en/2022/10/25/foeniculum-vulgare-en/ (accessed on 1 March 2024).
76. Available online: https://www.conifers.org/cu/Juniperus_communis.php (accessed on 1 March 2024).
77. Available online: https://antropocene.it/en/2022/11/26/juniperus-communis-2/ (accessed on 1 March 2024).
78. Available online: https://antropocene.it/en/2022/10/26/malva-sylvestris-en/ (accessed on 1 March 2024).
79. CABI Compendium: Wallingford, UK, 2021. Available online: https://www.cabidigitallibrary.org/doi/10.1079/cabicompendium.115485 (accessed on 1 March 2024).
80. Available online: https://antropocene.it/en/2022/12/13/gentiana-lutea-2/ (accessed on 1 March 2024).
81. CABI Compendium: Wallingford, UK, 2019. Available online: https://www.cabidigitallibrary.org/doi/10.1079/cabicompendium.114518 (accessed on 1 March 2024).
82. Alessi, N.; Wellstein, C.; Spada, F.; Zerbe, S. Population structure of *Laurus nobilis* L. in Central Italian forests: Evidence for its ongoing expansion. *Rend. Lincei. Sci. Fis. Nat.* **2021**, *32*, 365–376. [CrossRef]
83. Available online: https://antropocene.it/en/2022/10/29/laurus-nobilis-en/ (accessed on 1 March 2024).
84. Available online: https://antropocene.it/en/2022/12/20/satureja-montana-2/ (accessed on 1 March 2024).
85. Popay, I. *Silybum marianum (Variegated Thistle)*; CABI Compendium: Wallingford, UK, 2022. [CrossRef]
86. Available online: https://antropocene.it/en/2022/12/19/silybum-marianum-2/ (accessed on 1 March 2024).
87. Halder, S.; Sharma, A. A review on *Urtica dioica* L. *World J. Pharm. Pharm. Sci.* **2017**, *6*, 404–421. [CrossRef]
88. Available online: https://antropocene.it/en/2022/10/26/urtica-dioica-en/ (accessed on 1 March 2024).
89. Applequist, W.L.; Moerman, D.E. Yarrow (*Achillea millefolium* L.): A Neglected Panacea? A Review of Ethnobotany, Bioactivity, and Biomedical Research. *Econ. Bot.* **2011**, *65*, 209–225. [CrossRef]
90. Shah, R.; Peethambaran, B. Anti-inflammatory and Anti-microbial Properties of *Achillea millefolium* in Acne Treatment. In *Immunity and Inflammation in Health and Disease*; Academic Press: Cambridge, MA, USA, 2018; pp. 241–248.
91. Vitalini, S.; Beretta, G.; Iriti, M.; Orsenigo, S.; Basilico, N.; Dall'Acqua, S.; Iorizzi, M.; Fico, G. Phenolic compounds from *Achillea millefolium* L. and their bioactivity. *Acta Biochim. Pol.* **2011**, *58*, 203–209. [CrossRef] [PubMed]
92. Lazarević, J.; Radulović, N.; Zlatković, B.; Palić, R. Composition of *Achillea distans* Willd. subsp. distansroot essential oil. *Nat. Prod. Res.* **2010**, *24*, 718–731. [CrossRef] [PubMed]
93. Fierascu, I.; Ungureanu, C.; Avramescu, S.M.; Fierascu, R.C.; Ortan, A.; Soare, L.C.; Paunescu, A. In vitro antioxidant and antifungal properties of *Achillea millefolium* L. *Rom. Biotechnol. Lett.* **2015**, *20*, 10626–10636.
94. Gihan, F.A.; Alaa, Q.R.; Alaa, O.H.; Fatimah, Y.G.; Yousra, N.; Saeed, A. Potential Analgesic and Anti-Inflammatory Effect of *Cuminum Cyminum* and *Borago officinalis* in Rats and Mice. *Asian J. Pharm. Clin. Res.* **2019**, *13*, 216–218. [CrossRef]
95. Leporatti, M.L.; Ivancheva, S. Preliminary comparative analysis of medicinal plants used in the traditional medicine of Bulgaria and Italy. *J. Ethnopharmacol.* **2003**, *87*, 123–142. [CrossRef] [PubMed]
96. Lozano-Baena, M.-D.; Tasset, I.; Muñoz-Serrano, A.; Alonso-Moraga, Á.; de Haro-Bailón, A. Cancer Prevention and Health Benefices of Traditionally Consumed *Borago officinalis* Plants. *Nutrients* **2016**, *8*, 48. [CrossRef] [PubMed]
97. Michalak, M.; Zagórska-Dziok, M.; Klimek-Szczykutowicz, M.; Szopa, A. Phenolic Profile and Comparison of the Antioxidant, Anti-Ageing, Anti-Inflammatory, and Protective Activities of *Borago officinalis* Extracts on Skin Cells. *Molecules* **2023**, *28*, 868. [CrossRef] [PubMed]

98. De Natale, A.; Pollio, A. Plants species in the folk medicine of Montecorvino Rovella (inland Campania, Italy). *J. Ethnopharmacol.* **2007**, *109*, 295–303. [CrossRef] [PubMed]
99. Scherrer, A.M.; Motti, R.; Weckerle, C.S. Traditional plant use in the areas of Monte Vesole and Ascea, Cilento National Park (Campania, Southern Italy). *J. Ethnopharmacol.* **2005**, *97*, 129–143. [CrossRef] [PubMed]
100. Ramezani, M.; Amiri, M.S.; Zibartee, E.; Boghrati, Z.; Ayati, Z.; Sahebkar, A.; Emami, S.A. A Review on the Phytochemistry, Ethnobotanical Uses and Pharmacology of *Borago* Species. *Curr. Pharm. Des.* **2020**, *26*, 110–128. [CrossRef] [PubMed]
101. Zemmouri, H.; Ammar, S.; Boumendjel, A.; Messarah, M.; El Feki, A.; Bouaziz, M. Chemical composition and antioxidant activity of *Borago officinalis* L. leaf extract growing in Algeria. *Arab. J. Chem.* **2019**, *12*, 1954–1963. [CrossRef]
102. Fabrikov, D.; Guil-Guerrero, J.L.; González-Fernández, M.J.; Rodríguez-García, I.; Gómez-Mercado, F.; Urrestarazu, M.; Lao, M.T.; Rincón-Cervera, M.Á.; Álvaro, J.E.; Lyashenko, S. Borage oil: Tocopherols, sterols and squalene in farmed and endemic-wild *Borago* species. *J. Food Compos. Anal.* **2019**, *83*, 103299. [CrossRef]
103. Gilani, A.H.; Bashir, S.; Khan, A.-u. Pharmacological basis for the use of *Borago officinalis* in gastrointestinal, respiratory and cardiovascular disorders. *J. Ethnopharmacol.* **2007**, *114*, 393–399. [CrossRef] [PubMed]
104. Asadi-Samani, M.; Bahmani, M.; Rafieian-Kopaei, M. The chemical composition, botanical characteristic and biological activities of *Borago officinalis*: A review. *Asian Pac. J. Trop. Med.* **2014**, *7*, S22–S28. [CrossRef] [PubMed]
105. Mattalia, G.; Sõukand, R.; Corvo, P.; Pieroni, A. Scholarly vs. Traditional Knowledge: Effects of Sacred Natural Sites on Ethnobotanical Practices in Tuscany, Central Italy. *Hum. Ecol.* **2019**, *47*, 653–667. [CrossRef]
106. Motti, R.; Motti, P. An Ethnobotanical Survey of Useful Plants in the Agro Nocerino Sarnese (Campania, Southern Italy). *Hum. Ecol.* **2017**, *45*, 865–878. [CrossRef]
107. Menale, B.; De Castro, O.; Cascone, C.; Muoio, R. Ethnobotanical investigation on medicinal plants in the Vesuvio National Park (Campania, Southern Italy). *J. Ethnopharmacol.* **2016**, *192*, 320–349. [CrossRef] [PubMed]
108. Badgujar, S.B.; Patel, V.V.; Bandivdekar, A.H. *Foeniculum vulgare* Mill: A Review of Its Botany, Phytochemistry, Pharmacology, Contemporary Application, and Toxicology. *BioMed Res. Int.* **2014**, *2014*, 842674. [CrossRef] [PubMed]
109. Khammassi, M.; Mighri, H.; Ben Mansour, M.; Amri, I.; Jamoussi, B.; Khaldi, A. Metabolite profiling and potential antioxidant activity of sixteen fennel (*Foeniculum vulgare* Mill.) populations growing wild in Tunisia. *S. Afr. J. Bot.* **2022**, *148*, 407–414. [CrossRef]
110. Castaldo, L.; Izzo, L.; De Pascale, S.; Narváez, A.; Rodriguez-Carrasco, Y.; Ritieni, A. Chemical Composition, In Vitro Bioaccessibility and Antioxidant Activity of Polyphenolic Compounds from Nutraceutical Fennel Waste Extract. *Molecules* **2021**, *26*, 1968. [CrossRef] [PubMed]
111. Barakat, H.; Alkabeer, I.A.; Aljutaily, T.; Almujaydil, M.S.; Algheshairy, R.M.; Alhomaid, R.M.; Almutairi, A.S.; Mohamed, A. Phenolics and Volatile Compounds of Fennel (*Foeniculum vulgare*) Seeds and Their Sprouts Prevent Oxidative DNA Damage and Ameliorates CCl4-Induced Hepatotoxicity and Oxidative Stress in Rats. *Antioxidants* **2022**, *11*, 2318. [CrossRef] [PubMed]
112. Milenković, A.; Ilić, Z.; Stanojević, L.; Milenković, L.; Šunić, L.; Lalević, D.; Stanojević, J.; Danilović, B.; Cvetković, D. Essential Oil Yield, Composition, Antioxidant and Microbial Activity of Wild Fennel (*Foeniculum vulgare* Mill.) from Monte Negro Coast. *Horticulturae* **2022**, *8*, 1015. [CrossRef]
113. Di Napoli, M.; Castagliuolo, G.; Badalamenti, N.; Maresca, V.; Basile, A.; Bruno, M.; Varcamonti, M.; Zanfardino, A. Antimicrobial, Antibiofilm, and Antioxidant Properties of Essential Oil of *Foeniculum vulgare* Mill. Leaves. *Plants* **2022**, *11*, 3573. [CrossRef] [PubMed]
114. Crescenzi, M.A.; D'Urso, G.; Piacente, S.; Montoro, P. A Comparative UHPLC-Q-Trap-MS/MS-Based Metabolomics Analysis to Distinguish *Foeniculum vulgare* Cultivars' Antioxidant Extracts. *Molecules* **2023**, *28*, 900. [CrossRef]
115. Šunić, L.; Ilić, Z.S.; Stanojević, L.; Milenković, L.; Stanojević, J.; Kovač, R.; Milenković, A.; Cvetković, D. Comparison of the Essential Oil Content, Constituents and Antioxidant Activity from Different Plant Parts during Development Stages of Wild Fennel (*Foeniculum vulgare* Mill.). *Horticulturae* **2023**, *9*, 364. [CrossRef]
116. Križman, M.; Jakše, J. Chemical and Genetic Variability of Istrian *Foeniculum vulgare* Wild Populations. *Plants* **2022**, *11*, 2239. [CrossRef] [PubMed]
117. Petrović, M.; Vukosavljević, P.; Đurović, S.; Antić, M.; Gorjanović, S. New herbal bitter liqueur with high antioxidant activity and lower sugar content: Innovative approach to liqueurs formulations. *J. Food Sci. Technol.* **2019**, *56*, 4465–4473. [CrossRef] [PubMed]
118. Tonutti, I.; Liddle, P. Aromatic plants in alcoholic beverages. A review. *Flavour Fragr. J.* **2010**, *25*, 341–350. [CrossRef]
119. Motti, R.; Bonanomi, G.; de Falco, B. Wild and cultivated plants used in traditional alcoholic beverages in Italy: An ethnobotanical review. *Eur. Food Res. Technol.* **2022**, *248*, 1089–1106. [CrossRef]
120. Esquivel-Ferriño, P.C.; Favela-Hernández, J.M.J.; Garza-González, E.; Waksman, N.; Ríos, M.Y.; Camacho-Corona, M.d.R. Antimycobacterial Activity of Constituents from *Foeniculum vulgare* Var. Dulce Grown in Mexico. *Molecules* **2012**, *17*, 8471–8482. [CrossRef] [PubMed]
121. Pasdaran, A.; Naychov, Z.; Batovska, D.; Kerr, P.; Favre, A.; Dimitrov, V.; Aneva, I.; Hamedi, A.; Kozuharova, E. Some European Gentiana Species Are Used Traditionally to Cure Wounds: Bioactivity and Conservation Issues. *Diversity* **2023**, *15*, 467. [CrossRef]
122. Pieroni, A.; Giusti, M.E.; Quave, C.L. Cross-Cultural Ethnobiology in the Western Balkans: Medical Ethnobotany and Ethnozoology Among Albanians and Serbs in the Pešter Plateau, Sandžak, South-Western Serbia. *Hum. Ecol.* **2011**, *39*, 333–349. [CrossRef]
123. Marković, M.; Pljevljakušić, D.; Menković, N.; Matejić, J.; Papović, O.; Stankov-Jovanović, V. Traditional knowledge on the medicinal use of plants from genus *Gentiana* in the Pirot County (Serbia). *Lek. Sirovine* **2021**, *41*, 46–53. [CrossRef]
124. Menković, N.; Šavikin, K.; Tasić, S.; Zdunić, G.; Stešević, D.; Milosavljević, S.; Vincek, D. Ethnobotanical study on traditional uses of wild medicinal plants in Prokletije Mountains (Montenegro). *J. Ethnopharmacol.* **2011**, *133*, 97–107. [CrossRef] [PubMed]

125. Ponticelli, M.; Lela, L.; Moles, M.; Mangieri, C.; Bisaccia, D.; Faraone, I.; Falabella, R.; Milella, L. The healing bitterness of *Gentiana lutea* L., phytochemistry and biological activities: A systematic review. *Phytochemistry* **2023**, *206*, 113518. [CrossRef] [PubMed]
126. Łukasz Mikołajczak, P.; Kędzia, B.; Ożarowski, M.; Kujawski, R.; Bogacz, A.; Bartkowiak-Wieczorek, J.; Białas, W.; Gryszczyńska, A.; Buchwald, W.; Szulc, M.; et al. Evaluation of anti-inflammatory and analgesic activities of extracts from herb of *Chelidonium majus* L. *Cent. Eur. J. Immunol.* **2015**, *4*, 400–410. [CrossRef] [PubMed]
127. Mathew, A.; Taranalli, A.D.; Torgal, S.S. Evaluation of Anti-inflammatory and Wound Healing Activity of *Gentiana lutea* Rhizome Extracts in Animals. *Pharm. Biol.* **2008**, *42*, 8–12. [CrossRef]
128. Cioanca, O.; Hancianu, M.; Mihasan, M.; Hritcu, L. Anti-acetylcholinesterase and Antioxidant Activities of Inhaled Juniper Oil on Amyloid Beta (1–42)-Induced Oxidative Stress in the Rat Hippocampus. *Neurochem. Res.* **2015**, *40*, 952–960. [CrossRef] [PubMed]
129. Raina, R.; Verma, P.K.; Peshin, R.; Kour, H. Potential of *Juniperus communis* L as a nutraceutical in human and veterinary medicine. *Heliyon* **2019**, *5*, e02376. [CrossRef] [PubMed]
130. Gonçalves, A.; Flores-Félix, J.; Coutinho, P.; Alves, G.; Silva, L. Zimbro (*Juniperus communis* L.) as a Promising Source of Bioactive Compounds and Biomedical Activities: A Review on Recent Trends. *Int. J. Mol. Sci.* **2022**, *23*, 3197. [CrossRef] [PubMed]
131. Falasca, A.; Melck, D.; Paris, D.; Saviano, G.; Motta, A.; Iorizzi, M. Seasonal changes in the metabolic fingerprint of *Juniperus communis* L. berry extracts by 1H NMR-based metabolomics. *Metabolomics* **2013**, *10*, 165–174. [CrossRef]
132. Chiorcea-Paquim, A.M.; Enache, T.A.; De Souza Gil, E.; Oliveira-Brett, A.M. Natural phenolic antioxidants electrochemistry: Towards a new food science methodology. *Compr. Rev. Food Sci. Food Saf.* **2020**, *19*, 1680–1726. [CrossRef] [PubMed]
133. Gao, H.-W.; Huang, X.-F.; Yang, T.-P.; Chang, K.-F.; Yeh, L.-W.; Hsieh, M.-C.; Weng, J.-C.; Tsai, N.-M. *Juniperus communis* suppresses melanoma tumorigenesis by inhibiting tumor growth and inducing apoptosis. *Am. J. Chin. Med.* **2019**, *47*, 1171–1191. [CrossRef] [PubMed]
134. Falasca, A.; Caprari, C.; De Felice, V.; Fortini, P.; Saviano, G.; Zollo, F.; Iorizzi, M. GC-MS analysis of the essential oils of *Juniperus communis* L. berries growing wild in the Molise region: Seasonal variability and in vitro antifungal activity. *Biochem. Syst. Ecol.* **2016**, *69*, 166–175. [CrossRef]
135. Anzano, A.; de Falco, B.; Grauso, L.; Motti, R.; Lanzotti, V. Laurel, *Laurus nobilis* L.: A review of its botany, traditional uses, phytochemistry and pharmacology. *Phytochem. Rev.* **2022**, *21*, 565–615. [CrossRef]
136. Staub, P.O.; Geck, M.S.; Weckerle, C.S.; Casu, L.; Leonti, M. Classifying diseases and remedies in ethnomedicine and ethnopharmacology. *J. Ethnopharmacol.* **2015**, *174*, 514–519. [CrossRef] [PubMed]
137. Mohammed, R.R.; Omer, A.K.; Yener, Z.; Uyar, A.; Ahmed, A.K. Biomedical effects of *Laurus nobilis* L. leaf extract on vital organs in streptozotocin-induced diabetic rats: Experimental research. *Ann. Med. Surg.* **2021**, *61*, 188–197. [CrossRef]
138. Baydoun, S.A.; Kanj, D.; Raafat, K.; Aboul Ela, M.; Chalak, L.; Arnold-Apostolides, N. Ethnobotanical and Economic Importance of Wild Plant Species of Jabal Moussa Bioreserve, Lebanon. *J. Ecosyst. Ecography* **2017**, *7*, 1000245. [CrossRef]
139. Awada, F.; Hamade, K.; Kassir, M.; Hammoud, Z.; Mesnard, F.; Rammal, H.; Fliniaux, O. *Laurus nobilis* Leaves and Fruits: A Review of Metabolite Composition and Interest in Human Health. *Appl. Sci.* **2023**, *13*, 4606. [CrossRef]
140. Shukla, R.; Jain, N.; Rai, G.; Pandey, V.; Mourya, P.; Pal, B. Unlocking the potential of bay leaf: Exploring its role as a nutraceutical carrier through ethnomedicinal and pharmacological insights *Drug Pharm. Sci. Arch.* **2023**, *3*, 13–23.
141. Chahal, K.K.; Kaur, M.; Bhardwaj, U.; Singla, N.; Kaur, A. A review on chemistry and biological activities of *Laurus nobilis* L. essential oil. *J. Pharmacogn. Phytochem.* **2017**, *6*, 1153–1161.
142. De Marino, S.; Borbone, N.; Zollo, F.; Ianaro, A.; Di Meglio, P.; Iorizzi, M. Megastigmane and Phenolic Components from *Laurus nobilis* L. Leaves and Their Inhibitory Effects on Nitric Oxide Production. *J. Agric. Food Chem* **2004**, *52*, 7525–7531. [CrossRef] [PubMed]
143. Barros, L.; Carvalho, A.M.; Ferreira, I.C.F.R. Leaves, flowers, immature fruits and leafy flowered stems of *Malva sylvestris*: A comparative study of the nutraceutical potential and composition. *Food Chem. Toxicol.* **2010**, *48*, 1466–1472. [CrossRef] [PubMed]
144. Fathi, M.; Ghane, M.; Pishkar, L. Phytochemical Composition, Antibacterial, and Antibiofilm Activity of *Malva sylvestris* Against Human Pathogenic Bacteria. *Jundishapur J. Nat. Pharm. Prod.* **2021**, *17*, e114164. [CrossRef]
145. Billeter, M.; Meier, B.; Sticher, O. 8-hydroxyflavonoid glucuronides from *Malva sylvestris*. *Phytochemistry* **1991**, *30*, 987–990. [CrossRef]
146. Wu, Y.; Qiu, A.; Yang, Z.; Wu, J.; Li, X.; Bao, K.; Wang, M.; Wu, B. *Malva sylvestris* extract alleviates the astrogliosis and inflammatory stress in LPS-induced depression mice. *J. Neuroimmunol.* **2019**, *336*, 577029. [CrossRef] [PubMed]
147. Mohamadi Yarijani, Z.; Najafi, H.; Shackebaei, D.; Madani, S.H.; Modarresi, M.; Jassemi, S.V. Amelioration of renal and hepatic function, oxidative stress, inflammation and histopathologic damages by *Malva sylvestris* extract in gentamicin induced renal toxicity. *Biomed. Pharmacother.* **2019**, *112*, 108635. [CrossRef] [PubMed]
148. Chiclana, C.F.; Enrique, A.; Consolini, A.E. Actividad antiinflamatoria local de *Malva sylvestris* L. (Malvaceae) en el edema inducido por carragenina en ratas. *Lat. Am. J. Pharm.* **2009**, *28*, 275–278.
149. Camejo-Rodrigues, J.; Ascensão, L.; Bonet, M.À.; Vallès, J. An ethnobotanical study of medicinal and aromatic plants in the Natural Park of "Serra de São Mamede" (Portugal). *J. Ethnopharmacol.* **2003**, *89*, 199–209. [CrossRef] [PubMed]
150. Novais, M.H.; Santos, I.; Mendes, S.; Pinto-Gomes, C. Studies on pharmaceutical ethnobotany in Arrabida Natural Park (Portugal). *J. Ethnopharmacol.* **2004**, *93*, 183–195. [CrossRef] [PubMed]
151. Ferreira, A.; Proença, C.; Serralheiro, M.L.M.; Araújo, M.E.M. The in vitro screening for acetylcholinesterase inhibition and antioxidant activity of medicinal plants from Portugal. *J. Ethnopharmacol.* **2006**, *108*, 31–37. [CrossRef] [PubMed]
152. Guarrera, P.M.; Lucia, L.M. Ethnobotanical remarks on Central and Southern Italy. *J. Ethnobiol. Ethnomed.* **2007**, *3*, 23. [CrossRef] [PubMed]

153. Quave, C.L.; Pieroni, A.; Bennett, B.C. Dermatological remedies in the traditional pharmacopoeia of Vulture-Alto Bradano, inland southern Italy. *J. Ethnobiol. Ethnomed.* **2008**, *4*, 5. [CrossRef]
154. Leporatti, M.L.; Ghedira, K. Comparative analysis of medicinal plants used in traditional medicine in Italy and Tunisia. *J. Ethnobiol. Ethnomed.* **2009**, *5*, 31. [CrossRef] [PubMed]
155. Neves, J.M.; Matos, C.; Moutinho, C.; Queiroz, G.; Gomes, L.R. Ethnopharmacological notes about ancient uses of medicinal plants in Trás-os-Montes (northern of Portugal). *J. Ethnopharmacol.* **2009**, *124*, 270–283. [CrossRef]
156. Tepe, B.; Cilkiz, M. A pharmacological and phytochemical overview on *Satureja*. *Pharm. Biol.* **2015**, *54*, 375–412. [CrossRef] [PubMed]
157. Oalđe Pavlović, M.; Kolarević, S.; Đorđević, J.; Jovanović Marić, J.; Lunić, T.; Mandić, M.; Kračun Kolarević, M.; Živković, J.; Alimpić Aradski, A.; Marin, P.D.; et al. A Study of Phytochemistry, Genoprotective Activity, and Antitumor Effects of Extracts of the Selected Lamiaceae Species. *Plants* **2021**, *10*, 2306. [CrossRef] [PubMed]
158. Marković, M.; Pljevljakušić, D.; Matejić, J.; Nikolić, B.; Smiljić, M.; Đelić, G.; Papović, O.; Đokić, M.; Stankov-Jovanović, V. The plants traditionally used for the treatment of respiratory infections in the Balkan Peninsula (Southeast Europe). *Lek. Sirovine* **2022**, *42*, 68–88. [CrossRef]
159. Motti, R. Wild Plants Used as Herbs and Spices in Italy: An Ethnobotanical Review. *Plants* **2021**, *10*, 563. [CrossRef] [PubMed]
160. Grosso, C.; Figueiredo, A.C.; Burillo, J.; Mainar, A.M.; Urieta, J.S.; Barroso, J.G.; Coelho, J.A.; Palavra, A.M.F. Enrichment of the thymoquinone content in volatile oil from *Satureja montana* using supercritical fluid extraction. *J. Sep. Sci.* **2009**, *32*, 328–334. [CrossRef]
161. Cavalloro, V.; Robustelli della Cuna, F.S.; Quai, E.; Preda, S.; Bracco, F.; Martino, E.; Collina, S. Walking around the Autonomous Province of Trento (Italy): An Ethnobotanical Investigation. *Plants* **2022**, *11*, 2246. [CrossRef] [PubMed]
162. Capasso, R.; Aviello, G.; Capasso, F.; Savino, F.; Izzo, A.A.; Lembo, F.; Borrelli, F. Silymarin BIO-C®, an extract from *Silybum marianum* fruits, induces hyperprolactinemia in intact female rats. *Phytomedicine* **2009**, *16*, 839–844. [CrossRef] [PubMed]
163. Csupor, D.; Csorba, A.; Hohmann, J. Recent advances in the analysis of flavonolignans of *Silybum marianum*. *J. Pharm. Biomed. Anal.* **2016**, *130*, 301–317. [CrossRef] [PubMed]
164. Hanlidou, E.; Karousou, R.; Kleftoyanni, V.; Kokkini, S. The herbal market of Thessaloniki (N Greece) and its relation to the ethnobotanical tradition. *J. Ethnopharmacol.* **2004**, *91*, 281–299. [CrossRef] [PubMed]
165. Marmouzi, I.; Bouyahya, A.; Ezzat, S.M.; El Jemli, M.; Kharbach, M. The food plant *Silybum marianum* (L.) Gaertn.: Phytochemistry, Ethnopharmacology and clinical evidence. *J. Ethnopharmacol.* **2021**, *265*, 113303. [CrossRef]
166. Piluzza, G.; Virdis, S.; Serralutzu, F.; Bullitta, S. Uses of plants, animal and mineral substances in Mediterranean ethno-veterinary practices for the care of small ruminants. *J. Ethnopharmacol.* **2015**, *168*, 87–99. [CrossRef] [PubMed]
167. Kim, N.C.; Graf, T.N.; Sparacino, C.M.; Wani, M.C.; Wall, M.E. Complete Isolation and Characterization of Silybins and Isosilybins from Milk Thistle (*Silybum marianum*). *Org. Biomol. Chem.* **2003**, *1*, 1684–1689. [CrossRef] [PubMed]
168. Pieroni, A.; Quave, C.L.; Santoro, R.F. Folk pharmaceutical knowledge in the territory of the Dolomiti Lucane, inland southern Italy. *J. Ethnopharmacol.* **2004**, *95*, 373–384. [CrossRef] [PubMed]
169. Ahmed Kk, M.; Parsuraman, S. *Urtica dioica* L., (Urticaceae): A Stinging Nettle. *Syst. Rev. Pharm.* **2014**, *5*, 6–8. [CrossRef]
170. Taheri, Y.; Quispe, C.; Herrera-Bravo, J.; Sharifi-Rad, J.; Ezzat, S.M.; Merghany, R.M.; Shaheen, S.; Azmi, L.; Prakash Mishra, A.; Sener, B.; et al. *Urtica dioica* -Derived Phytochemicals for Pharmacological and Therapeutic Applications. *Evid.-Based Complement. Altern. Med.* **2022**, *2022*, 4024331. [CrossRef] [PubMed]
171. Carvalho, A.R.; Costa, G.; Figueirinha, A.; Liberal, J.; Prior, J.A.V.; Lopes, M.C.; Cruz, M.T.; Batista, M.T. *Urtica* spp.: Phenolic composition, safety, antioxidant and anti-inflammatory activities. *Food Res. Int.* **2017**, *99*, 485–494. [CrossRef] [PubMed]
172. Akram, M. Minireview on *Achillea millefolium* Linn. *J. Membr. Biol.* **2013**, *246*, 661–663. [CrossRef] [PubMed]
173. Orav, A.; Arak, E.; Raal, A. Phytochemical analysis of the essential oil of *Achillea millefolium* L. from various European Countries. *Nat. Prod. Res.* **2007**, *20*, 1082–1088. [CrossRef] [PubMed]
174. Ali, S.I.; Gopalakrishnan, B.; Venkatesalu, V. Pharmacognosy, Phytochemistry and Pharmacological Properties of *Achillea millefolium* L.: A Review. *Phytother. Res.* **2017**, *31*, 1140–1161. [CrossRef] [PubMed]
175. Motti, R.; Paura, B.; Cozzolino, A.; de Falco, B. Edible Flowers Used in Some Countries of the Mediterranean Basin: An Ethnobotanical Overview. *Plants* **2022**, *11*, 3272. [CrossRef]
176. Available online: https://www.atlantides.it/gentiana-lutea-subsp.-lutea.html (accessed on 1 March 2024).
177. Acta Plantarum, from 2007 on—"Lista Delle Schede Botaniche. Available online: https://www.actaplantarum.org/schede/schede.php (accessed on 1 March 2024).
178. Available online: http://www.lucolifloraefauna.it/Flora/Schede/Satureja_montana_montana.htm (accessed on 1 March 2024).
179. CABI Compendium: Wallingford, UK, 2013. Available online: https://www.cabidigitallibrary.org/doi/10.1079/cabicompendium.50304 (accessed on 1 March 2024).
180. Popay, I. *Urtica Dioica (Stinging Nettle)*; CABI Compendium: Wallingford, UK, 2014. [CrossRef]
181. Ahmed, A.F.; Shi, M.; Liu, C.; Kang, W. Comparative analysis of antioxidant activities of essential oils and extracts of fennel (*Foeniculum vulgare* Mill.) seeds from Egypt and China. *Food Sci. Hum. Wellness* **2019**, *8*, 67–72. [CrossRef]
182. Khammassi, M.; Loupassaki, S.; Tazarki, H.; Mezni, F.; Slama, A.; Tlili, N.; Zaouali, Y.; Mighri, H.; Jamoussi, B.; Khaldi, A. Variation in essential oil composition and biological activities of *Foeniculum vulgare* Mill. populations growing widely in Tunisia. *J. Food Biochem.* **2018**, *42*, e12532. [CrossRef]
183. Śliwińska, M.; Wiśniewska, P.; Dymerski, T.; Wardencki, W.; Namieśnik, J. The flavour of fruit spirits and fruit liqueurs: A review. *Flavour Fragr. J.* **2015**, *30*, 197–207. [CrossRef]

184. Prakash, O.; Singh, R.; Kumar, S.; Srivastava, S.; Ved, A. *Gentiana lutea* Linn. (Yellow Gentian): A comprehensive review. *J. Ayurvedic Herb. Med.* **2017**, *3*, 175–181. [CrossRef]
185. Pan, Y.; Zhao, Y.L.; Zhang, J.; Li, W.Y.; Wang, Y.Z. Phytochemistry and Pharmacological Activities of the Genus *Gentiana* (Gentianaceae). *Chem. Biodivers.* **2016**, *13*, 107–150. [CrossRef] [PubMed]
186. Mustafa, A.M.; Maggi, F.; Öztürk, N.; Öztürk, Y.; Sagratini, G.; Torregiani, E.; Vittori, S.; Caprioli, G. Chemical and biological analysis of the by-product obtained by processing *Gentiana lutea* L. and other herbs during production of bitter liqueurs. *Ind. Crops Prod.* **2016**, *80*, 131–140. [CrossRef]
187. Fejér, J.; Grul'ová, D.; Eliašová, A.; Kron, I.; De Feo, V. Influence of environmental factors on content and composition of essential oil from common juniper ripe berry cones (*Juniperus communis* L.). *Plant Biosyst.-Int. J. Deal. All Asp. Plant Biol.* **2018**, *152*, 1227–1235. [CrossRef]
188. Verheyen, K.; Adriaenssens, S.; Gruwez, R.; Michalczyk, I.M.; Ward, L.K.; Rosseel, Y.; Van den Broeck, A.; García, D. *Juniperus communis*: Victim of the combined action of climate warming and nitrogen deposition? *Plant Biol.* **2009**, *11*, 49–59. [CrossRef] [PubMed]
189. Bais, S.; Gill, N.S.; Kumar, N. Neuroprotective Effect of *Juniperus communis* on Chlorpromazine Induced Parkinson Disease in Animal Model. *Chin. J. Biol.* **2015**, *2015*, 542542. [CrossRef]
190. Guedri, M.M.; Romdhane, M.; Lebrihi, A.; Mathieu, F.; Bouajila, J. Chemical composition and antimicrobial and antioxidant activities of Tunisian, France and Austrian *Laurus nobilis* (Lauraceae) essential oils. *Not. Bot. Horti Agrobot. Cluj-Napoca.* **2020**, *48*, 1929–1940. [CrossRef]
191. Stefanova, G.; Girova, T.; Gochev, V.; Stoyanova, M.; Petkova, Z.; Stoyanova, A.; Zheljazkov, V.D.J.H. Comparative study on the chemical composition of laurel (*Laurus nobilis* L.) leaves from Greece and Georgia and the antibacterial activity of their essential oil. oil. *Heliyon* **2020**, *6*, e05491. [CrossRef] [PubMed]
192. Batiha, G.E.-S.; Tene, S.T.; Teibo, J.O.; Shaheen, H.M.; Oluwatoba, O.S.; Teibo, T.K.A.; Al-kuraishy, H.M.; Al-Garbee, A.I.; Alexiou, A.; Papadakis, M. The phytochemical profiling, pharmacological activities, and safety of *Malva sylvestris*: A review. *Naunyn-Schmiedeberg's Arch. Pharmacol.* **2022**, *396*, 421–440. [CrossRef] [PubMed]
193. Duggan, K.C.; Hermanson, D.J.; Musee, J.; Prusakiewicz, J.J.; Scheib, J.L.; Carter, B.D.; Banerjee, S.; Oates, J.A.; Marnett, L.J. (R)-Profens are substrate-selective inhibitors of endocannabinoid oxygenation by COX-2. *Nat. Chem. Biol.* **2011**, *7*, 803–809. [CrossRef] [PubMed]
194. Di Pietro, R. New Dry Grassland Associations from the Ausoni-Aurunci Mountains (Central Italy)—Syntaxonomical Updating and Discussion on the Higher Rank Syntaxa. *Hacquetia* **2011**, *10*, 183–231. [CrossRef]
195. Caprioli, G.; Lupidi, G.; Maggi, F. Comparison of chemical composition and antioxidant activities of two Winter savory subspecies (*Satureja montana* subsp. variegata and *Satureja montana* subsp. montana) cultivated in Northern Italy. *Nat. Prod. Res.* **2018**, *33*, 3143–3147. [CrossRef] [PubMed]
196. Tomaselli, V.; Silletti, G.; Forte, L. A new association of *Satureja montana* L. subsp. montana dominated garrigues in Puglia (SE Italy). *Plant Sociol.* **2021**, *58*, 1–14. [CrossRef]
197. Navarro-Rocha, J.; Andrés, M.F.; Díaz, C.E.; Burillo, J.; González-Coloma, A. Composition and biocidal properties of essential oil from pre-domesticated Spanish *Satureja montana*. *Ind. Crops Prod.* **2020**, *145*, 111958. [CrossRef]
198. Chizzola, R. Volatile Oil Composition of Four Populations of *Satureja montana* L. From Southern France. *Acta Hortic.* **2003**, *598*, 143–147. [CrossRef]
199. Bojović, D.; Šoškić, M.; Tadić, V. Comparative study of chemical composition of the essential oils from *Satureja cuneifolia* ten. and *Satureja montana* l., lamiaceae collected at national park Lovćen, Montenegro. *Stud. Univ. Babeș-Bolyai Chem.* **2018**, *63*, 167–180. [CrossRef]
200. Lumpert, M.; Kreft, S. Folk use of medicinal plants in Karst and Gorjanci, Slovenia. *J. Ethnobiol. Ethnomed.* **2017**, *13*, 16. [CrossRef] [PubMed]
201. Kremer, D.; Košir, I.; Končić, M.; Čerenak, A.; Potočnik, T.; Srečec, S.; Randić, M.; Kosalec, I. Antimicrobial and Antioxidant Properties of *Satureja montana* L. and S. Subspicata Vis. (Lamiaceae). *Curr. Drug Targets* **2015**, *16*, 1623–1633. [CrossRef] [PubMed]
202. Aćimović, M.; Todosijević, M.; Varga, A.; Kiprovski, B.; Tešević, V.; Čabarkapa, I.; Sikora, V. Bioactivity of essential oils from cultivated winter savory, sage and hyssop. *Lek. Sirovine* **2019**, *39*, 11–17. [CrossRef]
203. ČOpra-JanİĆİJevİĆ, A.; VİDİC, D.; MaksİMovİĆ, M. Chemical composition of the essential oil and headspace of *Satureja montana* L. *Nat. Volatiles Essent.* **2020**, *7*, 22–34. [CrossRef]
204. Souto-Maior, F.N.; Da Fonsêca, D.V.; Salgado, P.R.R.; de Oliveira Monte, L.; de Sousa, D.P.; de Almeida, R.N. Antinociceptive and anticonvulsant effects of the monoterpene linalool oxide. *Pharm. Biol.* **2016**, *55*, 63–67. [CrossRef] [PubMed]
205. Kustrak, D.; Kuftinec, J.; Blazevic, N.; Maffei, M. Comparison of the Essential Oil Composition of Two Subspecies of *Satureja montana*. *J. Essent. Oil Res.* **1996**, *8*, 7–13. [CrossRef]
206. Taoudiat, A.; Spigno, G.; Ferhat, Z.; Djenane, D. Bioenrichment using *Satureja montana* L. essential oil for the prevention against photooxidation of flavored extra virgin olive oil during light display. *N. Afr. J. Food Nutr. Res.* **2021**, *4*, 351–359. [CrossRef]
207. Babajafari, S.; Nikaein, F.; Mazloomi, S.M.; Zibaeenejad, M.J.; Zargaran, A. A Review of the Benefits of *Satureja* Species on Metabolic Syndrome and Their Possible Mechanisms of Action. *J. Evid.-Based Complement. Altern. Med.* **2015**, *20*, 212–223. [CrossRef] [PubMed]
208. Peschel, W. The use of community herbal monographs to facilitate registrations and authorisations of herbal medicinal products in the European Union 2004–2012. *J. Ethnopharmacol.* **2014**, *158*, 471–486. [CrossRef] [PubMed]

209. Dhouibi, R.; Affes, H.; Ben Salem, M.; Hammami, S.; Sahnoun, Z.; Zeghal, K.M.; Ksouda, K. Screening of pharmacological uses of *Urtica dioica* and others benefits. *Prog. Biophys. Mol. Biol.* **2020**, *150*, 67–77. [CrossRef] [PubMed]
210. Falconieri, D.; Piras, A.; Porcedda, S.; Marongiu, B.; Gonçalves, M.J.; Cabral, C.; Cavaleiro, C.; Salgueiro, L. Chemical Composition and Biological Activity of the Volatile Extracts of *Achillea millefolium*. *Nat. Prod. Commun.* **2011**, *6*, 1527–1530. [CrossRef] [PubMed]
211. Mohammed, H.A.; Abd-Elraouf, M.; Sulaiman, G.M.; Almahmoud, S.A.; Hamada, F.A.; Khan, R.A.; Hegazy, M.M.; Abd-El-Wahab, M.F.; Kedra, T.A.; Ismail, A. Variability in the volatile constituents and biological activities of *Achillea millefolium* L. essential oils obtained from different plant parts and by different solvents. *Arab. J. Chem.* **2023**, *16*, 105103. [CrossRef]
212. Chaouche, T.A.; Karim, A.; Mourad, B. Phytochemical screening of Algerian *Borago officinalis* L. and evaluation of its antioxidant and antimicrobial activities against respiratory pathogens. *Int. J. Phytomed.* **2014**, *6*, 369–376.
213. Samy, M.N.; Hamed, A.N.E.-S.; Sugimoto, S.; Otsuka, H.; Kamel, M.S.; Matsunami, K. Officinalioside, a new lignan glucoside from *Borago officinalis* L. *Nat. Prod. Res.* **2015**, *30*, 967–972. [CrossRef] [PubMed]
214. Karimi, E.; Oskoueian, E.; Karimi, A.; Noura, R.; Ebrahimi, M. *Borago officinalis* L. flower: A comprehensive study on bioactive compounds and its health-promoting properties. *J. Food Meas. Charact.* **2017**, *12*, 826–838. [CrossRef]
215. Herrmann, M.; Joppe, H.; Schmaus, G. Thesinine-4′-O-β-d-glucoside the first glycosylated plant pyrrolizidine alkaloid from *Borago officinalis*. *Phytochemistry* **2002**, *60*, 399–402. [CrossRef] [PubMed]
216. Montaner, C.; Zufiaurre, R.; Movila, M.; Mallor, C. Evaluation of Borage (*Borago officinalis* L.) Genotypes for Nutraceutical Value Based on Leaves Fatty Acids Composition. *Foods* **2021**, *11*, 16. [CrossRef]
217. Bowes, K.M.; Zheljazkov, V.D.J.H. Essential oil yields and quality of fennel grown in Nova Scotia. *HortScience* **2004**, *39*, 1640–1643. [CrossRef]
218. Shehata, A.M.; Elhafez, Z.A.A.; Ahmed, A.F. A Comparative Study of the Response of Fennel (*Foeniculum vulgare*, Mill) Plants from Egypt and China to Spraying with Benzyladenine (BA). *Eur. J. Med. Plants* **2022**, *33*, 1–12. [CrossRef]
219. Adefegha, S.A. Functional foods and nutraceuticals as dietary intervention in chronic diseases; novel perspectives for health promotion and disease prevention. *J. Diet. Suppl.* **2018**, *15*, 977–1009. [CrossRef] [PubMed]
220. Šavikin, K.; Menković, N.; Zdunić, G.; Stević, T.; Radanović, D.; Janković, T. Antimicrobial Activity of *Gentiana lutea* L. Extracts. *Z. Naturforschung C* **2009**, *64*, 339–342. [CrossRef] [PubMed]
221. Aberham, A.; Pieri, V.; Croom, E.M.; Ellmerer, E.; Stuppner, H. Analysis of iridoids, secoiridoids and xanthones in *Centaurium erythraea*, *Frasera caroliniensis* and *Gentiana lutea* using LC–MS and RP-HPLC. *J. Pharm. Biomed. Anal.* **2011**, *54*, 517–525. [CrossRef] [PubMed]
222. Mustafa, A.M.; Caprioli, G.; Dikmen, M.; Kaya, E.; Maggi, F.; Sagratini, G.; Vittori, S.; Öztürk, Y. Evaluation of neuritogenic activity of cultivated, wild and commercial roots of *Gentiana lutea* L. *J. Funct. Foods* **2015**, *19*, 164–173. [CrossRef]
223. Radanovic, D.; Antic-Mladenovic, S.; Jakovljevic, M.; Kresovic, M. Content of heavy metals in *Gentiana lutea* L. roots and galenic forms. *J. Serbian Chem. Soc.* **2007**, *72*, 133–138. [CrossRef]
224. Cafaro, T.; Carnicelli, V.; Caprioli, G.; Maggi, F.; Celenza, G.; Perilli, M.; Bozzi, A.; Amicosante, G.; Brisdelli, F. Anti-apoptotic and anti-inflammatory activity of *Gentiana lutea* root extract. *Adv. Tradit. Med.* **2020**, *20*, 619–630. [CrossRef]
225. Balijagić, J.; Janković, T.; Zdunić, G.; Bošković, J.; Šavikin, K.; Goćevac, D.; Stanojković, T.; Jovančević, M.; Menković, N. Chemical Profile, Radical Scavenging and Cytotoxic Activity of Yellow Gentian Leaves (*Genitaneae Luteae* Folium) Grown in Northern Regions of Montenegro. *Nat. Prod. Commun.* **2012**, *7*, 1487–1490. [CrossRef] [PubMed]
226. Vieira, L.M.M.; Kijjoa, A. Naturally-Occurring Xanthones: Recent Developments. *Curr. Med. Chem.* **2005**, *12*, 2413–2446. [CrossRef] [PubMed]
227. Citová, I.; Ganzera, M.; Stuppner, H.; Solich, P. Determination of gentisin, isogentisin, and amarogentin in *Gentiana lutea* L. by capillary electrophoresis. *J. Sep. Sci.* **2008**, *31*, 195–200. [CrossRef] [PubMed]
228. Rana, A.C.; Gulliya, B. Chemistry and Pharmacology of Flavonoids—A Review. *Indian J. Pharm. Educ. Res.* **2019**, *53*, 8–20. [CrossRef]
229. Živić, N.; Milošević, S.; Dekić, V.; Dekić, B.; Ristić, N.; Ristić, M.; Sretić, L. Phytochemical and antioxidant screening of some extracts of *Juniperus communis* L. and *Juniperus oxycedrus* L. *Czech J. Food Sci* **2019**, *37*, 351–358. [CrossRef]
230. Lima Reis, P.M.C.; Mezzomo, N.; Aguiar, G.P.S.; Hotza, D.; Baggio Ribeiro, D.H.; Salvador Ferreira, S.R.; Hense, H. Formation, stability and antimicrobial activity of laurel leaves essential oil (*Laurus nobilis* L.) particles in suspension obtained by SFEE. *J. Supercrit. Fluids* **2020**, *166*, 105032. [CrossRef]
231. Belasli, A.; Ben Miri, Y.; Aboudaou, M.; Aït Ouahioune, L.; Montañes, L.; Ariño, A.; Djenane, D. Antifungal, antitoxigenic, and antioxidant activities of the essential oil from laurel (*Laurus nobilis* L.): Potential use as wheat preservative. *Food Sci. Nutr.* **2020**, *8*, 4717–4729. [CrossRef] [PubMed]
232. Caputo, L.; Nazzaro, F.; Souza, L.; Aliberti, L.; De Martino, L.; Fratianni, F.; Coppola, R.; De Feo, V. *Laurus nobilis*: Composition of Essential Oil and Its Biological Activities. *Molecules* **2017**, *22*, 930. [CrossRef] [PubMed]
233. Řebíčková, K.; Bajer, T.; Šilha, D.; Ventura, K.; Bajerová, P. Comparison of Chemical Composition and Biological Properties of Essential Oils Obtained by Hydrodistillation and Steam Distillation of *Laurus nobilis* L. *Plant Foods Hum. Nutr.* **2020**, *75*, 495–504. [CrossRef] [PubMed]
234. Panza, E.; Tersigni, M.; Iorizzi, M.; Zollo, F.; De Marino, S.; Festa, C.; Napolitano, M.; Castello, G.; Ialenti, A.; Ianaro, A. Lauroside B, a Megastigmane Glycoside from *Laurus nobilis* (Bay Laurel) Leaves, Induces Apoptosis in Human Melanoma Cell Lines by Inhibiting NF-κB Activation. *J. Nat. Prod.* **2010**, *74*, 228–233. [CrossRef] [PubMed]
235. Nawwar, M.A.M.; El Dein, A.; El Sherbeiny, A.; El Ansari, M.A.; El Sissi, H.I. Two new sulphated flavonol glucosides from leaves of *Malva sylvestris*. *Phytochemistry* **1977**, *16*, 145–146. [CrossRef]

236. Nawwar, M.A.M.; Buddrus, J. A gossypetin glucuronide sulphate from the leaves of *Malva sylvestris*. *Phytochemistry* **1981**, *20*, 2446–2448. [CrossRef]
237. Brouillard, R.J.P. The in vivo expression of anthocyanin colour in plants. *Phytochemistry* **1983**, *22*, 1311–1323. [CrossRef]
238. Merlin, J.-C.; Statoua, A.; Brouillard, R. Investigation of the in vivo organization of anthocyanins using resonance raman microspectrometry. *Phytochemistry* **1985**, *24*, 1575–1581. [CrossRef]
239. Farina, A.; Doldo, A.; Cotichini, V.; Rajevic, M.; Quaglia, M.G.; Mulinacci, N.; Vincieri, F.F. HPTLC and reflectance mode densitometry of anthocyanins in *Malva Sylvestris* L.: A comparison with gradient-elution reversed-phase HPLC. *J. Pharm. Biomed. Anal.* **1995**, *14*, 203–211. [CrossRef] [PubMed]
240. Lewis, C. Effect of polysaccharides on the colour of anthocyanins. *Food Chem.* **1995**, *54*, 315–319. [CrossRef]
241. Sikorska, M.; Matławska, I.; Fra ski, R. 8-Hydroxyflavonoid glucuronides of *Malope trifida*. *Acta Physiol. Plant.* **2004**, *26*, 291–297. [CrossRef]
242. Takeda, K.; Enoki, S.; Harborne, J.B.; Eagles, J. Malonated anthocyanins in malvaceae: Malonylmalvin from *Malva sylvestris*. *Phytochemistry* **1989**, *28*, 499–500. [CrossRef]
243. Conforti, F.; Ioele, G.; Statti, G.A.; Marrelli, M.; Ragno, G.; Menichini, F. Antiproliferative activity against human tumor cell lines and toxicity test on Mediterranean dietary plants. *Food Chem. Toxicol.* **2008**, *46*, 3325–3332. [CrossRef]
244. Emets, T.I.; Steblyuk, M.V.; Klyuev, N.A.; Petrenko, V.V. Some components of the seed oil of *Malva sylvestris*. *Chem. Nat. Compd.* **1994**, *30*, 292–294. [CrossRef]
245. Mukarram, M.; Ahmad, I.; Ahmad, M. HBr-Reactive acids of *Malva sylvestris* seed oil. *J. Am. Oil Chem. Soc.* **1984**, *61*, 1060. [CrossRef]
246. Cutillo, F.; Dabrosca, B.; Dellagreca, M.; Fiorentino, A.; Zarrelli, A. Terpenoids and phenol derivatives from *Malva sylvestris*. *Phytochemistry* **2006**, *67*, 481–485. [CrossRef] [PubMed]
247. Paul, Z.A.; Malla, A.T.; Dar, M.A.; Masoodi, M.H. Phytochemistry and Pharmacological Activity of *Malva sylvestris* L: A Detailed Insight. *Comb. Chem. High Throughput Screen.* 2023; 26, *Online ahead of print*. [CrossRef]
248. Gladikostić, N.; Ikonić, B.; Teslić, N.; Zeković, Z.; Božović, D.; Putnik, P.; Bursać Kovačević, D.; Pavlić, B. Essential Oils from Apiaceae, Asteraceae, Cupressaceae and Lamiaceae Families Grown in Serbia: Comparative Chemical Profiling with In Vitro Antioxidant Activity. *Plants* **2023**, *12*, 745. [CrossRef]
249. Milojevic, S.; Radosavljevic, D.; Pavicevic, V.; Pejanovic, S.; Veljkovic, V. Modeling the kinetics of essential oil hydrodistillation from plant materials. *Hem. Ind.* **2013**, *67*, 843–859. [CrossRef]
250. Baydar, H.; Sağdiç, O.; Özkan, G.; Karadoğan, T. Antibacterial activity and composition of essential oils from *Origanum, Thymbra* and *Satureja* species with commercial importance in Turkey. *Food Control* **2004**, *15*, 169–172. [CrossRef]
251. Maccelli, A.; Vitanza, L.; Imbriano, A.; Fraschetti, C.; Filippi, A.; Goldoni, P.; Maurizi, L.; Ammendolia, M.G.; Crestoni, M.E.; Fornarini, S.; et al. *Satureja montana* L. Essential Oils: Chemical Profiles/Phytochemical Screening, Antimicrobial Activity and O/W NanoEmulsion Formulations. *Pharmaceutics* **2019**, *12*, 7. [CrossRef]
252. Semerdjieva, I.; Yankova-Tsvetkova, E.; Zheljazkov, V.D.; Koleva-Valkova, L.H.; Nikolova, R. Reproductive Capacity and Scanning Electron Microscopy (SEM) Analyses of the Micromorphological Surfaces of Three Endemic *Satureja* Species from Bulgaria. *Plants* **2023**, *12*, 2436. [CrossRef] [PubMed]
253. Magi, G.; Marini, E.; Facinelli, B. Antimicrobial activity of essential oils and carvacrol, and synergy of carvacrol and erythromycin, against clinical, erythromycin-resistant Group A Streptococci. *Front. Microbiol.* **2015**, *6*, 130454. [CrossRef] [PubMed]
254. Velicković, J.M.; Kostić, E.J. Comparative analysis of phenolic and mineral composition of traditionally used wild medicinal plants from Southeast Serbia. *Bulg. Chem. Commun.* **2020**, *52*, 197–202. [CrossRef]
255. Sucur, J.; Popovic, A.; Petrovic, M.; Anackov, G.; Bursic, V.; Kiprovski, B.; Prvulovic, D. Allelopathic effects and insecticidal activity of aqueous extracts of *Satureja montana* L. *J. Serbian Chem. Soc.* **2015**, *80*, 475–484. [CrossRef]
256. Arnone, A.; Merlini, L.; Zanarotti, A. Constituents of *Silybum marianum*. Structure of isosilybin and stereochemistry of silybin. *J. Chem. Soc. Chem. Commun.* **1979**, *16*, 696–697. [CrossRef]
257. Lucini, L.; Kane, D.; Pellizzoni, M.; Ferrari, A.; Trevisi, E.; Ruzickova, G.; Arslan, D. Phenolic profile and in vitro antioxidant power of different milk thistle [*Silybum marianum* (L.) Gaertn.] cultivars. *Ind. Crops Prod.* **2016**, *83*, 11–16. [CrossRef]
258. Orčić, D.; Francišković, M.; Bekvalac, K.; Svirčev, E.; Beara, I.; Lesjak, M.; Mimica-Dukić, N. Quantitative determination of plant phenolics in *Urtica dioica* extracts by high-performance liquid chromatography coupled with tandem mass spectrometric detection. *Food Chem.* **2014**, *143*, 48–53. [CrossRef] [PubMed]
259. Zeković, Z.; Cvetanović, A.; Švarc-Gajić, J.; Gorjanović, S.; Sužnjević, D.; Mašković, P.; Savić, S.; Radojković, M.; Đurović, S. Chemical and biological screening of stinging nettle leaves extracts obtained by modern extraction techniques. *Ind. Crops Prod.* **2017**, *108*, 423–430. [CrossRef]
260. Ilies, D.C.; Tudor, I.; Radulescu, V. Chemical composition of the essential oil of *Urtica dioica*. *Chem. Nat. Compd.* **2012**, *48*, 506–507. [CrossRef]
261. Gül, S.; Demirci, B.; Başer, K.H.C.; Akpulat, H.A.; Aksu, P. Chemical Composition and In Vitro Cytotoxic, Genotoxic Effects of Essential Oil from *Urtica dioica* L. *Bull. Environ. Contam. Toxicol.* **2012**, *88*, 666–671. [CrossRef] [PubMed]
262. Asgarpanah, J.; Mohajerani, R. Phytochemistry and pharmacologic properties of *Urtica dioica* L. *J. Med. Plants Res.* **2012**, *6*, 5714–5719.
263. Semwal, P.; Rauf, A.; Olatunde, A.; Singh, P.; Zaky, M.Y.; Islam, M.M.; Khalil, A.A.; Aljohani, A.S.M.; Al Abdulmonem, W.; Ribaudo, G. The medicinal chemistry of *Urtica dioica* L.: From preliminary evidence to clinical studies supporting its neuroprotective activity. *Nat. Prod. Bioprospect.* **2023**, *13*, 16. [CrossRef] [PubMed]

264. Dias, M.I.; Barros, L.; Dueñas, M.; Pereira, E.; Carvalho, A.M.; Alves, R.C.; Oliveira, M.B.P.P.; Santos-Buelga, C.; Ferreira, I.C.F.R. Chemical composition of wild and commercial *Achillea millefolium* L. and bioactivity of the methanolic extract, infusion and decoction. *Food Chem.* **2013**, *141*, 4152–4160. [CrossRef] [PubMed]
265. Farhadi, N.; Babaei, K.; Farsaraei, S.; Moghaddam, M.; Ghasemi Pirbalouti, A. Changes in essential oil compositions, total phenol, flavonoids and antioxidant capacity of *Achillea millefolium* at different growth stages. *Ind. Crops Prod.* **2020**, *152*, 112570. [CrossRef]
266. KapŁAn, M.; Borowy, A. Chemical composition and antioxidant activity of borage (*Borago officinalis* L.) seeds. *Acta Sci. Pol. Hortorum Cultus* **2020**, *19*, 79–90. [CrossRef]
267. Abdellaoui, M.; Bouhlali, E.d.T.; Kasrati, A.; El Rhaffari, L. The effect of domestication on seed yield, essential oil yield and antioxidant activities of fennel seed (*Foeniculum vulgare* Mill) grown in Moroccan oasis. *J. Assoc. Arab Univ. Basic Appl. Sci.* **2018**, *24*, 107–114. [CrossRef]
268. Sharopov, F.; Valiev, A.; Satyal, P.; Gulmurodov, I.; Yusufi, S.; Setzer, W.; Wink, M. Cytotoxicity of the Essential Oil of Fennel (*Foeniculum vulgare*) from Tajikistan. *Foods* **2017**, *6*, 73. [CrossRef] [PubMed]
269. Conforti, F.; Statti, G.; Uzunov, D.; Menichini, F. Comparative Chemical Composition and Antioxidant Activities of Wild and Cultivated *Laurus nobilis* L. Leaves and *Foeniculum vulgare* subsp. piperitum (Ucria) Coutinho Seeds. *Biol. Pharm. Bull.* **2006**, *29*, 2056–2064. [CrossRef] [PubMed]
270. Ighodaro, O.; Akinloye, O.A. First line defence antioxidants-superoxide dismutase (SOD), catalase (CAT) and glutathione peroxidase (GPX): Their fundamental role in the entire antioxidant defence grid. *Alex. J. Med.* **2018**, *54*, 287–293. [CrossRef]
271. Harris, C.S.; Cuerrier, A.; Lamont, E.; Haddad, P.S.; Arnason, J.T.; Bennett, S.A.L.; Johns, T. Investigating Wild Berries as a Dietary Approach to Reducing the Formation of Advanced Glycation Endproducts: Chemical Correlates of In Vitro Antiglycation Activity. *Plant Foods Hum. Nutr.* **2014**, *69*, 71–77. [CrossRef] [PubMed]
272. Emami, S.A.; Abedindo, B.F.; Hassanzadeh-Khayyat, M. Antioxidant activity of the essential oils of different parts of *Juniperus excelsa* M. Bieb. subsp. excelsa and *J. excelsa* M. Bieb. subsp. polycarpos (K. Koch) Takhtajan (Cupressaceae). *Iran. J. Pharm. Res.* **2011**, *10*, 799–810.
273. Petkova, N.; Popova, A.; Alexieva, I. Antioxidant properties and some phytochemical components of the edible medicinal *Malva sylvestris* L. *J. Med. Plants* **2019**, *7*, 96–99.
274. DellaGreca, M.; Cutillo, F.; Abrosca, B.D.; Fiorentino, A.; Pacifico, S.; Zarrelli, A. Antioxidant and Radical Scavenging Properties of *Malva sylvestris*. *Nat. Prod. Commun.* **2009**, *4*, 893–896. [CrossRef] [PubMed]
275. Conforti, F.; Sosa, S.; Marrelli, M.; Menichini, F.; Statti, G.A.; Uzunov, D.; Tubaro, A.; Menichini, F.; Loggia, R.D. In vivo anti-inflammatory and in vitro antioxidant activities of Mediterranean dietary plants. *J. Ethnopharmacol.* **2008**, *116*, 144–151. [CrossRef] [PubMed]
276. Nehir El, S.; Karakaya, S. Radical scavenging and iron-chelating activities of some greens used as traditional dishes in Mediterranean diet. *Int. J. Food Sci. Nutr.* **2009**, *55*, 67–74. [CrossRef] [PubMed]
277. Kumarasamy, Y.; Byres, M.; Cox, P.J.; Jaspars, M.; Nahar, L.; Sarker, S.D. Screening seeds of some Scottish plants for free radical scavenging activity. *Phytother. Res.* **2007**, *21*, 615–621. [CrossRef] [PubMed]
278. Jafari, F.; Ghavidel, F.; Zarshenas, M.M. A Critical Overview on the Pharmacological and Clinical Aspects of Popular *Satureja* Species. *J. Acupunct. Meridian Stud.* **2016**, *9*, 118–127. [CrossRef] [PubMed]
279. Vladić, J.; Ćebović, T.; Vidović, S.; Jokić, S. Evaluation of Anticancer Activity of *Satureja montana* Supercritical and Spray-Dried Extracts on Ehrlich's Ascites Carcinoma Bearing Mice. *Plants* **2020**, *9*, 1532. [CrossRef] [PubMed]
280. Bourgeois, C.; Leclerc, É.A.; Corbin, C.; Doussot, J.; Serrano, V.; Vanier, J.-R.; Seigneuret, J.-M.; Auguin, D.; Pichon, C.; Lainé, É.; et al. Nettle (*Urtica dioica* L.) as a source of antioxidant and anti-aging phytochemicals for cosmetic applications. *Comptes Rendus Chimie* **2016**, *19*, 1090–1100. [CrossRef]
281. Choi, E.-M.; Hwang, J.-K. Antiinflammatory, analgesic and antioxidant activities of the fruit of *Foeniculum vulgare*. *Fitoterapia* **2004**, *75*, 557–565. [CrossRef]
282. De Marino, S.; Gala, F.; Borbone, N.; Zollo, F.; Vitalini, S.; Visioli, F.; Iorizzi, M. Phenolic glycosides from *Foeniculum vulgare* fruit and evaluation of antioxidative activity. *Phytochemistry* **2007**, *68*, 1805–1812. [CrossRef] [PubMed]
283. De Vita, S.; Chini, M.G.; Saviano, G.; Finamore, C.; Festa, C.; Lauro, G.; De Marino, S.; Russo, R.; Avagliano, C.; Casapullo, A.; et al. Biological Profile of Two *Gentiana lutea* L. Metabolites Using Computational Approaches and In Vitro Tests. *Biomolecules* **2021**, *11*, 1490. [CrossRef] [PubMed]
284. Sleiman, N.H.; Daher, C.F. *Malva sylvestris* water extract: A potential anti-Inflammatory and anti-ulcerogenic remedy. *Planta Med.* **2009**, *75*, PH10. [CrossRef]
285. Bach, H.; Benso, B.; Franchin, M.; Massarioli, A.P.; Paschoal, J.A.R.; Alencar, S.M.; Franco, G.C.N.; Rosalen, P.L. Anti-Inflammatory, Anti-Osteoclastogenic and Antioxidant Effects of *Malva sylvestris* Extract and Fractions: In Vitro and In Vivo Studies. *PLoS ONE* **2016**, *11*, e0162728. [CrossRef]
286. Choi, K.-S.; Kundu, J.K.; Chun, K.-S.; Na, H.-K.; Surh, Y.-J. Rutin inhibits UVB radiation-induced expression of COX-2 and iNOS in hairless mouse skin: p38 MAP kinase and JNK as potential targets. *Arch. Biochem. Biophys.* **2014**, *559*, 38–45. [CrossRef] [PubMed]
287. Abdelshafeek, K.A.; Osman, A.F.; Mouneir, S.M.; Elhenawy, A.A.; Abdallah, W.E. Phytochemical profile, comparative evaluation of *Satureja montana* alcoholic extract for antioxidants, anti-inflammatory and molecular docking studies. *BMC Complement. Med. Ther.* **2023**, *23*, 108. [CrossRef] [PubMed]
288. Wang, X.; Zhang, Z.; Wu, S.-C. Health Benefits of *Silybum marianum*: Phytochemistry, Pharmacology, and Applications. *J. Agric. Food Chem.* **2020**, *68*, 11644–11664. [CrossRef] [PubMed]

289. Collard, M.; Camenzuli, L.; Saunders, D.; Vallotton, N.; Curtis-Jackson, P. Persistence and Mobility (Defined as Organic-Carbon Partitioning) Do Not Correlate to the Detection of Substances Found in Surface and Groundwater: Criticism of the Regulatory Concept of Persistent and Mobile Substances. *Sci. Total Environ.* **2023**, *865*, 161228. [CrossRef] [PubMed]
290. Khan, R.U.; Fatima, A.; Naz, S.; Ragni, M.; Tarricone, S.; Tufarelli, V. Perspective, Opportunities and Challenges in Using Fennel (*Foeniculum vulgare*) in Poultry Health and Production as an Eco-Friendly Alternative to Antibiotics: A Review. *Antibiotics* **2022**, *11*, 278. [CrossRef] [PubMed]
291. Anwar, F.; Ali, M.; Hussain, A.I.; Shahid, M. Antioxidant and antimicrobial activities of essential oil and extracts of fennel (*Foeniculum vulgare* Mill.) seeds from Pakistan. *Flavour Fragr. J.* **2009**, *24*, 170–176. [CrossRef]
292. Marín, I.; Sayas-Barberá, E.; Viuda-Martos, M.; Navarro, C.; Sendra, E. Chemical Composition, Antioxidant and Antimicrobial Activity of Essential Oils from Organic Fennel, Parsley, and Lavender from Spain. *Foods* **2016**, *5*, 18. [CrossRef] [PubMed]
293. Domínguez-Vigil, I.G.; Mata-Cárdenas, B.D.; Esquivel-Ferriño, P.C.; Avalos-Alanís, F.G.; Vargas-Villarreal, J.; Camacho-Corona, M.d.R. Antigiardial Activity of *Foeniculum vulgare* Hexane Extract and Some of Its Constituents. *Plants* **2022**, *11*, 2212. [CrossRef] [PubMed]
294. Burt, S. Essential oils: Their antibacterial properties and potential applications in foods—A review. *Int. J. Food Microbiol.* **2004**, *94*, 223–253. [CrossRef] [PubMed]
295. Malin, V.; Elez Garofulić, I.; Repajić, M.; Zorić, Z.; Pedisić, S.; Sterniša, M.; Smole Možina, S.; Dragović-Uzelac, V. Phenolic Characterization and Bioactivity of Fennel Seed (*Foeniculum vulgare* Mill.) Extracts Isolated by Microwave-Assisted and Conventional Extraction. *Processes* **2022**, *10*, 510. [CrossRef]
296. Khaleil, M.M.; Alnoman, M.M.; Elrazik, E.S.A.; Zagloul, H.; Khalil, A.M.A. Essential Oil of *Foeniculum vulgare* Mill. as a Green Fungicide and Defense-Inducing Agent against Fusarium Root Rot Disease in *Vicia faba* L. *Biology* **2021**, *10*, 696. [CrossRef] [PubMed]
297. Kartal, M. Intellectual property protection in the natural product drug discovery, traditional herbal medicine and herbal medicinal products. *Phytother. Res.* **2006**, *21*, 113–119. [CrossRef] [PubMed]
298. Makola, D.; Peura, D.A.; Crowe, S.E. *Helicobacter pylori* Infection and Related Gastrointestinal Diseases. *J. Clin. Gastroenterol.* **2007**, *41*, 548–558. [CrossRef] [PubMed]
299. Xavier, V.; Finimundy, T.C.; Heleno, S.A.; Amaral, J.S.; Calhelha, R.C.; Vaz, J.; Pires, T.C.S.P.; Mediavilla, I.; Esteban, L.S.; Ferreira, I.C.F.R.; et al. Chemical and Bioactive Characterization of the Essential Oils Obtained from Three Mediterranean Plants. *Molecules* **2021**, *26*, 7472. [CrossRef] [PubMed]
300. Cabral, C.; Francisco, V.; Cavaleiro, C.; Gonçalves, M.J.; Cruz, M.T.; Sales, F.; Batista, M.T.; Salgueiro, L. Essential Oil of *Juniperus communis* subsp. alpina (Suter) Čelak Needles: Chemical Composition, Antifungal Activity and Cytotoxicity. *Phytother. Res.* **2012**, *26*, 1352–1357. [CrossRef] [PubMed]
301. Paparella, A.; Nawade, B.; Shaltiel-Harpaz, L.; Ibdah, M. A Review of the Botany, Volatile Composition, Biochemical and Molecular Aspects, and Traditional Uses of *Laurus nobilis*. *Plants* **2022**, *11*, 1209. [CrossRef] [PubMed]
302. Batool, S.; Khera, R.A.; Hanif, M.A.; Ayub, M.A. Bay Leaf. In *Medicinal Plants of South Asia*; Elsevier: Amsterdam, The Netherlands, 2020; pp. 63–74.
303. Salima, B.; Yasmina Tlili Ait, K.; Abdelghani, D.; Youcef, H.; Azzedine, C. Antibiotic Activity of the Essential Oil of Laurel (*Laurus nobilis* L.) on Eight Bacterial Strains. *J. Life Sci.* **2013**, *8*, 814. [CrossRef]
304. Quave, C.L.; Plano, L.R.W.; Pantuso, T.; Bennett, B.C. Effects of extracts from Italian medicinal plants on planktonic growth, biofilm formation and adherence of methicillin-resistant *Staphylococcus aureus*. *J. Ethnopharmacol.* **2008**, *118*, 418–428. [CrossRef] [PubMed]
305. Coelho de Souza, G.; Haas, A.P.S.; von Poser, G.L.; Schapoval, E.E.S.; Elisabetsky, E. Ethnopharmacological studies of antimicrobial remedies in the south of Brazil. *J. Ethnopharmacol.* **2004**, *90*, 135–143. [CrossRef]
306. Cogo, L.L.; Monteiro, C.L.B.; Miguel, M.D.; Miguel, O.G.; Cunico, M.M.; Ribeiro, M.L.; de Camargo, E.R.; Kussen, G.M.B.; da Silva Nogueira, K.; Dalla Costa, L.M. Anti-*Helicobacter pylori* activity of plant extracts traditionally used for the treatment of gastrointestinal disorders. *Braz. J. Microbiol.* **2010**, *41*, 304–309. [CrossRef] [PubMed]
307. Delfine, S.; Marrelli, M.; Conforti, F.; Formisano, C.; Rigano, D.; Menichini, F.; Senatore, F. Variation of *Malva sylvestris* essential oil yield, chemical composition and biological activity in response to different environments across Southern Italy. *Ind. Crops Prod.* **2017**, *98*, 29–37. [CrossRef]
308. Magro, A.; Carolino, M.; Bastos, M.; Mexia, A. Efficacy of plant extracts against stored products fungi. *Rev. Iberoam. Micol.* **2006**, *23*, 176–178. [CrossRef] [PubMed]
309. Ebani, V.V.; Pieracci, Y.; Cagnoli, G.; Bertelloni, F.; Munafò, C.; Nardoni, S.; Pistelli, L.; Mancianti, F. In Vitro Antimicrobial Activity of *Thymus vulgaris, Origanum vulgare, Satureja montana* and Their Mixture against Clinical Isolates Responsible for Canine Otitis Externa. *Vet. Sci.* **2023**, *10*, 30. [CrossRef] [PubMed]
310. Kovačević, Z.; Kladar, N.; Čabarkapa, I.; Radinović, M.; Maletić, M.; Erdeljan, M.; Božin, B. New Perspective of *Origanum vulgare* L. and *Satureja montana* L. Essential Oils as Bovine Mastitis Treatment Alternatives. *Antibiotics* **2021**, *10*, 1460. [CrossRef] [PubMed]
311. Oliveira-Pinto, P.R.; Mariz-Ponte, N.; Gil, R.L.; Cunha, E.; Amorim, C.G.; Montenegro, M.C.B.S.M.; Fernandes-Ferreira, M.; Sousa, R.M.O.F.; Santos, C. Montmorillonite Nanoclay and Formulation with *Satureja montana* Essential Oil as a Tool to Alleviate *Xanthomonas euvesicatoria* Load on *Solanum lycopersicum*. *Appl. Nano* **2022**, *3*, 126–142. [CrossRef]
312. Gomes, F.; Dias, M.I.; Lima, Â.; Barros, L.; Rodrigues, M.E.; Ferreira, I.C.F.R.; Henriques, M. *Satureja montana* L. and *Origanum majorana* L. Decoctions: Antimicrobial Activity, Mode of Action and Phenolic Characterization. *Antibiotics* **2020**, *9*, 294. [CrossRef] [PubMed]
313. El-Hagrassi, A.M.; Abdallah, W.E.; Osman, A.F.; Abdelshafeek, K.A. Phytochemical Study of Bioactive Constituents from *Satureja montana* L. Growing in Egypt and Their Antimicrobial and Antioxidant Activities. *Asian J. Pharm. Clin. Res.* **2018**, *11*, 142–148. [CrossRef]

314. Pino-Otín, M.R.; Gan, C.; Terrado, E.; Sanz, M.A.; Ballestero, D.; Langa, E. Antibiotic properties of *Satureja montana* L. hydrolate in bacteria and fungus of clinical interest and its impact in non-target environmental microorganisms. *Sci. Rep.* **2022**, *12*, 18460. [CrossRef] [PubMed]
315. Vitanza, L.; Maccelli, A.; Marazzato, M.; Scazzocchio, F.; Comanducci, A.; Fornarini, S.; Crestoni, M.E.; Filippi, A.; Fraschetti, C.; Rinaldi, F.; et al. *Satureja montana* L. essential oil and its antimicrobial activity alone or in combination with gentamicin. *Microb. Pathog.* **2019**, *126*, 323–331. [CrossRef]
316. Dunkić, V.; Bezić, N.; Vuko, E.; Cukrov, D. Antiphytoviral Activity of *Satureja montana* L. ssp. variegata (Host) P. W. Ball Essential Oil and Phenol Compounds on CMV and TMV. *Molecules* **2010**, *15*, 6713–6721. [CrossRef]
317. Lee, D.G.; Kim, H.K.; Park, Y.; Park, S.-C.; Woo, E.-R.; Jeong, H.G.; Hahm, K.-S. Gram-positive bacteria specific properties of silybin derived from *Silybum marianum*. *Arch. Pharmacal Res.* **2003**, *26*, 597–600. [CrossRef] [PubMed]
318. Rakelly de Oliveira, D.; Relison Tintino, S.; Morais Braga, M.F.B.; Boligon, A.A.; Linde Athayde, M.; Douglas Melo Coutinho, H.; de Menezes, I.R.A.; Fachinetto, R. In Vitro Antimicrobial and Modulatory Activity of the Natural Products Silymarin and Silibinin. *BioMed Res. Int.* **2015**, *2015*, 292797. [CrossRef] [PubMed]
319. Yun, D.G.; Lee, D.G. Silibinin triggers yeast apoptosis related to mitochondrial Ca^{2+} influx in *Candida albicans*. *The Int. J. Biochem. Cell Biol* **2016**, *80*, 1–9. [CrossRef] [PubMed]
320. Singh, R.; Hussain, S.; Verma, R.; Sharma, P. Anti-mycobacterial screening of five Indian medicinal plants and partial purification of active extracts of *Cassia sophera* and *Urtica dioica*. *Asian Pac. J. Trop. Med.* **2013**, *6*, 366–371. [CrossRef] [PubMed]
321. Zenão, S.; Aires, A.; Dias, C.; Saavedra, M.J.; Fernandes, C. Antibacterial potential of *Urtica dioica* and *Lavandula angustifolia* extracts against methicillin resistant *Staphylococcus aureus* isolated from diabetic foot ulcers. *J. Herb. Med.* **2017**, *10*, 53–58. [CrossRef]
322. Abdul-Ghani, A.S.; Amin, R. The vascular action of aqueous extracts of *Foeniculum vulgare* leaves. *J. Ethnopharmacol.* **1988**, *24*, 213–218. [CrossRef] [PubMed]
323. Khan, A.; Zaman, G.; Anderson, R.A. Bay Leaves Improve Glucose and Lipid Profile of People with Type 2 Diabetes. *J. Clin. Biochem. Nutr.* **2009**, *44*, 52–56. [CrossRef] [PubMed]
324. Hamidpour, R.; Hamidpour, S.; Hamidpour, M.; Shahlari, M.; Sohraby, M. Summer Savory: From the Selection of Traditional Applications to the Novel Effect in Relief, Prevention, and Treatment of a Number of Serious Illnesses such as Diabetes, Cardiovascular Disease, Alzheimer's Disease, and Cancer. *J. Tradit. Complement. Med.* **2014**, *4*, 140–144. [CrossRef] [PubMed]
325. Chkhikvishvili, I.; Sanikidze, T.; Gogia, N.; McHedlishvili, T.; Enukidze, M.; Machavariani, M.; Vinokur, Y.; Rodov, V. Rosmarinic Acid-Rich Extracts of Summer Savory (*Satureja hortensis* L.) Protect Jurkat T Cells against Oxidative Stress. *Oxidative Med. Cell. Longev.* **2013**, *2013*, 456253. [CrossRef] [PubMed]
326. El Haouari, M.; Bnouham, M.; Bendahou, M.; Aziz, M.; Ziyyat, A.; Legssyer, A.; Mekhfi, H. Inhibition of Rat Platelet Aggregation by *Urtica dioica* Leaves Extracts. *Phytother. Res.* **2006**, *20*, 568–572. [CrossRef]
327. Guil-Guerrero, J.L.; Gómez-Mercado, F.; Ramos-Bueno, R.P.; González-Fernández, M.J.; Urrestarazu, M.; Jiménez-Becker, S.; de Bélair, G. Fatty acid profiles and sn -2 fatty acid distribution of γ-linolenic acid-rich *Borago* species. *J. Food Compos. Anal.* **2018**, *66*, 74–80. [CrossRef]
328. Ke, W.; Zhao, X.; Lu, Z. *Foeniculum vulgare* seed extract induces apoptosis in lung cancer cells partly through the down-regulation of Bcl-2. *Biomed. Pharmacother.* **2021**, *135*, 111213. [CrossRef] [PubMed]
329. Mohamad, R.H.; El-Bastawesy, A.M.; Abdel-Monem, M.G.; Noor, A.M.; Al-Mehdar, H.A.R.; Sharawy, S.M.; El-Merzabani, M.M. Antioxidant and Anticarcinogenic Effects of Methanolic Extract and Volatile Oil of Fennel Seeds (*Foeniculum vulgare*). *J. Med. Food* **2011**, *14*, 986–1001. [CrossRef] [PubMed]
330. Huang, N.-C.; Huang, R.-L.; Huang, X.-F.; Chang, K.-F.; Lee, C.-J.; Hsiao, C.-Y.; Lee, S.-C.; Tsai, N.-M. Evaluation of anticancer effects of *Juniperus communis* extract on hepatocellular carcinoma cells in vitro and in vivo. *Biosci. Rep.* **2021**, *41*, 1–14. [CrossRef] [PubMed]
331. Tsai, N.-M.; Chang, K.-F.; Wang, J.-C. *Juniperus communis* Extract Exerts Antitumor Effects in Human Glioblastomas Through Blood-Brain Barrier. *Cell. Physiol. Biochem.* **2018**, *49*, 2443–2462. [CrossRef]
332. Daniela, A.; Pichichero, E.; Canuti, L.; Cicconi, R.; Karou, D.; D'Arcangelo, G.; Canini, A. Identification of phenolic compounds from medicinal and melliferous plants and their cytotoxic activity in cancer cells. *Caryologia* **2013**, *60*, 90–95. [CrossRef]
333. Jayakumar, S.; Madankumar, A.; Asokkumar, S.; Raghunandhakumar, S.; Gokuladhas, K.; Kamaraj, S.; Josephine Divya, M.G.; Devaki, T. Potential preventive effect of carvacrol against diethylnitrosamine-induced hepatocellular carcinoma in rats. *Mol. Cell. Biochem.* **2011**, *360*, 51–60. [CrossRef] [PubMed]
334. Yin, Q.-H.; Yan, F.-X.; Zu, X.-Y.; Wu, Y.-H.; Wu, X.-P.; Liao, M.-C.; Deng, S.-W.; Yin, L.-L.; Zhuang, Y.-Z. Anti-proliferative and pro-apoptotic effect of carvacrol on human hepatocellular carcinoma cell line HepG-2. *Cytotechnology* **2011**, *64*, 43–51. [CrossRef] [PubMed]
335. Arunasree, K.M. Anti-proliferative effects of carvacrol on a human metastatic breast cancer cell line, MDA-MB 231. *Phytomedicine* **2010**, *17*, 581–588. [CrossRef] [PubMed]
336. Durak, I.; Biri, H.; Devrim, E.; Sözen, S.; Avcı, A. Aqueous extract of *Urtica dioica* makes significant inhibition on adenosine deaminase activity in prostate tissue from patients with prostate cancer. *Cancer Biol. Ther.* **2014**, *3*, 855–857. [CrossRef] [PubMed]
337. Fattahi, S.; Ghadami, E.; Asouri, M.; Motevalizadeh Ardekanid, A.; Akhavan-Niaki, H. *Urtica dioica* inhibits cell growth and induces apoptosis by targeting Ornithine decarboxylase and Adenosine deaminase as key regulatory enzymes in adenosine and polyamines homeostasis in human breast cancer cell lines. *Cell. Mol. Biol.* **2018**, *64*, 97–102. [CrossRef] [PubMed]
338. Jenabi, E.; Fereidoony, B. Effect of *Achillea millefolium* on Relief of Primary Dysmenorrhea: A Double-Blind Randomized Clinical Trial. *J. Pediatr. Adolesc. Gynecol.* **2015**, *28*, 402–404. [CrossRef] [PubMed]

339. Basati, G.; Abbaszadeh, S.; Zebardast, A.; Teimouri, H. Analgesic Medicinal Plants in Shahrekord, Southwest of Iran: An Ethnobotanical Study. *Galen Med. J.* **2019**, *8*, e1593. [CrossRef] [PubMed]
340. Imran, A.; Xiao, L.; Ahmad, W.; Anwar, H.; Rasul, A.; Imran, M.; Aziz, N.; Razzaq, A.; Arshad, M.U.; Shabbir, A.; et al. *Foeniculum vulgare* (Fennel) promotes functional recovery and ameliorates oxidative stress following a lesion to the sciatic nerve in mouse model. *J. Food Biochem.* **2019**, *43*, e12983. [CrossRef] [PubMed]
341. Maqbool, J.; Anwar, H.; Iqbal, J.; Rasul, A.; Imran, A.; Ahmad Malik, S.; Shabbir, A.; Ijaz, F.; Sajid, F.; Akram, R.; et al. Methanolic extract of Fennel (*Foeniculum vulgare*) escalates functional restoration following a compression injury to the sciatic nerve in a mouse model. *Food Sci. Nutr.* **2020**, *9*, 701–710. [CrossRef] [PubMed]
342. Raman, S.; Asle-Rousta, M.; Rahnema, M. Protective effect of fennel, and its major component trans-anethole against social isolation induced behavioral deficits in rats. *Physiol. Int.* **2020**, *107*, 30–39. [CrossRef] [PubMed]
343. Yamazaki, M.; Chiba, K.; Mohri, T.J.B.; Bulletin, P. Neuritogenic effect of natural iridoid compounds on PC12h cells and its possible relation to signaling protein kinases. *Biol. Pharm. Bull.* **1996**, *19*, 791–795. [CrossRef] [PubMed]
344. More, S.V.; Koppula, S.; Kim, I.-S.; Kumar, H.; Kim, B.-W.; Choi, D.-K. The Role of Bioactive Compounds on the Promotion of Neurite Outgrowth. *Molecules* **2012**, *17*, 6728–6753. [CrossRef] [PubMed]
345. Roszek, K.; Czarnecka, J. Is Ecto-nucleoside Triphosphate Diphosphohydrolase (NTPDase)-based Therapy of Central Nervous System Disorders Possible? *Mini-Rev. Med. Chem.* **2015**, *15*, 5–20. [CrossRef] [PubMed]
346. Brinza, I.; Boiangiu, R.S.; Hancianu, M.; Cioanca, O.; Erdogan Orhan, I.; Hritcu, L. Bay Leaf (*Laurus nobilis* L.) Incense Improved Scopolamine-Induced Amnesic Rats by Restoring Cholinergic Dysfunction and Brain Antioxidant Status. *Antioxidants* **2021**, *10*, 259. [CrossRef] [PubMed]
347. Kostadinova, I.; Kandilarov, I.; Kotetarova, M.; Zlatanova, H.; Kostadinov, I.; Delev, D.; Vilmosh, N. Anxiolytic Effect of *Satureja montana* Dry Extract and its Active Compounds Rosmarinic Acid and Carvacrol in Acute Stress Experimental Model. *J. Integr. Neurosci.* **2022**, *21*, 124. [CrossRef]
348. Silva, F.V.M.; Martins, A.; Salta, J.; Neng, N.R.; Nogueira, J.M.F.; Mira, D.; Gaspar, N.l.; Justino, J.; Grosso, C.; Urieta, J.S.; et al. Phytochemical Profile and Anticholinesterase and Antimicrobial Activities of Supercritical versus Conventional Extracts of *Satureja montana*. *J. Agric. Food Chem.* **2009**, *57*, 11557–11563. [CrossRef] [PubMed]
349. Rodríguez-Magaña, M.P.; Cordero-Pérez, P.; Rivas-Morales, C.; Oranday-Cárdenas, M.A.; Moreno-Peña, D.P.; García-Hernández, D.G.; Leos-Rivas, C. Hypoglycemic Activity of *Tilia americana*, *Borago officinalis*, *Chenopodium nuttalliae*, and *Piper sanctumon* Wistar Rats. *J. Diabetes Res.* **2019**, *2019*, 7836820. [CrossRef] [PubMed]
350. El-Ouady, F.; Lahrach, N.; Ajebli, M.; Haidani, A.E.; Eddouks, M. Antihyperglycemic Effect of the Aqueous Extract of *Foeniculum vulgare* in Normal and Streptozotocin-induced Diabetic Rats. *Cardiovasc. Hematol. Disord.-Drug Targets* **2020**, *20*, 54–63. [CrossRef] [PubMed]
351. Samadi-Noshahr, Z.; Hadjzadeh, M.A.R.; Moradi-Marjaneh, R.; Khajavi-Rad, A. The hepatoprotective effects of fennel seeds extract and trans-Anethole in streptozotocin-induced liver injury in rats. *Food Sci. Nutr.* **2020**, *9*, 1121–1131. [CrossRef]
352. Shahat, A.; Radwan, H.A.; Elkholy, Y.; Ghanem, M.M.; Mahdy, E.-S.; Hassanein, H. Phenolic compounds from *Foeniculum vulgare* (Subsp. Piperitum) (Apiaceae) herb and evaluation of hepatoprotective antioxidant activity. *Pharmacogn. Res.* **2012**, *4*, 104. [CrossRef] [PubMed]
353. Xiao, J.; Wang, F.; Wong, N.-K.; He, J.; Zhang, R.; Sun, R.; Xu, Y.; Liu, Y.; Li, W.; Koike, K.; et al. Global liver disease burdens and research trends: Analysis from a Chinese perspective. *J. Hepatol.* **2019**, *71*, 212–221. [CrossRef] [PubMed]
354. Han, H.; Xu, L.; Xiong, K.; Zhang, T.; Wang, Z. Exploration of Hepatoprotective Effect of Gentiopicroside on Alpha-Naphthylisothiocyanate-Induced Cholestatic Liver Injury in Rats by Comprehensive Proteomic and Metabolomic Signatures. *Cell. Physiol. Biochem.* **2018**, *49*, 1304–1319. [CrossRef] [PubMed]
355. Kandis, H.; Karapolat, S.; Yildirim, U.; Saritas, A.; Gezer, S.; Memisogullari, R. Effects of *Urtica dioica* on hepatic ischemia-reperfusion injury in rats. *Clinics* **2010**, *65*, 1357–1361. [CrossRef]
356. Ved, A.; Gupta, A.; Rawat, A.K.S. Antioxidant and hepatoprotective potential of phenol-rich fraction of *Juniperus communis* Linn. leaves. *Pharmacogn. Mag.* **2017**, *13*, 108–113. [CrossRef] [PubMed]
357. Mohi-ud-din, R.; Mir, R.H.; Sawhney, G.; Dar, M.A.; Bhat, Z.A. Possible Pathways of Hepatotoxicity Caused by Chemical Agents. *Curr. Drug Metab.* **2019**, *20*, 867–879. [CrossRef] [PubMed]
358. Pashtetsky, V.; Ostapchuk, P.; Usmanova, E.; Zyablitskaya, E.; Makalish, T.; Danilova, I.; Kuevda, T.; Zubochenko, D.; Uppe, V.; Pashtetskaia, A.; et al. *Satureja montana* L. essential oil various dosages effect on the main rats' biological features. *Potravin. Slovak J. Food Sci.* **2021**, *15*, 799–809. [CrossRef] [PubMed]
359. Yang, H.; Mehrabi Nasab, D.; Athari, S.S. Effects of Silymarin and Baicalein on Glycogen Storage in the Hepatocytes of Rat Models of Hepatic Injury. *Hepat. Mon.* **2021**, *21*, e113114. [CrossRef]
360. Federico, A.; Dallio, M.; Loguercio, C. Silymarin/Silybin and Chronic Liver Disease: A Marriage of Many Years. *Molecules* **2017**, *22*, 191. [CrossRef] [PubMed]
361. Jalali, S.M.; Najafzadeh, H.; Bahmei, S. Protective role of silymarin and D-penicillamine against lead-induced liver toxicity and oxidative stress. *Toxicol. Ind. Health* **2017**, *33*, 512–518. [CrossRef] [PubMed]
362. Türkdoğan, M.K.; Ozbek, H.; Yener, Z.; Tuncer, I.; Uygan, I.; Ceylan, E. The role of *Urtica dioica* and *Nigella sativa* in the prevention of carbon tetrachloride-induced hepatotoxicity in rats. *Phytother. Res.* **2003**, *17*, 942–946. [CrossRef] [PubMed]
363. Foster, R.H.; Hardy, G.; Alany, R.G. Borage oil in the treatment of atopic dermatitis. *Nutrition* **2010**, *26*, 708–718. [CrossRef] [PubMed]

364. Leech, B.; Schloss, J.; Steel, A. Treatment Interventions for the Management of Intestinal Permeability: A Cross-Sectional Survey of Complementary and Integrative Medicine Practitioners. *J. Altern. Complement. Med.* **2019**, *25*, 623–636. [CrossRef] [PubMed]
365. Pathak, S.; Tewari, R.; Prakash, A.O. Hormonal properties of ethanolic extract of *Juniperus communis* Linn. *Anc. Sci. Life* **1990**, *10*, 106–113. [PubMed]
366. Turgut, F.; Bayrak, O.; Catal, F.; Bayrak, R.; Atmaca, A.F.; Koc, A.; Akbas, A.; Akcay, A.; Unal, D. Antioxidant and protective effects of silymarin on ischemia and reperfusion injury in the kidney tissues of rats. *Int. Urol. Nephrol.* **2008**, *40*, 453–460. [CrossRef] [PubMed]
367. Wilasrusmee, C.; Kittur, S.; Shah, G.; Siddiqui, J.; Bruch, D.; Wilasrusmee, S.; Kittur, D.S. Immunostimulatory effect of *Silybum marianum* (milk thistle) extract. *Med. Sci. Monit.* **2002**, *8*, BR439–BR443. [PubMed]
368. Khayyal, M.; El-Ghazaly, M.; Kenawy, S.; Seif-El-Nasr, M.; Mahran, L.; Kafafi, Y.; Okpanyi, S. Antiulcerogenic Effect of Some Gastrointestinally Acting Plant Extracts and their Combination. *Arzneimittelforschung* **2011**, *51*, 545–553. [CrossRef] [PubMed]

nutraceuticals

Review

Effects of Ursolic Acid on Colorectal Cancer: A Review of Recent Evidence

Amanda Kornel [1] and Evangelia Tsiani [1,2,*]

[1] Department of Health Sciences, Faculty of Applied Health Sciences, Brock University, St. Catharines, ON L2S 3A1, Canada
[2] Centre for Bone and Muscle Health, Applied Health Sciences, Brock University, St. Catharines, ON L2S 3A1, Canada
* Correspondence: etsiani@brocku.ca

Abstract: Colorectal cancer is the third leading cause of cancer-related deaths, and the five-year survival rate of the metastatic disease is less than 15%. Treatment approaches include surgery, systemic chemotherapy and radiotherapy. The aggressive nature and low five-year survival rate of metastatic colorectal cancer indicate a need for new treatment options to help combat this disease. Ursolic acid is a pentacyclic triterpenoid naturally occurring in many plants, with high concentrations found in cranberries. This review summarizes evidence from the last ten years of the effects of ursolic acid on colorectal cancer. Overall, the available studies indicate that the treatment of colon cancer cells with ursolic acid results in a significant inhibition of proliferation and induction of apoptosis. In addition, the limited in vivo studies indicate a significant reduction in tumor volume and tumor angiogenesis in animal models of colorectal cancer administered ursolic acid. More in vivo animal studies are required to better understand the potential anticancer properties of ursolic acid and to form the basis for human clinical trials.

Keywords: colorectal cancer; ursolic acid; proliferation; metastasis; xenograft

Citation: Kornel, A.; Tsiani, E. Effects of Ursolic Acid on Colorectal Cancer: A Review of Recent Evidence. *Nutraceuticals* **2024**, *4*, 373–394. https://doi.org/10.3390/nutraceuticals4030022

Academic Editors: Ivan Cruz-Chamorro and Guillermo Santos Sánchez

Received: 23 May 2024
Revised: 28 June 2024
Accepted: 4 July 2024
Published: 8 July 2024

1. Introduction

Colorectal cancer (CRC) accounts for approximately 10% of yearly diagnosed cancer cases, causing nearly 900,000 deaths worldwide annually [1], and it is the fourth most common cancer globally [2]. In the United States, it is the third most common cause of cancer-related deaths, causing approximately 53,200 deaths every year [2,3]. The 5-year survival for localized CRC is approximately 91% but only 14% for metastatic CRC [2,4]. Many of the cases and deaths from CRC are due to modifiable risk factors including smoking, high alcohol consumption, excess body weight, unhealthy diet and low physical activity [1–3]. Age and gender are factors in developing CRC. Although more than 50% of diagnoses occur in people over the age of 65, incidence rates in people under 50 years of age are on the rise and men are 33% more likely to develop this disease compared with women [2]. Heritable factors account for only 12–35% of CRC cases, therefore the majority of cases are sporadic [5].

Colorectal cancer arises from the glandular epithelial cells of the large intestine. The sporadic growth of CRC is typically due to the accumulation of genetic mutations and/or epigenetic modifications such as methylation in key cellular signaling pathways (resulting in hyper-proliferative cells [1,3,6,7]) and the transformation of normal glandular epithelial cells to adenocarcinomas [8]. CRC pathogenesis includes the following main steps: from a normal colon epithelium to aberrant crypt focus formation to polyp/adenoma formation to adenocarcinoma [8]. The crypts and polyps are benign but are considered to be pre-cancerous.

CRC is divided into four unique molecular subtypes known as the consensus molecular subtypes or CMSs. CMS1 is associated with immune evasion mechanisms, CMS2 is the largest subtype and is known as the canonical subtype, CMS3 is a metabolic subtype and CMS4 is known as the mesenchymal subtype [9].

A person's diet and lifestyle are key risk factors in the development of colorectal cancer as they influence gut microbiota, which has been established to play a pivotal role in colorectal carcinogenesis [10–12]. Bacteria in the gut play an important role in regulating digestion, absorbing nutrients and providing critical immune functions [12]. The wide diversity in microorganisms that make up an individual's gut microbiome contributes to making colorectal cancer a heterogenous cancer type [4,13,14].

The heterogeneity seen in CRC [14] makes its treatment challenging. Another cause of difficulty in treating colorectal cancer is that CRC cells are prone to developing resistance to chemotherapy drugs [14,15].

The main function of the colon in normal physiology is to reabsorb water, minerals and nutrients for use throughout the body, and malfunctions in these processes, which happen in CRC, lead to dehydration and malnourishment [12,16]. Symptoms of CRC can vary depending on the region of the colon that is effected but typically include dehydration, fatigue, anemia, abdominal pain, altered bowel movement (diarrhea or constipation), weight loss and blood in the stool [17,18].

Surgery is often the first treatment option when CRC is diagnosed early as it is relatively easy to remove the primary tumor. However, once the tumor develops metastatic properties, surgery is no longer the best/most effective method of treatment. Systemic chemotherapy is used most often in patients with CRC. There are many (approximately 35) Federal Drug Administration (FDA)-approved drugs currently in use to treat CRC. The five most commonly used FDA-approved drugs to treat CRC are 5-fluorouracil, capecitabine, irinotecan, oxaliplatin and trifluridine/tipiracil [3]. Many of the approved chemotherapy drugs work by interrupting DNA synthesis or causing DNA breaks [3]. Chemotherapy drugs that are currently available and in use are associated with severe side effects including anemia, diarrhea, gastrointestinal perforations and cardiac ischemia among others. Novel compounds with less side effects and high efficacy are needed to treat metastatic and drug-resistant CRC.

Traditionally, plant-derived chemicals have been developed into agents used to treat cancer. Some examples of plant-derived chemotherapy agents include paclitaxel, derived from the bark of the Pacific yew tree, *Taxus brevifolia*, and the semi-synthetic docetaxel, a taxoid derived from a precursor extracted from the leaves of the European yew tree, *Taxus baccata* [19,20]. These chemotherapy medications are being used in treatments against prostate, breast and lung cancers. The search for novel plant-derived chemicals with strong anticancer potential is on-going and is an important field of research. Ursolic acid (UA), a pentacyclic triterpenoid (Figure 1) (chemical formula $C_{30}H_{48}O_3$), is found in the leaves and fruits of more than 120 plant species. Originally, it was isolated and identified in the 1920s in the epicuticular waxes of apples [21].

Fruits such as cranberries, black elderberries, apples and pears contain substantial levels of UA [22,23], as does olive oil [24]. Many flowering plants also contain UA, with high levels found in lavender, white deadnettle, marigold, rosinweed, basil, rosemary, daylily and olive tree leaves [24–27] (Table 1).

Table 1. Ursolic acid (UA) concentrations in different plants and fruits.

	Source Common and Botanical Name	Concentration of UA (FW = Fresh Weight) (DW = Dry Weight)	Reference
Fruits	Apple (peel) *Malus*	1.52 mg/g DW	[28]
	Apple (whole fruit) *Malus*	0.77 ± 0.1 mg/g to 1.85 ± 0.17 mg/g	[22]
	Cranberry *Vaccinium macrocarpon*	0.46–1.09 mg/g FW	[23]
	Pear *pyrus*	0.3481 mg/g (mature fruit) 0.1293 mg/g FW (young fruit)	[29,30]

Table 1. *Cont.*

	Source Common and Botanical Name	Concentration of UA (FW = Fresh Weight) (DW = Dry Weight)	Reference
Herbs	Basil *Ocimum tenuiflorum*	20.2 mg/g DW	[27]
	Rosemary *Rosmarinus officinalis*	15.8–29.5 mg/g	[31]
	Thyme *Thymus vulgaris*	9.4 mg/g DW	[31,32]
	Oregano *Origanum vulgare*	2.8 mg/g DW	[31]
	Sage *Salvia officinalus*	18 mg/g DW	[31]
Flowering plants	Lavender *Lavandula*	106.7–153.1 mg/g F.W. 3.463–6.484 mg/g D.W. 10.5 mg/g (flowers)	[26,31]
	White deadnettle *Lamii albi flos*	39.1–110.4 mg/g D.W.	[33]
	Oleander leaves *Nerium oleander*	12.7 mg/g DW	[31]
	Rosinweed *Silphium* sp. *Flowers*	17.95–22.05 mg/g D.W.	[34]
	Olive leaves *Olea europeae*	1.8 mg/g DW	[31]
Other	Arabica coffee leaves *Coffea arabic*	18 mg/g DW	[31]

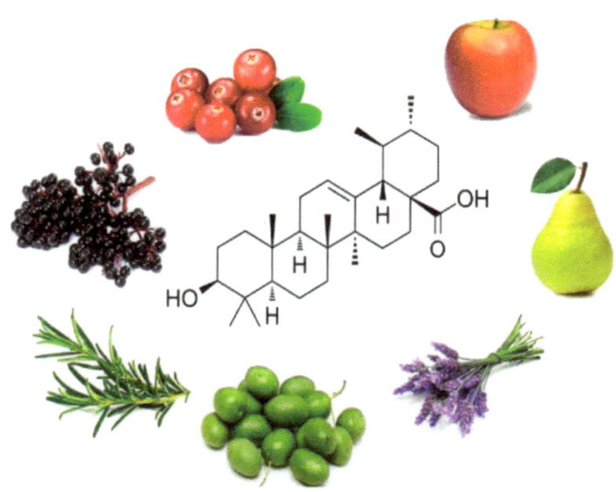

Figure 1. Chemical structure of ursolic acid and some of the plants (apples, pears, cranberries, elderberries, lavender, olives and rosemary) that contain high concentrations of UA. This chemical structure was created in BioRender.com (https://www.biorender.com; accessed 23 May 2024) and the image was created using Microsoft PowerPoint 2024.

There is substantial evidence showing that UA exhibits a wide range of biological activities [35,36], including anti-inflammatory [37], neuroprotective [38,39], antidiabetic [40] and anticancer properties [41–44].

This review article provides a summary of research data published in the last 30 years that examine the effects of UA against CRC in vitro and in vivo. Although there are a number of published reviews examining UA and its effects against cancer (PubMed search listed 125), only one focused specifically on colorectal cancer [41]. A PubMed search was performed using the keywords ursolic acid and colorectal cancer (63 results) and ursolic acid and colon cancer (66 results). Articles that were specific to colorectal cancer or colon cancer and ursolic acid were included in this review. Articles were excluded if they were not specific to these cancer types, did not use ursolic acid in their treatments, were review papers or were not available in English.

2. Effects of Ursolic Acid against Colorectal/Colon Cancer

2.1. Effects of Ursolic Acid against Colorectal/Colon Cancer: In Vitro Evidence

A significant number of studies (Table 2) have examined the effects of ursolic acid in colorectal cells in vitro.

Table 2. Effects of ursolic acid against colorectal cancer: in vitro studies.

Cell Type	Dose/Duration	Effects	Mechanism	Reference
HCT15	UA 30 μM	↓ Cell viability Cell cycle arrest (G0/G1 phase)	Not examined	[45]
HT-29	UA 10, 20 and 40 μM	↓ Cell proliferation ↑ Apoptosis	↓ p-EGFR ↓ p-ERK $\frac{1}{2}$ ↓ p-p38 ↓ p-JNK ↓ Bcl-2 ↓ Bcl-xL ↑ Cleaved caspase 3 ↑ Cleaved caspase 9	[46]
HCT15 CO115	UA 2.5 and 4 μM (HCT15) and UA 10 and 15 μM (CO115)	↓ Cell proliferation ↑ Apoptosis	↓ p-AKT ↓ KRAS	[47]
HT-29	UA 20 and 30 μM	↓ Cell proliferation ↑ Apoptosis	↑ Caspase 3 activity ↑ DNA fragmentation ↑ Cleaved Parp ↑ PGE2 concentration ↓ p-ERK ↑ p-p38 ↑COX-2	[48]
HT-29	UA 25 μM		↑ ATP in cytosol ↑ P2Y2 mRNA ↑ COX-2 protein ↑ DNA fragmentation ↑p-p38↑ p-Src protein	[49]
HCT-116	UA 5, 34.7 and 50 μM	↓ Cell viability ↑ Apoptosis ↑ ROS ↓ Cell migration	↓ BCL-2 protein ↓ Survivin protein ↓ NFkB ↓ SP1 protein ↑ BAX mRNA ↑ P21 mRNA ↑ P53 mRNA ↓ FN1 mRNA ↓CDH2 ↓↓CTNNB1 ↓ Twist	[50]
HT-29	UA 20, 40 and 80 μM		↓ SHH protein and mRNA ↓ Gli-1 protein and mRNA ↓ VEGF-A protein and mRNA ↓ bFGF protein and mRNA	[51]
HUVEC	UA 20, 40 and 80 μM	↓ Cell viability ↓ Cell migration	Not examined	
HT-29	UA 20, 40 and 80 μM	↓ Cell viability ↓ Cell survival ↓ Cell cycle progression ↑ Apoptosis	↓ Cells in s-phase ↓ Cyclin D1 protein and mRNA ↓ CDK4 protein and mRNA ↑ p21 protein and mRNA ↑ DNA fragmentation ↓ Bcl-2 protein and mRNA ↑ Bax protein and mRNA ↓ p-Erk1/2 protein ↓ p-JNK protein ↓ p-p38 protein	[52]
HCT116 HT29 SW480	UA 25 μM	↓ Cell viability ↓ Tumor sphere formation	↓ p-STAT3 protein ↑ Cleaved caspase 3	[53]

Table 2. *Cont.*

Cell Type	Dose/Duration	Effects	Mechanism	Reference
SW480 LoVo	UA 20, 40 and 60 μM	↓ Cell viability ↓ Colony formation ↓ Cell migration ↑ Apoptosis	↓ MMP9 mRNA ↑ CDH1 mRNA ↓ p-Akt ↓ p-mTOR ↑ p-PTEN ↓ p-JNK ↓ p-ERK ↓ COX-2 protein and mRNA ↓ PGE2 ↑ NF-kB translocation ↑ p300 translocation ↑ Cleaved PARP ↑ Cleaved caspase -3, -8 and -9	[54]
HCT116 HT29	UA 20, 40, 60 and 80 μM	↑ Apoptosis ↓ Cell viability	↑ TUNEL positive cells ↑ Cleaved PARP ↑ Cleaved caspase-3 ↓ p-JAK2 ↓ p-STAT3 ↓ STAT3 nuclear translocation ↓ miR-4500 mRNA expression	[55]
HCT116 HCT-8	UA 40 μM	↓ Cell viability ↓ Cell migration ↓ Cell invasion	↓ TGF-β1 protein ↓ p-Smad2/3 ↓ p-FAK ↓ ZEB1 ↓ N-cadherin ↑ miR-200a mRNA ↑ miR-200c mRNA	[56]
SW620 HCT116	UA 10, 30 and 60 μM	↓ Cell viability ↓ Clone formation ↓ Cell migration ↓ EMT	↑ Caspase 3 activity ↓ Mesenchymal phenotype ↑ E-cad protein ↓ Integrin protein ↓ Vimentin protein ↓ Twist protein ↓ Zeb1 protein	[57]
SW-480 HCT116	UA 10 μM	↓ Cell viability ↑ Cell injury Cell cycle arrest (S phase)	↓ CCNB1 mRNA and protein ↓ CDK1 mRNA and protein ↓ CDK2 mRNA ↓ CCND1 mRNA and protein ↓ CCNA2 mRNA and protein ↓ CDC20 mRNA and protein ↓ CKS2 mRNA ↓ CCNB2 mRNA	[58]
HT-29 HCT116	UA 2.5–40 μM	↓ Cell viability ↓ Cell number ↓ Colony formation ↑ Apoptosis	↓ NUFIP1 mRNA	[59]
RKO	UA 14, 17 and 20 μM Conventional conditions	↓ Cell viability ↑ Apoptosis ↑ ROS Cell cycle arrest (G0/G1 phase)	↑ Casp-3, -8 and -9 activity ↑ Bax protein ↓ Bcl-2 protein	[60]
	UA 25, 28, 31 μM 24 h Poly-HEMA coated plates	↓ Cell viability ↑ Apoptosis ↑ Anoikis	↓ p-FAK ↓ p-PI3K ↓ p-Akt ↓ N-cadherin ↑ E-cadherin	
HCT-116^{hSMO-}	UA 20 μM	↓ Cell proliferation ↓ Migration ↑ Apoptosis	↓ Bcl-2 protein and mRNA ↑ Bax protein and mRNA ↑ Caspase -3 and -9 mRNA ↓ c-Myc protein and mRNA ↓ GLI1 protein and mRNA ↓ SHH protein and mRNA ↓ SUFU protein and mRNA ↓ p-Akt protein	[61]

Table 2. *Cont.*

Cell Type	Dose/Duration	Effects	Mechanism	Reference
SW620	UA 7.5, 15 and 30 μM	↓ Cell proliferation ↓ Migration ↑ Apoptosis Cell cycle arrest–G0/G1 phase	↓ c-Myc protein ↓ Cyclin D1 protein ↓ Wnt4 mRNA and protein ↓ TCF4 mRNA and protein ↓ LEF1 mRNA and protein ↑ GSK3β mRNA and protein ↓ p-GSK3-β protein ↓ β-catenin mRNA and protein ↑ p-β-catenin	[62]
HCT-116 and SW480	UA 15 μM	↓ Migration ↓ Invasion	↓ p-Akt ↓ p-mTOR ↓ ARL4C ↓ MMP2	[63]
N/A	UA–computational model	↓ Cell proliferation ↑ Apoptosis ↓ Angiogenesis		[64]

Table legend: ↑ increased, ↓ reduced, p—phosphorylated.

Human colorectal cancer cells (HCT15) treated with UA (24 to 72 h) had increased cell fragmentation and death, with an IC_{50} value of 30 μM UA. There was an accumulation of cells in the G0/G1 phase of the cell cycle. These data showed significant anticancer activity of UA through the induction of colorectal cancer cell cycle arrest [45].

UA inhibited the growth of HT-29 colon cancer cells and increased apoptosis, as indicated by the decreased levels of the antiapoptotic proteins Bcl-2 and Bcl-xL and the increased activation of the apoptotic markers caspases -3 and -9 [46]. UA treatment decreased the phosphorylation of EGFR, ERK1/2, p38 MAPK and JNK signaling molecules. The use of the EGFR inhibitor AG1478 or the MEK inhibitor U0126 in combination with UA enhanced the inhibitory cell proliferation response. Together, these data indicate the inhibition of colorectal cancer cell and the suppression of the EGFR/MAPK signaling pathway [46].

Treatment of HCT15 and CO115 human colon cancer cells with UA decreased proliferation and increased apoptosis [47]. UA decreased phosphorylation of Akt in CO115 cells compared with the controls. In HCT15 cells, UA significantly reduced the protein level of KRAS. Based on these results, the authors concluded UA may act on the PI3K and Ras pathways.

UA induced apoptosis in HT-29 colorectal cancer cells, an effect that was associated with increased p38 phosphorylation and COX-2 levels [48]. The elevation of COX-2 was suggested to play a role in UA-induced apoptosis resistance, as the use of a p38-specific inhibitor or COX-2 siRNA resulted in an increase in UA-mediated apoptosis [48]. In subsequent studies, the same team of researchers examined signaling upstream of p38 and found that treatment of UA HT-29 colorectal cancer cells with UA increased cytosolic ATP and expression of the ATP receptor P2Y2 [49]. The activation of P2Y2 by UA treatment resulted in the activation of Src, downstream phosphorylation/activation of p38, induction of COX-2 and apoptosis resistance [49].

Human HTC-116 colon cancer cells treated with UA had decreased cell viability and increased apoptosis, as indicated by the increased nuclear fragmentation and reduced cell migration [50]. UA treatment increased the levels of reactive oxygen species (ROS) and the increase was attenuated with the addition of NAC, a ROS inhibitor. BCL-2, survivin, NFkB and SP1 mRNA levels were decreased, while BAX, P21 and P53 mRNA levels were increased with UA treatment. In addition, mRNA levels of FN1, CDH2, CTNNB1 and TWIST (cell migration-associated genes) were decreased following UA treatment. HT-29 colorectal cancer cells treated with UA (20, 40 and 80 μM) had a significant decrease in phosphorylation of STAT3, Akt and p70S6K [51] (Table 2). Treated cells had a decrease in mRNA and protein levels of sonic hedgehog (SHH) and Gli-1, both proteins associated with angiogenesis. This treatment also inhibited protein and mRNA levels of vascular endothelial growth factor (VEGF-1) and basic fibroblast growth factor (bFGF), two more angiogenesis markers. Human umbilical vein endothelial cells (HUVECs) treated with the same concentrations of UA had significantly reduced viability and migration [51]. Although these data suggest a role for UA to act as an anti-angiogenesis agent in CRC cells, the researchers did not examine the functional effects of UA.

The same HT-29 colorectal cancer cells were found in another study to have had reduced viability and survival with UA treatment [52]. Fluorescence-activated cell sorting (FACS) analysis following UA treatment revealed less cells in the S phase of the cell cycle and an increased number of apoptotic cells. UA-induced apoptosis was also seen by an increase in DNA fragmentation. Treatment with UA (20, 40 and 80 μM) reduced protein and mRNA levels of cyclin D1 and CDK4 and increased p21. The protein and mRNA levels of anti-apoptotic Bcl-2 was decreased with UA while expression of pro-apoptotic Bax was increased. UA reduced phosphorylation of mitogen-activated protein (MAP) kinases Erk1/2, JNK and p38 [52].

Ursolic acid (25 μM) inhibited phosphorylation of signal transducer and activator of transcription-3 (STAT3) and increased cleavage of caspase 3 in human HT29, HCT116 and SW480 colorectal cancer cells [53] (Table 2). Inhibition of p-STAT3 was associated with a significantly reduced viability in these cells. The IC_{50} for UA-induced inhibition of viability in these cells was lower than the IC_{50} of resveratrol or capsaicin, two other common dietary compounds reported to have antiproliferative properties, suggesting that UA is a more potent inhibitor. In these three CRC cell lines, UA completely inhibited the formation of tumor spheres when grown in anchorage-independent conditions [53].

Colorectal cancer cells SW480 and LoVo had reduced viability and survival when exposed to UA [54], while the same concentration of UA did not significantly affect CCD841 normal epithelial-like cells, indicating a sparing of normal/healthy cells. In addition, treatment with UA inhibited cell migration, an effect that was associated with a decreased mRNA level of MMP-9 and an increased mRNA level of CDH1 (key molecules in invasion and migration). Treatment of SW480 cells with UA reduced phosphorylation of Akt, mTOR and ERK yet increased phosphorylation of PTEN. Pre-treating cells with Akt inhibitor (LY294002, 5 μM) or ERK inhibitor (U0126, 20 μM) prior to UA treatment abrogated the UA-induced reduction in viability. UA (20 and 40 μM) inhibited protein and mRNA levels of COX-2 and decreased prostaglandin E2 (PGE2) production. UA induced translocation of NF-kB and p300 from the nuclei to the cytoplasm and attenuated p300-mediated NF-kB and CREB2 acetylation. Finally, UA induced apoptosis and caused an increase in cleavage of apoptosis signaling molecules PARP (caspases -3, -8 and -9) [54].

Human HCT116 and HT29 CRC cells treated with UA (20, 40, 60 or 80 μM) had increased apoptosis, as seen by TUNEL staining (Table 2) [55]. UA treatment increased the cleavage of apoptotic proteins PARP and caspase-3 and reduced phosphorylation levels of signaling molecules Janus kinase 2 (JAK2) and signal transducer and activator of transcription 3 (STAT3). Nuclear translocation of STAT3 was also blocked with UA treatment. qRT-PCR analysis showed that UA increased the mRNA level of miR-4500 in the HCT116 cells. Application of an miR-4500 inhibitor reversed the UA-induced cytotoxicity and apoptosis. These findings provide evidence that, in CRC cells, UA induces apoptosis by upregulating miR-4500 and inhibiting the JAK2/STAT3 signaling pathway [55].

Ursolic acid treatment of HCT116 and HCT-8 colon cancer cell lines resulted in significantly decreased cell viability, altered cell morphology and decreased cell migration and invasion [56]. These effects were associated with downregulation of proteins in the TGF-β1 signaling pathway (total protein levels of TGF-β1, p-Smad/2/3, p-FAK, ZEB1 and N-cadherin). qt-PCR analysis showed that UA increased mRNA levels of miR-200a and miR-200c in both cell lines, and that HCT-8 cells also had an increase in miR-200b. Together, these results suggest that UA acts to inhibit CRC cell viability, migration and invasion through modulating the TGF-β1/ZEB1/miR200 signaling pathway [56].

Treatment of human colon cancer cells SW620 and HCT116 with UA (10–60 μM) resulted in decreased cell viability, clone formation and inducement of caspase-3-mediated apoptosis [57]. UA induced a decrease in cell migration and epithelial–mesenchymal transition (EMT), which was associated with an increased protein level of E-cadherin, while protein levels of vimentin, integrin, twist and Zeb1 (metastasis biomarkers) were decreased. These findings suggest that UA has antiproliferative and antimetastatic properties via the inhibition of EMT in colorectal cancer [57].

SW-480 and HCT-116 human colorectal cancer cells had a decreased viability when treated with UA (5–100 μM) [58]. Flow cytometry analysis of the cell cycle showed that UA treatment increased the number of cells in the S phase in both CRC cell lines. UA significantly decreased mRNA and protein levels of genes associated with cell cycle progression (CCNB1, CDK1, CCND1, CCNA2 and CDC20). mRNA levels of CDK2, CKS2 and CCNB2 were also decreased with UA. Fluorescence staining with Ethd-1 showed an increased number of injured cells following treatment with 30 μM UA and a significant increase in damaged cells when treated with the CCNB1 inhibitor (R0-3306; 10 μM), suggesting that CCNB1 and its associated targets are involved in the anticancer effects of UA. These results suggest that UA inhibits the proliferation of colon adenocarcinoma cells by downregulating CCNB1, thus blocking the division of cells [58].

Human CRC cells (HT-29) had reduced viability and colony formation and increased apoptosis following treatment with UA (2.5–40 μM) [59]. Nuclear fragile X mental retardation interacting protein 1 (NUFIP1) mRNA and protein levels were reduced following UA treatment and this effect was increased in sh-NUFIP1 knockdown models. When NUFIP1 was knocked down, HCT116 cells had increased protein levels of tumor suppressors p53 and p21. These findings suggest that NUFIP1 is an oncogene that drives CRC development and can be counteracted with UA treatment. Further studies are needed to explore the interaction between UA and NUFIP1 in more detail [59].

Human colorectal cancer RKO cells treated with UA had reduced cell viability [60]. Flow cytometry revealed a G0/G1 cell cycle phase arrest. Further analysis showed increased activity of caspases -3, -8 and -9. UA reduced the protein level of Bcl-2 yet increased the Bax protein level. UA-induced apoptosis was associated with an increase in levels of ROS. When RKO cells were grown in suspension (a detached condition), there was again a UA-induced decrease in cell growth. In detached conditions, UA induced anoikis (another form of programmed cell death that occurs in anchorage-dependent cells when they are detached from their surrounding extracellular matrix). Increased anoikis was associated with a decrease in phosphorylation of anoikis-related proteins FAK, PI3K and Akt. UA treatment of RKO cells also inhibited epithelial–mesenchymal transition. UA downregulated N-cadherin expression and upregulated E-cadherin. These findings together show that, in colorectal cancer cells, UA is able to induce caspase-dependent apoptosis and FAK/PI3K/Akt signaling-related anoikis, suggesting it has potential as an effective treatment against anchorage-dependent and anchorage-independent cancers [60]. The authors suggest that they saw a UA-induced time dependent effect; however, the data presented here do not appear to support that, and more detail is required to show that there was a true time dependent effect. While this paper begins to examine anoikis, it does not provide strong enough evidence that this event is what is occurring or being induced with their UA treatment protocol over other forms of cell death; more evidence for the initiation of anoikis would make the claim stronger.

Ursolic acid inhibited the proliferation and migration of CRC cells HCT-116^{hSMO-} (smoothened (SMO) gene knockdown cell line) [61]. Smoothened is a g-protein coupled receptor important in the canonical hedgehog signaling pathway; however, the non-canonical hedgehog pathway can be activated without the involvement of SMO. In this study, UA (10, 20, 30 and 40 μM) reduced cell proliferation in SMO knockdown cells and HCT-116 cells at the same concentrations. Normal human colon mucosa epithelial cells (NCM460) did not show any significant changes in proliferation with UA (15–300 μM), suggesting cancer cell specific effects. Wound healing assays showed significantly decreased cell migration with UA treatment in both the normal and knockdown cells. Apoptosis was induced in the HCT-116^{hSMO-} cells, as examined by flow cytometry. Furthermore, the apoptosis-related protein BCL-2 level was reduced, while the BAX level was increased. qRT-PCR analysis showed a decreased mRNA level of BCL-2 and increased mRNA levels for BAX, caspase-9 and caspase-3. Signaling molecules important in the hedgehog pathway were examined with UA treatment and showed decreased protein and mRNA levels of MYC (c-Myc), glioma-associated oncogene (GLI1) and sonic hedgehog (SHH), while protein and mRNA levels of suppressor of fused (SUFU) were increased. Finally, this research group found that UA decreased the phosphorylation of Akt and suggested that the suppression of Akt signaling inhibited the hedgehog signaling cascade [61]. Together, these results show that UA exerted its anticancer effects against CRC cells through the suppression of Akt signaling-dependent activation of SMO-independent hedgehog signaling. Importantly, this paper provides evidence that UA did not exert cytotoxic effects against normal colon tissue, suggesting that it may exert its anticancer effects with lower overall side effects.

Treatment of SW620 human colorectal cancer cells with UA reduced proliferation and colony formation and increased apoptosis [62]. UA inhibited the rate of cell migration in a wound healing assay. UA at the same concentrations showed no effect on NCM460 (normal human colonic cells), indicating a cancer cell specific effect. UA caused cell cycle arrest in the G0/G1 phase, which was associated with a decrease in mRNA levels of c-Myc and cyclin D1 (common cell cycle markers). These effects were associated with a decrease in Wnt/β-catenin signaling. UA increased mRNA and protein levels of GSK-3β and decreased mRNA levels of β-catenin, Wnt4 and TCF4. Phosphorylation of GSK3β and Wnt4 was decreased, while the phosphorylation level of β-catenin increased [62].

Zhang et al. [63] found that treatment of HCT-116 and SW480 cells with UA resulted in significant inhibition of migration and invasion that was associated with reduced ADP-ribosylation factor like GTPase 4C (ARL4C) and MMP2 levels and reduced phosphorylation of Akt and mTOR. The inhibitory effect of UA on cell migration and invasion and on MMP2 were reversed by ARL4C overexpression. In addition, UA, the Akt inhibitor LY294002 and the mTOR inhibitor rapamycin increased ARL4C ubiquitination, an effect reversed by the proteasome inhibitor MG-132 [63]. Altogether, these data indicate that UA inhibits CRC cell migration and invasion by inhibiting Akt-mTOR signaling and increasing ubiquitination of ARL4C, resulting in its degradation.

Using a series of database and computational methods, a protein interaction (PPI) network was created. Using this, PPI Zhao et al. (2021) concluded that UA can target multiple signaling pathways [64]. Analysis revealed 113 potential targets of UA in colon cancer cells; the core targets were interleukin-6 (IL-6), mitogen-activated protein kinase 3 (MAPK3), vascular endothelial growth factor receptor (VEGFA), caspase-3, mitogen-activated protein kinase 8, tumor necrosis factor, cyclin D1 and STAT3 [64]. The computational analysis performed in this paper provided strong evidence for UA to act against colon cancer cells through a wide variety of signaling pathways; however, experiments based off these data need to be performed using in vitro and in vivo models to confirm that UA does have these effects in a physiological model.

Overall, these studies indicate that the treatment of colorectal cancer cells with ursolic acid leads to decreased cell viability, survival, migration and invasion and increased apoptosis and cell cycle arrest. These effects are associated with decreased phosphorylation of signaling proteins Akt, mTOR, JNK and ERK and increased levels of cleaved PARP, casp-3 and casp-9 (Figure 2).

2.2. *Effects of UA in Combination with Chemotherapy Agents and Radiation*

Colorectal cancer can be treated systemically with chemotherapy agents, and there are several drugs with FDA approval for use against CRC; however, their efficacy is often reduced due to patients developing drug resistance. To help overcome drug resistance and harmful side effects experienced with many chemotherapy agents, researchers are examining other compounds that can be given in combination to enhance the efficacy of the chemo drugs, reducing the required dosage and lowering the severity of the side effects. A number of studies have examined the effects of UA treatment in combination with other agents.

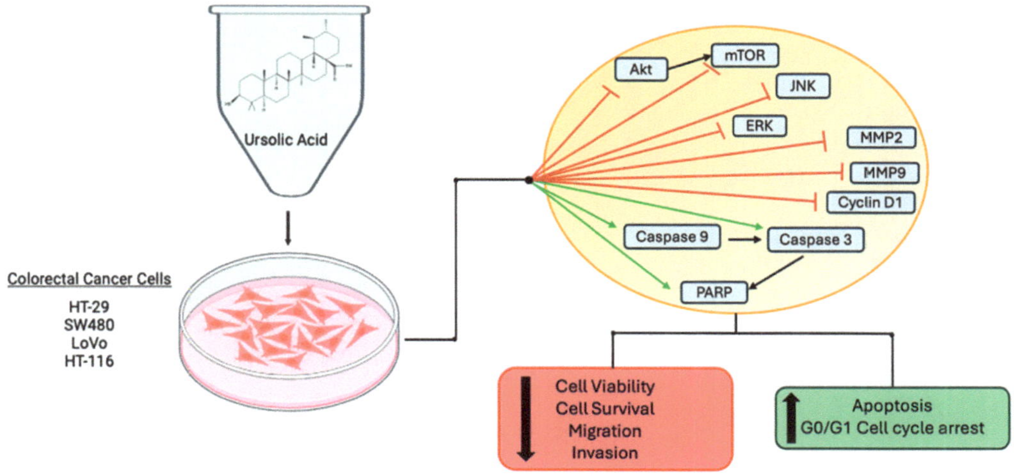

Figure 2. Summary of the effects of ursolic acid in human colorectal cancer cells. This figure was created using BioRender.com and Microsoft PowerPoint based on studies presented in Table 2 above. Green arrows show stimulation; red arrows show inhibition; black arrows show established pathways.

Colon cancer cells SW480 and LoVo had reduced cell viability and migration when treated with UA in combination with 1 mM melatonin (MT) [65] (Table 3). This treatment induced apoptosis, caused changes in cell morphology and modulated cytochrome c/caspase, MMP9/COX-2 and p300/NF-kB signaling. UA combined with MT triggered the release of cytochrome c from the mitochondria, inducing cleavage of caspase (-3 and -9) and PARP proteins. UA plus MT treatment inhibited MMP-9 and COX-2 protein levels and promoted p300 and NF-kB translocation from the nucleus to the cytoplasm. The effects of the combined treatment were greater than UA or MT treatment alone [65].

Table 3. Effects of UA in combination with chemotherapy agents.

Cell Type	Dose/Duration	Effects	Mechanism	Reference
SW480 LoVo	UA 20, 40 and 60 μM Melatonin 1 mM	↓ Cell viability ↓ Cell migration ↑ Apoptosis	↓ MMP9 mRNA expression ↑ Cleaved PARP ↑ Cleaved caspase -3 and -9 ↓ COX-2 protein and mRNA ↑ p300 cytoplasmic translocation ↑ NF-kB cytoplasmic translocation	[65]
HCT15	UA 4 μM 5-FU 100 μM	↑ Apoptosis	↑ p-JNK p46 protein ↓ p-mTOR protein ↑ LC3-I protein ↑ LC3-II protein ↑ p62 protein ↑ p53 levels	[66]
SW480 SW620 LoVo RKO	UA 10 μmol/L Oxaliplatin 0.4 μmol/L	↓ Cell viability ↑ Apoptosis	↓ Mitochondrial membrane potential ↑ Cleaved caspase -3, -8 and -9 ↓ p-B-Raf ↓ p-MEK1/2 ↓ p-ERK1/2 ↓ p-Akt ↓ p-p38 ↓ p-JNK ↓ p-IKKα ↓ p-IkBα ↓ p-p65 ↓ p-NF-kB (plasma and nucleus)	[67]

Table 3. *Cont.*

Cell Type	Dose/Duration	Effects	Mechanism	Reference
RKO LoVo SW480	UA 20 and 40 μM/L 5-FU 4 and 8 μM/L Oxaliplatin 0.5, 1 and 1.5 μM/L	↓ Cell viability ↑ Chemosensitivity (hypoxia) ↑ Apoptosis	↓ MDR1 protein and mRNA expression ↓ HIF-1α protein and mRNA ↓ VEGF	[68]
HCT8 SW480	UA 20 μmo/L Oxa 0.4 μmol/L	↓ Cell viability ↑ Apoptosis	↑ Cleaved caspase -3 ↑ ROS ↑ NAPDH protein ↓ P-gp mRNA and protein ↓ MRP mRNA and protein ↓ BCRP mRNA and protein	[69]
RKO	UA 15 μM Oxa 2.5 μM 48 h	↓ Cell survival ↑ Apoptosis	↑ Caspase -3, -8 and -9 activity ↑ Cleaved PARP ↓ Survivin protein ↓ XIAP protein	[70]
HT-29 SW 620	UA 5 μg/mL OA 100 μg/mL CPT-11 0.075 μg/ml	↓ Cell viability ↓ Migration	N/A	[71]
HCT116 HT-29	UA 15 μM DOX 1.5 μM	↓ Cell proliferation ↑ Apoptosis ↓ Colony formation ↓ Cell migration G1 cell cycle arrest	↑ Cleaved caspase -9 ↑ Cleaved PARP ↑ E-cadherin ↓ MMP-9 ↓ uPA ↓ CDK4 and CDK6 ↓ cyclin D1 ↓ p-Akt ↓ p-GSK-3β ↓ c-Myc ↑ Rassf1A ↑ Mst1 and Mst2 ↑ Sav1 ↑ p-Mob1 ↑ p-Yap ↓ CTGF	[72]
LoVo HCT116	UA 10 mM Sorafenib 10 mM	↓ Cell viability ↓ Colony formation ↑ Apoptosis ↑ ROS	↑ Cleaved PARP ↑ Cleaved caspases 9 and 8 ↑ LC3 I and II ↓ Mcl-1 ↑ Bim ↑ MDA ↓ GSH	[73]
CT26 Mouse colon cancer cells	UA 15 μM 15 Gy radiation	↓ Cell survival ↑ Apoptosis ↑ Ros ↓ GSH	↓ Casp 3 ↓ Bcl2 ↑ Cleaved PARP	[74]

Table legend: ↑ increased, ↓ reduced, p—phosphorylated.

HCT15 cells treated with a combination of UA and 5-fluorouracil (5-FU, 100 μM) (a chemotherapy agent used in CRC) showed a significant increase in TUNEL positive cells compared with either compound alone [66]. The combined treatment increased the phosphorylation level of JNK while decreasing the phosphorylation level of mTOR (p46), and increased protein levels of autophagy markers LC3-I, LC3-II and p62. The combined treatment also increased p53 protein levels. Based on these results, the authors concluded that UA is able to enhance the apoptotic effects of 5-FU by activating JNK signaling [66]. This study provides evidence that UA has the potential to act synergistically to enhance the effects of 5-FU (a chemotherapy drug already in use).

Ursolic acid's ability to enhance the efficacy of oxaliplatin (OXL) treatment was examined in human CRC cells (SW480, SW620, LoVo and RKO) [67]. The combination treatment of SW620 cells reduced viability, induced apoptosis and reduced mitochondrial membrane potential. Cleaved caspase -3, -8 and -9 were increased, while levels of anti-apoptotic markers Bcl-xL, Bcl-2 and survivin were decreased. A drug combination index analysis revealed a synergistic effect between UA and oxaliplatin for the inhibition of proliferation in all cell lines tested; RKO cells had the greater interaction, with 0.68 synergy. Multiple signaling pathways appeared to be involved in the anticancer effects, seen as the treatment causing changes in levels of proteins associated with MAPK, PI3K/Akt and NF-kB signaling [67].

Treatment with UA caused colon cancer cells (RKO, LoVo and SW480) to become more sensitive to treatment with chemotherapy agents 5-FU or oxaliplatin [68] (Table 3).

This occurred by the inhibition of MDR1 via HIF-1α under hypoxic conditions; UA alone and the combinations also inhibited HIF-1α in normoxic conditions. Additionally, UA downregulated VEGF protein levels and inhibited angiogenesis. These findings indicate that UA could act as a chemosensitizer in colon cancer by inhibiting HIF-1α, MDR1 and VEGF [68].

Human CRC cells HCT8 and SW480 treated with UA (20 μmol/L) and oxaliplatin (Oxa) (0.4 μmol/L) had reduced proliferation, increased apoptosis and ROS production [69]. The combination treatment significantly increased these effects compared with either parent compound alone. The protein level of apoptotic marker cleaved caspase-3 was increased. ROS levels, detected by DCFH-DA assay, were significantly increased with the combination treatment compared with either drug alone. Similarly, the combination treatment increased the protein level of NADPH (a primary resource for ROS production). The combination treatment significantly reduced mRNA and protein levels of permeability glycoprotein (p-gp), MRP and BCRP (genes associated with drug resistance) [69]. Based on these results, the authors conclude that UA enhances the anticancer effects of oxaliplatin in CRC cells via the ROS-mediated inhibition of drug resistance.

Another research group, Zheng et al. 2020, also examined the synergistic effect of ursolic acid and oxaliplatin (Oxa), using RKO cells [70]. The combination treatment (UA and Oxa) enhanced apoptosis and reduced survival when compared with treatment with either compound alone. Increased activity of caspase -3, -8 and -9 and a significant increase in cleaved PARP protein levels was seen. In RKO cells, the combination treatment decreased the protein levels of the X-linked inhibitor of apoptosis (XIAP) and survivin. These results provide evidence that UA and Oxa have a synergistic effect and may be used together in the treatment of CRC [70].

Human HT-29 colon cancer cells treated with UA (5 μg/mL) and oleanolic acid (OA, 100 μg/mL) had reduced viability [71]. UA combined with camptohecin-11 (CPT-11, 0.075 μg/mL) had no cytotoxic effect. Normal healthy cells (CCD 841) did not show cytotoxicity with either treatment protocol. UA combined with OA or CPT-11 inhibited the migration of HT-29 and SW620 CRC cells. HT-29 cells had reduced protein levels of MMP2 with both treatment protocols, while SW620 cells had reduced protein levels of MMP9 with both treatments. Both HT-29 and SW620 cells treated with UA combined with CPT-11 had reduced levels of urokinase-type plasminogen activator receptor (uPAR) immunofluorescence [71].

Ursolic acid in combination with chemotherapy agent doxorubicin (DOX) was examined in human colon cancer cell line HCT116 and human colorectal cancer cell line HT-29; the results showed a synergistic effect [72]. UA combined with DOX significantly decreased cell proliferation and colony formation, a response that was greater than with either compound alone. The combination treatment significantly increased the number of apoptotic cells and levels of cleaved caspase 9 and cleaved PARP. A wound healing assay revealed the significant inhibition of migration associated with an increase in protein levels of E-cadherin, MMP-9 and uPA (metastasis markers with the combined treatment). Cell cycle arrest in the G1 phase was seen and protein levels of cell cycle markers CDK4, CDK6 and cyclin D1 were all decreased. The combined treatment decreased phosphorylation levels of Akt and GSK3β and reduced c-Myc levels. UA combined with DOX increased the protein levels of Hippo signaling pathway proteins Rassf1A, Mst1, Mst2, Sav1 and CTGF and increased phosphorylation levels of Mob1 and Yap. The inhibition of Akt signaling with PI3K inhibitor (LY2940002) further increased the protein levels of Hippo pathway proteins, and the synergistic anticancer effects seen with the combined treatment occurred through the Akt/Hippo signaling pathway [72].

The combination treatment of ursolic acid and sorafenib (an FDA-approved tyrosine kinase inhibitor drug) showed synergistic anticancer effects in HCT-116 and LoVo colon cancer cells. The combined treatment significantly decreased cell viability and survival and increased apoptosis compared with either compound alone [73]. The protein levels of cleaved PARP, cleaved caspases -9 and -8 and LC3 I and II increased with the combined treatment.

Mouse colon cancer cells (CT26) exposed to 15 Gy of gamma irradiation and then treated with UA had significantly decreased survival and increased apoptosis compared with each treatment alone [74]. Irradiation and UA combined decreased caspase3 and Bcl2 protein levels and increased the level of cleaved PARP. The combination treatment significantly increased the level of cellular ROS (DCF assay) and increased mitochondrial ROS (DHR 123 assay), while levels of GSH were decreased [74].

The above studies provide evidence that the combination of ursolic acid with the chemotherapy drugs 5-fluorouracil, oxaliplatin and doxorubicin and the tyrosine kinase inhibitor sorafenib or in combination with irradiation result in significant enhanced responses. The combination treatment increased the level of apoptosis in CRC cells above the level of either treatment alone (Figure 3).

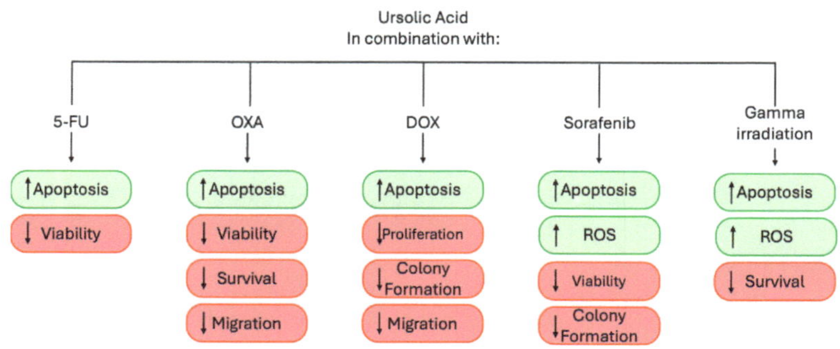

Figure 3. Ursolic acid has synergistic effects with 5-fluorouracil (5-FU), oxaliplatin (OXA), doxorubicin (DOX), sorafenib and gamma irradiation. This figure was created with Microsoft PowerPoint 2024 based on the studies presented in Table 3. (↑: increased, ↓: reduced).

2.3. Effects of Ursolic Acid on Animal Models of Colorectal Cancer

This section summarizes articles that examined the effects of UA utilizing animal models of colorectal cancer.

The daily intraperitoneal administration of UA inhibited the growth of tumors in mice xenografted with human CRC cells (HT-29), evidenced by reduced tumor volume and weight [52] (Table 4). No change in the body weight of the animals was seen. Tumor samples from UA-treated mice had reduced PCNA cell staining. Consistent with the in vitro results reported in Section 2.1, tumor tissue samples from mice treated with UA had reduced mRNA and protein levels of cyclin D1 and CDK4, while p21 mRNA and protein levels were increased. Tumor samples had an increase in the percent of TUNEL positive cells. UA treatment decreased the protein level of Bcl-2 in tumor samples yet increased the protein and mRNA levels of Bax. Reduced phosphorylation of the MAP kinases, Erk1/2, JNK and p38 and reduced phosphorylation of STAT3 were seen in the tumor samples, suggesting that UA acts to reduce tumor growth in CRC xenografts via the inhibition of these signaling pathways [52].

Table 4. Effects of ursolic acid on animal models of colorectal cancer.

Model	Dose/Duration	Effects	Mechanism	Reference
Male BALB/c athymic mice xenografted with HT-29 cells (1.5×10^6)	UA 12.5 mg/kg Daily intraperitoneal injection	↓ Tumor volume ↓ Tumor weight	↓ PCNA ↓ cyclin D1 protein and mRNA ↓ CDK4 protein and mRNA ↑ p21 protein and mRNA ↑ TUNEL ↓ Bcl-2 protein and mRNA ↑ Bax protein and mRNA ↓ p-STAT3 ↓ p-Erk1/2 ↓ p-JNK ↓ p-p38	[52]

Table 4. *Cont.*

Model	Dose/Duration	Effects	Mechanism	Reference
Male BALB/c athymic mice xenografted with HT-29 cells (1.5×10^6)	UA 12.5 mg/kg Daily intraperitoneal injection	↓ Tumor volume	↓ CD31 positive cells ↓ p-STAT3 ↓ p-Akt ↓ p-p70S6K ↓ SHH positive cells and mRNA ↓ Gli-1 positive cells and mRNA ↓ VEGF-A positive and mRNA ↓ bFGF positive cells and mRNA	[51]
Chick chorioallantoic membrane	UA 0.25 mg 72 h	↓ Number of blood vessels		
Female athymic nude mice xenografted with HCT116 cells (1×10^7)	UA 10 mg/mg Daily intraperitoneal injection	↓ Tumor volume	N/A	[53]
Female nude mice xenografted with HCT15 cells (10^6 cells)	UA 75 mg/kg daily Orally in Nutella	↓ Tumor size	↑ p62 (ns) ↑ p-JNK (ns)	[66]
Male BALB/c nude mice HCT-116^{hSMO-} cells (1×10^7)	UA 10, 20 or 40 mg/kg Intraperitoneal injection 12 consecutive days	↓ Tumor weight ↓ Tumor volume	↓ BCL-2 protein and mRNA ↑ BAX protein and mRNA ↑ Caspase -3 and -9 mRNA ↓ c-Myc protein and mRNA ↓ GLI1 ↓ SHH ↓ SUFU ↓ p-Akt	[61]
Nude mice xenografted with SW620 cells (1×10^7)	UA 15, 30 or 60 mg/kg Intragastrical	↑ Body weight ↓ Tumor weight ↓ Tumor volume ↑ Apoptosis	↑ GSK3β mRNA and protein ↓ β-catenin mRNA and protein ↓ WNT4 mRNA and protein ↓ TCF4 mRNA and protein ↓ LEF1 mRNA and protein ↑ p-β-catenin ↓ p-GSK3β ↓ Nuclear β-catenin	[62]
Male BALB/c-nude mice HCT-116 cells (5×10^6) injected in tail vein	UA 20 mg/kg Intraperitoneal Daily/42 days	↓ Lung metastasis	↓ ARL4C	[63]
Female nude mice xenografted with SW620 cells	UA 20 mg/kg Oxaliplatin 10 mg/kg	↓ Tumor weight ↓ Tumor volume	↓ p-ERK1/2 ↓ p-Akt ↓ p-IKKα ↓ Ki-67 pos cells ↑ TUNEL pos cells	[67]
Female nude mice xenografted with HCT8 or SW480 cells (1×10^5)	UA 10 mg/kg Oxaliplatin 10 mg/kg	↑ Animal survival time ↓ Tumor volume	N/A	[69]
Athymic nude mice xenografted with HCT116 cells	UA 10 mg/kg/day DOX 2 mg/kg/twice weekly	↓ Tumor weight ↓ Tumor volume	↓ Ki67 ↓ p-Akt ↑ Rassf1A ↑ Mst1 ↑ Mst2 ↑ Sav1 ↑ p-Mob1 ↑ p-Yap ↓ CTGF	[72]
Male athymic nu/nu mice Luciferase-transfected HCT116 cells	UA 250 mg/kg Orally, daily Capecitabine 60 mg/kg Orally, 2/week combination	↓ Tumor growth ↓ Tumor volume ↓ Tumor weight ↓ Metastasis	↓ Ki67 ↓ CD31 ↓ Nuclear p65 ↓ β-catenin ↓ p-STAT3 ↓ Cyclin D1 protein ↓ cMyc ↓ p-EGFR protein ↓ Bcl-2 protein ↓ Bcl-xl protein ↓ Survivin protein ↓ ICAM-1 ↓ VEGF ↓ MMP9 ↓ p53 ↓ p21	[75]

Table legend: ↑ increased, ↓ reduced, p—phosphorylated.

Ursolic acid (12.5 mg/kg, daily) inhibited CRC tumor volume yet did not affect the body weight of mice xenografted with human HT-29 colorectal cancer cells, suggesting that UA acts as an anticancer agent in vivo, with minimal toxicity to the animal [51]. UA treatment in these mice caused reduced intratumoral microvessel density and decreased the protein level of CD31 (a vascular differential marker). This research group used a chick embryo chorioallantoic membrane model treated with UA (0.25 mg for 72 h), which resulted in a reduced total number of blood vessels. A bio-plex phosphoprotein assay showed that UA decreased phosphorylation of STAT3, Akt and p70S6K. The immunohistochemistry of tumor samples showed decreased protein levels of sonic hedgehog (SHH) and Gli-1 with the UA treatment, suggesting that UA suppresses the sonic hedgehog signaling pathway. Immunohistochemistry analysis further showed a decrease in levels of VEGF-A and bFGF (important markers of angiogenesis). Taken together, the findings from this study support a role for UA as an anticancer treatment through its ability to inhibit angiogenesis in tumors through multiple signaling pathways [51].

Ursolic acid (10 mg/kg) treatment (by daily intraperitoneal injection for 13 days) suppressed the growth of tumors in mice xenografted with human colon cancer cells (HCT116) [53] (Table 4). The tumor volume was calculated by caliper measurements and found to be reduced in the UA-treated mice compared with the controls [53]. Unfortunately, no tumor tissue examination or other measurements were performed in this study.

Female nude mice xenografted with human CRC (HCT15) cells had reduced tumor size when treated with UA (75 mg/kg orally daily for 14 days) [66]. Immunohistochemical analysis of excised tumor tissues showed a slight increase in the protein level of p62 and the phosphorylation level of JNK following UA treatment, while levels of LC3 and Ki67 were not changed. These results suggest that UA reduces tumor growth in vivo, possibly through JNK signaling and induced autophagy [66].

Nude mice xenografted with smoothed knockdown HCT-116^{hSMO-} cells and treated with UA (10, 20 or 40 mg/kg) for 12 consecutive days had decreased tumor volume and weight compared with no-treatment mice [61]. Analysis of the extracted tumors showed that UA induced apoptosis in the tumor cells through the inhibition of Bcl-2 and increased BAX (protein and mRNA). Tumor tissues had increased mRNA levels of caspases -9 and -3, increased mRNA and protein levels of SUFU and reduced mRNA and protein levels of c-Myc, GLO1 and SHH. These results were consistent with the cell culture data from the same research group (reported above, Section 2.1) [61]. The combined results from this study provide evidence that UA acts through a complex signaling mechanism involving non-canonical hedgehog signaling and inhibited Akt signaling.

Human SW620 CRC cell-xenografted mice treated with UA (15, 30 and 60 mg/kg) by intragastric administration had a significantly higher body weight after 16 days compared with the controls, and a significantly decreased tumor weight and volume [62]. Analysis of the tumors showed that UA treatment induced apoptosis; mRNA and protein levels of Bcl-2 were decreased, while mRNA and protein levels of Bax were increased. UA treatment increased mRNA levels of caspases -3 and -9. UA-treated mice had increased mRNA and protein levels of GSK3β and decreased mRNA and protein levels of β-catenin, WNT4, TCF4 and LEF1. The phosphorylation level of β-catenin increased, while the level of GSK3β decreased. There was a decrease in the level of nuclear β-catenin protein [62].

Zhang et al. [63] injected HCT-116 cells into the tail vein of male BALB/c-nude mice, establishing a lung metastatic model of colon cancer and found that daily administration of UA (20 mg/kg) for 42 days resulted in a significant decrease in lung metastases (accompanied by prevention of weight loss). Lung tissue immunohistochemical analysis revealed that the expression of ARL4C was significantly decreased in animals administered UA compared with the control untreated animals. These data indicate that UA could inhibit the metastasis of colon cancer to the lungs.

Mice xenografted with human CRC SW620 cells were treated with UA (20 mg), OXL (10 mg) or a combination for five consecutive days and sacrificed on day 50 for tumor and tissue examination [67]. The combination of UA and OXL caused a decrease in tumor volume and tumor weight above that seen with either compound alone. Immunohistochemistry showed that the combination treatment led to a decrease in Ki-67 protein levels and an increase in TUNEL positive cells. Serum concentrations of ALT and AST showed that OXA caused impaired liver function, which was attenuated with the addition of UA to the treatment. Tumor tissues examined with immunohistochemistry and Western blot showed that the antitumor effects were associated with the inhibition of Akt and ERK1/2 signaling. Phosphorylation levels of ERK1/2, Akt and IKKα were reduced following the combined treatment [67]. These results suggest that UA enhances the effect of the chemotherapy agent OXL and contributes to reducing the OXL-associated side effects.

The combination of ursolic acid and oxaliplatin increased the survival time and reduced the tumor volume of mice xenografted with human CRC (HCT8 or SW480) cells [69]. The anticancer effects with the combined treatment were enhanced compared with either compound alone.

Mice xenografted with HCT116 human colon cancer cells were treated with UA (10 mg/kg/day), DOX (2 mg/kg/twice weekly) or combined UA and DOX by intraperitoneal injection [72] (Table 4). Following treatment, the mice were sacrificed and tumor and serum samples were collected. Serum samples showed no significant change in biochemical markers ALT, AST, BUN and creatinine, suggesting that the treatments did not impair liver or kidney function of the animals. The body weight of the animals did not differ between the treatment groups. Tumor volume was significantly decreased with UA or DOX treatment, and the decrease was enhanced with the combined UA and DOX treatment. Cell proliferation marker Ki67 was reduced with the combination treatment when examined with immunohistochemistry. Tumors from UA- and DOX-treated animals had reduced protein levels of CTGF, reduced phosphorylation levels of Akt, increased protein levels of Rassf1A, Mst1, Mst2 and Sav1 and increased phosphorylation levels of Mob1 and Yap [72]. These results are consistent with in vitro findings previously reported by this group (Section 2.2 above). The results from this study, both in vitro and in vivo, show that the combination of UA and DOX have strong anticancer effects over either treatment alone.

Nude mice were orthotopically implanted with luciferase-transfected HCT116 human colorectal cancer cells and treated via oral gavage with UA (250 mg/kg daily), capecitabine (CAP) (60 mg/kg twice/week) or both in combination. Tumor growth was measured on days 7, 14, 21 and 28 by bioluminescence imaging and showed reduced growth with UA treatment. Importantly, greater growth inhibition of the tumor was seen with the combination treatment over 28 days [75]. The excised tumors had significantly decreased tumor volume and weight only with the combination treatment. The sacrificed animals were examined for metastatic growths; UA, CAP and combination treatment mice all had reduced metastases in the liver, intestine, lung and spleen. Significant levels of UA (measured by HPLC) were detected in serum (480 ng/mL) and tissue (356 mg/mL) samples that were collected four hours after UA oral administration. These data suggest that UA reached the animals' circulation and infiltrated the tumor. Extracted tumor tissues were further analyzed by immunohistochemistry and Western blot and it was found that UA enhanced the effects of capecitabine. UA inhibited levels of nuclear p65 and β-catenin, reduced protein levels of cyclin D1, cMyc, Bcl2, Bcl-xl and survivin and reduced phosphorylation levels of STAT3 and EGFR. The metastasis markers ICAM-1, VEGF and MMP-9 were reduced in tumor tissues with the UA treatment of animals, and expression of tumor suppressor genes p53 and p21 were increased with either UA or the combined treatment. Immunohistochemical analysis also showed a reduced expression of proliferation markers Ki67 and CD31 [75]. This paper provides strong evidence that UA acts as an anticancer drug against colorectal cancer and also acts as a capecitabine chemosensitizing agent. Importantly, this paper shows that the oral administration of UA results in significant UA serum levels and that UA is able to enter the tumor tissue, achieving detectable levels in the tumor.

Although limited in vivo animal studies exist, the evidence indicates that oral or intraperitoneal administration of UA in mice xenografted with human CRC cells causes reduced tumor volume and tumor weight. Oral administration also reduces cancer metastasis (Figure 4).

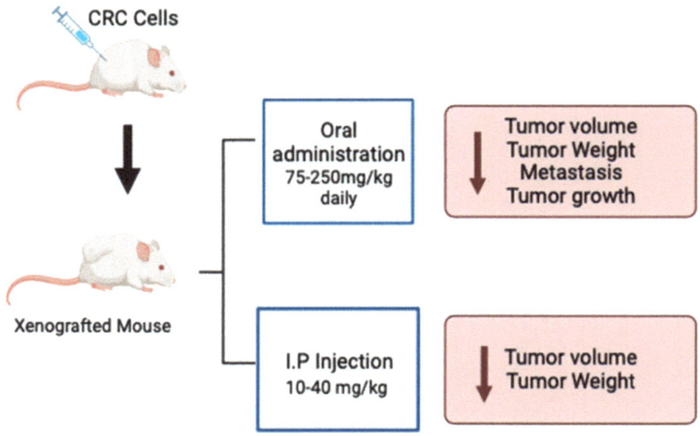

Figure 4. Ursolic acid reduced tumor growth in animals xenografted with human CRC cells. This figure was created with BioRender.com. ↓: reduced.

A search of ClinicalTrials.gov (accessed on 22 June 2024) for the disease "colorectal cancer" or "colon cancer" and the term "ursolic acid" found no results. A search for "cancer" and "ursolic acid" also had no results at this time. Overall, there are currently no clinical trials examining UA against CRC or any other cancer.

2.4. Bioavailability and Potential Toxicity of UA

Similar to many other polyphenols, UA has limited bioavailability due to its low water solubility and high molecular weight (456.7 g/mol) [76,77]. Studies examining the bioavailability of UA are limited. Yin et al. administered UA in mice (0.05% UA added to the diet) for 8 weeks, then measured UA plasma levels and examined UA tissue distribution. The plasma level of UA was 580 ng/mL or 1.26 μM. Tissues that showed UA accumulation included the liver (9.7 μg/g), colon (6.4 μg/g), kidney (5.9 μg/g), heart (3.9 μg/g), bladder (2.9 μg/g) and brain (1.6 μg/g) [77]. In another study, oral administration of UA (100 mg/kg body weight) in rats resulted in a plasma concentration of 300 ng/mL (656.89 nM) within one hour. The concentration decreased after 4 h, but the animals maintained a plasma concentration of about 100 ng/mL (218.96 nM) for 12 h [76]. Although there are currently no studies examining UA plasma levels in humans after oral administration of the parent UA compound, a study by Xia et al. [78] administered in humans UA nanoliposomes via intravenous infusion at a dose of 98 mg/m^2. These subjects had peak plasma concentrations of 3404.6 ng/mL (7454.78 nM) four hours after infusion. This peak was followed by a rapid decline, about 10-fold, and the concentration 6 h post-infusion was approximately 300 mg/mL (656.89 nM). There was another 10-fold decline and, at 16 h post-infusion, the subjects had approximately 30 ng/mL (65.69 nM) UA plasma concentration [78]. The above studies suggest that UA administration results in plasma UA levels in the nano- to micromolar range. The concentrations found in the plasma were comparable to concentrations of UA used in many of the in vitro studies showing potent anticancer effects.

A 2013 study by Wang et al. administered ursolic acid liposomes via a 4 h intravenous infusion in healthy adult volunteers and patients with advanced solid tumors. The maximum tolerated dose was found to be 98 mg/m^2 and the dose-limiting toxicity presented as diarrhea and hepatotoxicity. UA liposomes showed a linear pharmacokinetic profile [79].

Mice that were orally given one dose of 10 mg/kg of body weight UA had an elevated neutrophil count, serum glutamic oxaloacetic transaminase (SGOT), alkaline phosphatase (ALP) and urea levels and a decrease in white blood cells, lymphocytes and platelets. Some alterations were also seen in the architecture of the liver, kidney and spleen tissue that were recoverable, indicating that the toxic effects caused with oral administration are reversable [80]. An acute oral toxicity study in which mice were given a single dose of UA and then observed for 2 weeks for signs of toxicity or mortality showed the oral LD_{50} of UA was >2000 mg/kg in mice [80].

3. Discussion/Conclusions

Colorectal cancer is the fourth most diagnosed cancer worldwide [4]. It is the second leading cause of cancer-related deaths among men and third leading cause of cancer-related deaths among women. The incidence rate is continuing to rise, and new treatment options are needed to overcome treatment resistance commonly seen in CRCs.

Many medications, including current chemotherapy drugs such as paclitaxel and docetaxel [19,20], were originally derived from plants. Other plant-derived compounds may have strong anticancer effects and they need to be studied. In this paper, studies examining the pentacyclic triterpenoid ursolic acid and its effects against colorectal cancer were reviewed.

The in vitro studies all showed reduced proliferation and survival and increased levels of apoptosis of colorectal cancer cells treated with UA (Table 2 and Figure 2). Most of the research groups used concentrations of UA ranging from 20 to 80 μM. Only one study with low (2.5 μM) UA concentration found significant inhibitory effects [59].

Some studies examined the effects of ursolic acid when it was given in combination with compounds that are already known to inhibit CRC. Oxaliplatin, a platinum compound that inhibits DNA synthesis and is therefore cytotoxic, is currently used in the treatment of CRC but is associated with severe side effects. 5-fluorouracil is another currently used chemotherapy agent given for CRC; this drug works systemically by inhibiting thymidylate synthase thus blocking DNA replication. Doxorubicin is an antibiotic, originally derived from *Streptomyces peucetius* bacterium, that interacts with DNA by intercalation and inhibits macromolecule biosynthesis. All these currently accepted treatments act on DNA on a systemic level that causes many severe systemic side effects. Results of the articles summarized in this review (Table 3) suggest that the doses of some of these compounds may be reduced when combined with UA, thereby reducing the occurrence and severity of side effects.

The administration of UA in mice xenografted with human colorectal cancer cells showed decreased tumor growth and reduced tumor volumes, weights and size. Importantly, none of the animal studies reported reduced weight or premature death of the treated animals, suggesting that UA is safe and has limited side effects.

The evidence from the studies reviewed here indicates that UA is active against colorectal cancer cells. UA inhibits human colorectal cancer cell proliferation, survival and migration and induces apoptosis. In addition, the evidence indicates that UA may act as a chemosensitizer to enhance the effectiveness of currently approved drugs in the treatment of CRC. The limited evidence from in vivo animal studies indicates that UA administration reduces CRC tumor growth without causing side effects.

More studies should be performed to examine the anticancer potential of UA in animals before considering human clinical trials.

Author Contributions: A.K. formulated the review topic, performed searches and created figures A.K. and E.T. composed the entirety of the manuscript. All authors have read and agreed to the published version of the manuscript.

Funding: This review received no external funding.

Institutional Review Board Statement: Not applicable.

Informed Consent Statement: Not applicable.

Data Availability Statement: Not applicable.

Conflicts of Interest: The authors declare no conflicts of interest.

References

1. Dekker, E.; Tanis, P.J.; Vleugels, J.L.A.; Kasi, P.M.; Wallace, M.B. Colorectal Cancer. *Lancet* **2019**, *394*, 1467–1480. [CrossRef] [PubMed]
2. Siegel, R.L.; Wagle, N.S.; Cercek, A.; Smith, R.A.; Jemal, A. Colorectal Cancer Statistics, 2023. *CA A Cancer J. Clin.* **2023**, *73*, 233–254. [CrossRef] [PubMed]
3. Biller, L.H.; Schrag, D. Diagnosis and Treatment of Metastatic Colorectal Cancer: A Review. *JAMA* **2021**, *325*, 669–685. [CrossRef] [PubMed]
4. Bray, F.; Ferlay, J.; Soerjomataram, I.; Siegel, R.L.; Torre, L.A.; Jemal, A. Global Cancer Statistics 2018: GLOBOCAN Estimates of Incidence and Mortality Worldwide for 36 Cancers in 185 Countries. *CA A Cancer J. Clin.* **2018**, *68*, 394–424. [CrossRef] [PubMed]
5. Schubert, S.A.; Morreau, H.; de Miranda, N.F.C.C.; van Wezel, T. The Missing Heritability of Familial Colorectal Cancer. *Mutagenesis* **2020**, *35*, 221–231. [CrossRef]
6. Li, J.; Ma, X.; Chakravarti, D.; Shalapour, S.; DePinho, R.A. Genetic and Biological Hallmarks of Colorectal Cancer. *Genes Dev.* **2021**, *35*, 787–820. [CrossRef] [PubMed]
7. Dariya, B.; Aliya, S.; Merchant, N.; Alam, A.; Nagaraju, G.P. Colorectal Cancer Biology, Diagnosis, and Therapeutic Approaches. *Crit. Rev. Oncog.* **2020**, *25*, 71–94. [CrossRef] [PubMed]
8. Grady, W.M.; Markowitz, S.D. The Molecular Pathogenesis of Colorectal Cancer and Its Potential Application to Colorectal Cancer Screening. *Dig. Dis. Sci.* **2015**, *60*, 762–772. [CrossRef]
9. Kasi, A.; Handa, S.; Bhatti, S.; Umar, S.; Bansal, A.; Sun, W. Molecular Pathogenesis and Classification of Colorectal Carcinoma. *Curr. Color. Cancer Rep.* **2020**, *16*, 97–106. [CrossRef]
10. Pandey, H.; Tang, D.W.T.; Wong, S.H.; Lal, D. Gut Microbiota in Colorectal Cancer: Biological Role and Therapeutic Opportunities. *Cancers* **2023**, *15*, 866. [CrossRef]
11. Liu, Y.; Lau, H.C.-H.; Cheng, W.Y.; Yu, J. Gut Microbiome in Colorectal Cancer: Clinical Diagnosis and Treatment. *Genom. Proteom. Bioinform.* **2023**, *21*, 84–96. [CrossRef] [PubMed]
12. Rinninella, E.; Raoul, P.; Cintoni, M.; Franceschi, F.; Miggiano, G.A.D.; Gasbarrini, A.; Mele, M.C. What Is the Healthy Gut Microbiota Composition? A Changing Ecosystem across Age, Environment, Diet, and Diseases. *Microorganisms* **2019**, *7*, 14. [CrossRef] [PubMed]
13. Rothschild, D.; Weissbrod, O.; Barkan, E.; Kurilshikov, A.; Korem, T.; Zeevi, D.; Costea, P.I.; Godneva, A.; Kalka, I.N.; Bar, N.; et al. Environment Dominates over Host Genetics in Shaping Human Gut Microbiota. *Nature* **2018**, *555*, 210–215. [CrossRef] [PubMed]
14. Saoudi González, N.; Salvà, F.; Ros, J.; Baraibar, I.; Rodríguez-Castells, M.; García, A.; Alcaráz, A.; Vega, S.; Bueno, S.; Tabernero, J.; et al. Unravelling the Complexity of Colorectal Cancer: Heterogeneity, Clonal Evolution, and Clinical Implications. *Cancers* **2023**, *15*, 4020. [CrossRef]
15. Ashique, S.; Bhowmick, M.; Pal, R.; Khatoon, H.; Kumar, P.; Sharma, H.; Garg, A.; Kumar, S.; Das, U. Multi Drug Resistance in Colorectal Cancer- Approaches to Overcome, Advancements and Future Success. *Adv. Cancer Biol.-Metastasis* **2024**, *10*, 100114. [CrossRef]
16. Sellers, R.S.; Morton, D. The Colon: From Banal to Brilliant. *Toxicol. Pathol.* **2014**, *42*, 67–81. [CrossRef] [PubMed]
17. Fritz, C.D.L.; Otegbeye, E.E.; Zong, X.; Demb, J.; Nickel, K.B.; Olsen, M.A.; Mutch, M.; Davidson, N.O.; Gupta, S.; Cao, Y. Red-Flag Signs and Symptoms for Earlier Diagnosis of Early-Onset Colorectal Cancer. *JNCI J. Natl. Cancer Inst.* **2023**, *115*, 909–916. [CrossRef] [PubMed]
18. Briggs, N.L.; Ton, M.; Malen, R.C.; Reedy, A.M.; Cohen, S.A.; Phipps, A.I.; Burnett-Hartman, A.N.; Newcomb, P.A. Colorectal Cancer Pre-Diagnostic Symptoms Are Associated with Anatomic Cancer Site. *BMC Gastroenterol.* **2024**, *24*, 65. [CrossRef]
19. Bissery, M.C.; Nohynek, G.; Sanderink, G.J.; Lavelle, F. Docetaxel (Taxotere): A Review of Preclinical and Clinical Experience. Part I: Preclinical Experience. *Anticancer Drugs* **1995**, *6*, 339–355. [CrossRef]
20. Weaver, B.A. How Taxol/Paclitaxel Kills Cancer Cells. *Mol. Biol. Cell* **2014**, *25*, 2677–2681. [CrossRef]
21. Panda, S.S.; Thangaraju, M.; Lokeshwar, B.L. Ursolic Acid Analogs as Potential Therapeutics for Cancer. *Molecules* **2022**, *27*, 8981. [CrossRef] [PubMed]
22. Butkeviciute, A.; Viskelis, J.; Liaudanskas, M.; Viskelis, P.; Bobinas, C.; Janulis, V. Variation of Triterpenes in Apples Stored in a Controlled Atmosphere. *Molecules* **2021**, *26*, 3639. [CrossRef] [PubMed]
23. Kondo, M.; MacKinnon, S.L.; Craft, C.C.; Matchett, M.D.; Hurta, R.A.R.; Neto, C.C. Ursolic Acid and Its Esters: Occurrence in Cranberries and Other Vaccinium Fruit and Effects on Matrix Metalloproteinase Activity in DU145 Prostate Tumor Cells: Anti-Tumor Activity and Content of Ursolic Acid from Vaccinium Fruit. *J. Sci. Food Agric.* **2011**, *91*, 789–796. [CrossRef] [PubMed]
24. Allouche, Y.; Jiménez, A.; Uceda, M.; Aguilera, M.P.; Gaforio, J.J.; Beltrán, G. Triterpenic Content and Chemometric Analysis of Virgin Olive Oils from Forty Olive Cultivars. *J. Agric. Food Chem.* **2009**, *57*, 3604–3610. [CrossRef]
25. Guinda, Á.; Rada, M.; Delgado, T.; Gutiérrez-Adánez, P.; Castellano, J.M. Pentacyclic Triterpenoids from Olive Fruit and Leaf. *J. Agric. Food Chem.* **2010**, *58*, 9685–9691. [CrossRef] [PubMed]

26. Ivanov, I.; Petkova, N.; Tumbarski, Y.; Vrancheva, R.; Stoyanova, M. Lavender Waste-Promising Source of Triterpenoids and Polyphenols with Antioxidant and Antimicrobial Activity. *Ind. Technol.* **2018**, *5*, 26–32.
27. Silva, M.G.V.; Vieira, I.G.P.; Mendes, F.N.P.; Albuquerque, I.L.; dos Santos, R.N.; Silva, F.O.; Morais, S.M. Variation of Ursolic Acid Content in Eight Ocimum Species from Northeastern Brazil. *Molecules* **2008**, *13*, 2482–2487. [CrossRef] [PubMed]
28. He, X.; Liu, R.H. Triterpenoids Isolated from Apple Peels Have Potent Antiproliferative Activity and May Be Partially Responsible for Apple's Anticancer Activity. *J. Agric. Food Chem.* **2007**, *55*, 4366–4370. [CrossRef] [PubMed]
29. Li, X.; Wang, T.; Zhou, B.; Gao, W.; Cao, J.; Huang, L. Chemical Composition and Antioxidant and Anti-Inflammatory Potential of Peels and Flesh from 10 Different Pear Varieties (*Pyrus* Spp.). *Food Chem.* **2014**, *152*, 531–538. [CrossRef]
30. Sun, L.; Tao, S.; Zhang, S. Characterization and Quantification of Polyphenols and Triterpenoids in Thinned Young Fruits of Ten Pear Varieties by UPLC-Q TRAP-MS/MS. *Molecules* **2019**, *24*, 159. [CrossRef]
31. Jäger, S.; Trojan, H.; Kopp, T.; Laszczyk, M.; Scheffler, A. Pentacyclic Triterpene Distribution in Various Plants–Rich Sources for a New Group of Multi-Potent Plant Extracts. *Molecules* **2009**, *14*, 2016–2031. [CrossRef] [PubMed]
32. Zannou, O.; İPekci, B.; Koca, İ.; Odabaş, H.İ. Assessing Ursolic Acid Contents of Some Commonly Consumed Herbs Grown in Turkey. *Gümüşhane Üniversitesi Fen Bilim. Enstitüsü Derg.* **2022**, *12*, 301–308. [CrossRef]
33. Wójciak-Kosior, M.; Sowa, I.; Kocjan, R.; Nowak, R. Effect of Different Extraction Techniques on Quantification of Oleanolic and Ursolic Acid in Lamii Albi Flos. *Ind. Crops Prod.* **2013**, *44*, 373–377. [CrossRef]
34. Kowalski, R. Studies of Selected Plant Raw Materials as Alternative Sources of Triterpenes of Oleanolic and Ursolic Acid Types. *J. Agric. Food Chem.* **2007**, *55*, 656–662. [CrossRef] [PubMed]
35. Mlala, S.; Oyedeji, A.O.; Gondwe, M.; Oyedeji, O.O. Ursolic Acid and Its Derivatives as Bioactive Agents. *Molecules* **2019**, *24*, 2751. [CrossRef] [PubMed]
36. Woźniak, Ł.; Skąpska, S.; Marszałek, K. Ursolic Acid—A Pentacyclic Triterpenoid with a Wide Spectrum of Pharmacological Activities. *Molecules* **2015**, *20*, 20614–20641. [CrossRef] [PubMed]
37. Zhao, J.; Zheng, H.; Sui, Z.; Jing, F.; Quan, X.; Zhao, W.; Liu, G. Ursolic Acid Exhibits Anti-Inflammatory Effects through Blocking TLR4-MyD88 Pathway Mediated by Autophagy. *Cytokine* **2019**, *123*, 154726. [CrossRef] [PubMed]
38. Gudoityte, E.; Arandarcikaite, O.; Mazeikiene, I.; Bendokas, V.; Liobikas, J. Ursolic and Oleanolic Acids: Plant Metabolites with Neuroprotective Potential. *Int. J. Mol. Sci.* **2021**, *22*, 4599. [CrossRef]
39. Alam, M.; Ali, S.; Ahmed, S.; Elasbali, A.M.; Adnan, M.; Islam, A.; Hassan, M.I.; Yadav, D.K. Therapeutic Potential of Ursolic Acid in Cancer and Diabetic Neuropathy Diseases. *Int. J. Mol. Sci.* **2021**, *22*, 12162. [CrossRef]
40. Dai, Y.; Sun, L.; Tan, Y.; Xu, W.; Liu, S.; Zhou, J.; Hu, Y.; Lin, J.; Yao, X.; Mi, P.; et al. Recent Progress in the Development of Ursolic Acid Derivatives as Anti-Diabetes and Anti-Cardiovascular Agents. *Chem. Biol. Drug Des.* **2023**, *102*, 1643–1657. [CrossRef]
41. Chan, E.W.C.; Soon, C.Y.; Tan, J.B.L.; Wong, S.K.; Hui, Y.W. Ursolic Acid: An Overview on Its Cytotoxic Activities against Breast and Colorectal Cancer Cells. *J. Integr. Med.* **2019**, *17*, 155–160. [CrossRef]
42. Kornel, A.; Nadile, M.; Tsiani, E. Evidence of the Beneficial Effects of Ursolic Acid against Lung Cancer. *Molecules* **2022**, *27*, 7466. [CrossRef] [PubMed]
43. Kornel, A.; Nadile, M.; Retsidou, M.I.; Sakellakis, M.; Gioti, K.; Beloukas, A.; Sze, N.S.K.; Klentrou, P.; Tsiani, E. Ursolic Acid against Prostate and Urogenital Cancers: A Review of In Vitro and In Vivo Studies. *Int. J. Mol. Sci.* **2023**, *24*, 7414. [CrossRef] [PubMed]
44. Yin, R.; Li, T.; Tian, J.X.; Xi, P.; Liu, R.H. Ursolic Acid, a Potential Anticancer Compound for Breast Cancer Therapy. *Crit. Rev. Food Sci. Nutr.* **2018**, *58*, 568–574. [CrossRef] [PubMed]
45. Li, J.; Guo, W.-J.; Yang, Q.-Y. Effects of Ursolic Acid and Oleanolic Acid on Human Colon Carcinoma Cell Line HCT15. *World J. Gastroenterol.* **2002**, *8*, 493–495. [CrossRef] [PubMed]
46. Shan, J.; Xuan, Y.; Zheng, S.; Dong, Q.; Zhang, S. Ursolic Acid Inhibits Proliferation and Induces Apoptosis of HT-29 Colon Cancer Cells by Inhibiting the EGFR/MAPK Pathway. *J. Zhejiang Univ. Sci. B* **2009**, *10*, 668–674. [CrossRef]
47. Xavier, C.P.R.; Lima, C.F.; Preto, A.; Seruca, R.; Fernandes-Ferreira, M.; Pereira-Wilson, C. Luteolin, Quercetin and Ursolic Acid Are Potent Inhibitors of Proliferation and Inducers of Apoptosis in Both KRAS and BRAF Mutated Human Colorectal Cancer Cells. *Cancer Lett.* **2009**, *281*, 162–170. [CrossRef] [PubMed]
48. Limami, Y.; Pinon, A.; Leger, D.Y.; Mousseau, Y.; Cook-Moreau, J.; Beneytout, J.-L.; Delage, C.; Liagre, B.; Simon, A. HT-29 Colorectal Cancer Cells Undergoing Apoptosis Overexpress COX-2 to Delay Ursolic Acid-Induced Cell Death. *Biochimie* **2011**, *93*, 749–757. [CrossRef]
49. Limami, Y.; Pinon, A.; Leger, D.Y.; Pinault, E.; Delage, C.; Beneytout, J.-L.; Simon, A.; Liagre, B. The P2Y2/Src/P38/COX-2 Pathway Is Involved in the Resistance to Ursolic Acid-Induced Apoptosis in Colorectal and Prostate Cancer Cells. *Biochimie* **2012**, *94*, 1754–1763. [CrossRef]
50. Rawat, L.; Nayak, V. Ursolic Acid Disturbs ROS Homeostasis and Regulates Survival-Associated Gene Expression to Induce Apoptosis in Intestinal Cancer Cells. *Toxicol. Res.* **2021**, *10*, 369–375. [CrossRef]
51. Lin, J.; Chen, Y.; Wei, L.; Hong, Z.; Sferra, T.J.; Peng, J. Ursolic Acid Inhibits Colorectal Cancer Angiogenesis through Suppression of Multiple Signaling Pathways. *Int. J. Oncol.* **2013**, *43*, 1666–1674. [CrossRef] [PubMed]
52. Lin, J.; Chen, Y.; Wei, L.; Shen, A.; Sferra, T.J.; Hong, Z.; Peng, J. Ursolic Acid Promotes Colorectal Cancer Cell Apoptosis and Inhibits Cell Proliferation via Modulation of Multiple Signaling Pathways. *Int. J. Oncol.* **2013**, *43*, 1235–1243. [CrossRef] [PubMed]

53. Wang, W.; Zhao, C.; Jou, D.; Lü, J.; Zhang, C.; Lin, L.; Lin, J. Ursolic Acid Inhibits the Growth of Colon Cancer-Initiating Cells by Targeting STAT3. *Anticancer Res.* **2013**, *33*, 4279–4284.
54. Wang, J.; Liu, L.; Qiu, H.; Zhang, X.; Guo, W.; Chen, W.; Tian, Y.; Fu, L.; Shi, D.; Cheng, J.; et al. Ursolic Acid Simultaneously Targets Multiple Signaling Pathways to Suppress Proliferation and Induce Apoptosis in Colon Cancer Cells. *PLoS ONE* **2013**, *8*, e63872. [CrossRef] [PubMed]
55. Kim, K.; Shin, E.A.; Jung, J.H.; Park, J.E.; Kim, D.S.; Shim, B.S.; Kim, S.-H. Ursolic Acid Induces Apoptosis in Colorectal Cancer Cells Partially via Upregulation of MicroRNA-4500 and Inhibition of JAK2/STAT3 Phosphorylation. *Int. J. Mol. Sci.* **2018**, *20*, 114. [CrossRef] [PubMed]
56. Zhang, L.; Cai, Q.-Y.; Liu, J.; Peng, J.; Chen, Y.-Q.; Sferra, T.J.; Lin, J.-M. Ursolic Acid Suppresses the Invasive Potential of Colorectal Cancer Cells by Regulating the TGF-B1/ZEB1/miR-200c Signaling Pathway. *Oncol. Lett.* **2019**, *18*, 3274–3282. [CrossRef] [PubMed]
57. Wang, X.; Wang, T.; Yi, F.; Duan, C.; Wang, Q.; He, N.; Zhu, L.; Li, Q.; Deng, W. Ursolic Acid Inhibits Tumor Growth via Epithelial-to-Mesenchymal Transition in Colorectal Cancer Cells. *Biol. Pharm. Bull.* **2019**, *42*, 685–691. [CrossRef]
58. Yang, M.; Hu, C.; Cao, Y.; Liang, W.; Yang, X.; Xiao, T. Ursolic Acid Regulates Cell Cycle and Proliferation in Colon Adenocarcinoma by Suppressing Cyclin B1. *Front. Pharmacol.* **2020**, *11*, 622212. [CrossRef] [PubMed]
59. Shen, A.; Wu, M.; Liu, L.; Chen, Y.; Chen, X.; Zhuang, M.; Xie, Q.; Cheng, Y.; Li, J.; Shen, Z.; et al. Targeting NUFIP1 Suppresses Growth and Induces Senescence of Colorectal Cancer Cells. *Front. Oncol.* **2021**, *11*, 681425. [CrossRef]
60. Zheng, J.-L.; Wang, S.-S.; Shen, K.-P.; Chen, L.; Peng, X.; Chen, J.-F.; An, H.-M.; Hu, B. Ursolic Acid Induces Apoptosis and Anoikis in Colorectal Carcinoma RKO Cells. *BMC Complement. Med. Ther.* **2021**, *21*, 52. [CrossRef]
61. Chen, L.; Liu, M.; Yang, H.; Ren, S.; Sun, Q.; Zhao, H.; Ming, T.; Tang, S.; Tao, Q.; Zeng, S.; et al. Ursolic Acid Inhibits the Activation of Smoothened-Independent Non-Canonical Hedgehog Pathway in Colorectal Cancer by Suppressing AKT Signaling Cascade. *Phytother. Res.* **2022**, *36*, 3555–3570. [CrossRef]
62. Zhao, H.; Tang, S.; Tao, Q.; Ming, T.; Lei, J.; Liang, Y.; Peng, Y.; Wang, M.; Liu, M.; Yang, H.; et al. Ursolic Acid Suppresses Colorectal Cancer by Down-Regulation of Wnt/β-Catenin Signaling Pathway Activity. *J. Agric. Food Chem.* **2023**, *71*, 3981–3993. [CrossRef]
63. Zhang, M.; Xiang, F.; Sun, Y.; Liu, R.; Li, Q.; Gu, Q.; Kang, X.; Wu, R. Ursolic Acid Inhibits the Metastasis of Colon Cancer by Downregulating ARL4C Expression. *Oncol. Rep.* **2024**, *51*, 27. [CrossRef] [PubMed]
64. Zhao, J.; Leng, P.; Xu, W.; Sun, J.-L.; Ni, B.-B.; Liu, G.-W. Investigating the Multitarget Pharmacological Mechanism of Ursolic Acid Acting on Colon Cancer: A Network Pharmacology Approach. *Evid. Based Complement. Altern. Med.* **2021**, *2021*, 9980949. [CrossRef]
65. Wang, J.; Guo, W.; Chen, W.; Yu, W.; Tian, Y.; Fu, L.; Shi, D.; Tong, B.; Xiao, X.; Huang, W.; et al. Melatonin Potentiates the Antiproliferative and Pro-Apoptotic Effects of Ursolic Acid in Colon Cancer Cells by Modulating Multiple Signaling Pathways. *J. Pineal. Res.* **2013**, *54*, 406–416. [CrossRef] [PubMed]
66. Xavier, C.P.R.; Lima, C.F.; Pedro, D.F.N.; Wilson, J.M.; Kristiansen, K.; Pereira-Wilson, C. Ursolic Acid Induces Cell Death and Modulates Autophagy through JNK Pathway in Apoptosis-Resistant Colorectal Cancer Cells. *J. Nutr. Biochem.* **2013**, *24*, 706–712. [CrossRef]
67. Shan, J.; Xuan, Y.; Zhang, Q.; Zhu, C.; Liu, Z.; Zhang, S. Ursolic Acid Synergistically Enhances the Therapeutic Effects of Oxaliplatin in Colorectal Cancer. *Protein. Cell* **2016**, *7*, 571–585. [CrossRef] [PubMed]
68. Shan, J.-Z.; Xuan, Y.-Y.; Zhang, Q.; Huang, J.-J. Ursolic Acid Sensitized Colon Cancer Cells to Chemotherapy under Hypoxia by Inhibiting MDR1 through HIF-1α. *J. Zhejiang Univ. Sci. B* **2016**, *17*, 672–682. [CrossRef]
69. Zhang, Y.; Huang, L.; Shi, H.; Chen, H.; Tao, J.; Shen, R.; Wang, T. Ursolic Acid Enhances the Therapeutic Effects of Oxaliplatin in Colorectal Cancer by Inhibition of Drug Resistance. *Cancer Sci.* **2018**, *109*, 94–102. [CrossRef]
70. Zheng, J.-L.; Wang, S.-S.; Shen, K.-P.; Huang, X.-W.; Li, M.; Chen, L.; Peng, X.; An, H.-M.; Hu, B. Ursolic Acid Potentiated Oxaliplatin to Induce Apoptosis in Colorectal Cancer RKO Cells. *Pharmazie* **2020**, *75*, 246–249.
71. Piet, M.; Paduch, R. Ursolic and Oleanolic Acids in Combination Therapy Inhibit Migration of Colon Cancer Cells through Down-Regulation of the uPA/uPAR-Dependent MMPs Pathway. *Chem. Biol. Interact.* **2022**, *368*, 110202. [CrossRef]
72. Hu, D.; Meng, R.Y.; Nguyen, T.V.; Chai, O.H.; Park, B.H.; Lee, J.-S.; Kim, S.M. Inhibition of Colorectal Cancer Tumorigenesis by Ursolic Acid and Doxorubicin Is Mediated by Targeting the Akt Signaling Pathway and Activating the Hippo Signaling Pathway. *Mol. Med. Rep.* **2023**, *27*, 11. [CrossRef] [PubMed]
73. Li, H.; Yu, Y.; Liu, Y.; Luo, Z.; Law, B.Y.K.; Zheng, Y.; Huang, X.; Li, W. Ursolic Acid Enhances the Antitumor Effects of Sorafenib Associated with Mcl-1-Related Apoptosis and SLC7A11-Dependent Ferroptosis in Human Cancer. *Pharmacol. Res.* **2022**, *182*, 106306. [CrossRef] [PubMed]
74. Koh, S.J.; Tak, J.K.; Kim, S.T.; Nam, W.S.; Kim, S.Y.; Park, K.M.; Park, J.-W. Sensitization of Ionizing Radiation-Induced Apoptosis by Ursolic Acid. *Free Radic. Res.* **2012**, *46*, 339–345. [CrossRef]
75. Prasad, S.; Yadav, V.R.; Sung, B.; Reuter, S.; Kannappan, R.; Deorukhkar, A.; Diagaradjane, P.; Wei, C.; Baladandayuthapani, V.; Krishnan, S.; et al. Ursolic Acid Inhibits Growth and Metastasis of Human Colorectal Cancer in an Orthotopic Nude Mouse Model by Targeting Multiple Cell Signaling Pathways: Chemosensitization with Capecitabine. *Clin. Cancer Res.* **2012**, *18*, 4942–4953. [CrossRef]

76. Yang, L.; Sun, Z.; Zu, Y.; Zhao, C.; Sun, X.; Zhang, Z.; Zhang, L. Physicochemical Properties and Oral Bioavailability of Ursolic Acid Nanoparticles Using Supercritical Anti-Solvent (SAS) Process. *Food Chem.* **2012**, *132*, 319–325. [CrossRef]
77. Yin, M.-C.; Lin, M.-C.; Mong, M.-C.; Lin, C.-Y. Bioavailability, Distribution, and Antioxidative Effects of Selected Triterpenes in Mice. *J. Agric. Food Chem.* **2012**, *60*, 7697–7701. [CrossRef] [PubMed]
78. Xia, Y.; Wei, G.; Si, D.; Liu, C. Quantitation of Ursolic Acid in Human Plasma by Ultra Performance Liquid Chromatography Tandem Mass Spectrometry and Its Pharmacokinetic Study. *J. Chromatogr. B* **2011**, *879*, 219–224. [CrossRef]
79. Wang, X.-H.; Zhou, S.-Y.; Qian, Z.-Z.; Zhang, H.-L.; Qiu, L.-H.; Song, Z.; Zhao, J.; Wang, P.; Hao, X.-S.; Wang, H.-Q. Evaluation of Toxicity and Single-Dose Pharmacokinetics of Intravenous Ursolic Acid Liposomes in Healthy Adult Volunteers and Patients with Advanced Solid Tumors. *Expert Opin. Drug Metab. Toxicol.* **2013**, *9*, 117–125. [CrossRef]
80. Mishra, V.; Soren, A.D.; Yadav, A.K. Toxicological Evaluations of Betulinic Acid and Ursolic Acid; Common Constituents of Houttuynia Cordata Used as an Anthelmintic by the Naga Tribes in North-East India. *Future J. Pharm. Sci.* **2021**, *7*, 39. [CrossRef]

nutraceuticals

Article

Protection against Microglia Senescence by the Dietary Supplement Dekosilhue® in BV2 Cells: A New Perspective for Obesity and Related Complications

Vittoria Borgonetti [1], Chiara Sasia [1], Martina Morozzi [1], Lorenzo Cenci [2] and Nicoletta Galeotti [1,*]

1 Department of Neuroscience, Psychology, Drug Research and Child Health (NEUROFARBA), Section of Pharmacology and Toxicology, University of Florence, Viale G. Pieraccini 6, 50139 Florence, Italy; vittoria.borgonetti@unifi.it (V.B.); chiara.sasia@unifi.it (C.S.); martina.morozzi@unifi.it (M.M.)
2 Department of Biomedical Sciences, University of Padua, Via Marzolo, 3, 35131 Padua, Italy; lorenzo.cenci@studenti.unipd.it
* Correspondence: nicoletta.galeotti@unifi.it

Abstract: Growing evidence indicates chronic low-grade systemic inflammation as a major pathophysiological mechanism of obesity. Systemic inflammation provokes an immune response in the brain through the activation of microglia that results in the development of neuroinflammation, cellular senescence, and occurrence of neurological dysfunction. In the efforts to identify an innovative intervention with potential efficacy on obesity and associated complications, our aim was to study the capability of the dietary supplement Dekosilhue® (DKS), successfully used for improving the control of body weight, to attenuate microglia senescent phenotype. Microglia senescence was induced by intermittent stimulation of BV2 cells with LPS 500 ng/mL every 72 h for 4 h/day, over a period of 10 days. DKS (100 µg/mL) treatment reduced ß-galactosidase activity and expression, the formation of senescence-associated heterochromatin foci to control levels, and increased cell viability of senescent BV2 (2 folds of control). DSK reduced the expression of Nuclear Factor-kB (NF-kB) (20% lower than control), a key molecule involved in the acquisition of the senescence-associated secretory phenotype (SASP). DKS promoted a neuroprotective effect by increasing the cell viability of SH-SY5Y neuronal cells exposed to the senescent BV2 conditioned medium to values of non-senescent cells. In conclusion, DKS attenuated the senescent microglia phenotype, showing senotherapeutic activity that might be further investigated as adjunctive intervention for obesity and obesity-related neurological disturbances.

Keywords: microglia; cellular senescence; inflammation; neuroinflammation; phytocomplex

Citation: Borgonetti, V.; Sasia, C.; Morozzi, M.; Cenci, L.; Galeotti, N. Protection against Microglia Senescence by the Dietary Supplement Dekosilhue® in BV2 Cells: A New Perspective for Obesity and Related Complications. *Nutraceuticals* **2023**, *3*, 250–261. https://doi.org/10.3390/nutraceuticals3020020

Academic Editors: Ivan Cruz-Chamorro and Guillermo Santos Sánchez

Received: 20 March 2023
Revised: 17 May 2023
Accepted: 19 May 2023
Published: 23 May 2023

1. Introduction

Obesity affects nearly one-third of the entire population; its incidence continues to rise [1] and is not effectively controlled by current interventions [2]. Obesity, rather than being simply the result of an energy imbalance between calorie intake and expenditure, is a complex, chronic multifactorial disorder associated with a wide range of metabolic and neurological disorders. Growing evidence indicates chronic low-grade systemic inflammation as a major pathophysiological mechanism underlying obesity [3]. Obesity causes a phenotypic modification of adipocytes which become inflamed and secrete proinflammatory cytokines, leading to an elevation of the peripheral inflammatory response [4,5]. In spite of its low-grade nature, inflammation of the adipose tissue is mechanistically linked to metabolic disease and remote organ complications of obese individuals [6,7]. Since the peripheral and central innate immune systems are in constant communication [8], systemic inflammation provokes an immune response in the brain [9], leading obesity to promote the development of neuroinflammation [10,11] and the occurrence of neurological disturbances [12].

On a cellular level, a main trait of neuroinflammation is the activation of the resident immune cells in the brain, the microglia, whose primary function is to maintain the CNS homeostasis. Upon stimulation, these cells rapidly become activated and undergo morphological and molecular changes [13]. Microglia alter their morphology and functional activity in response to the increased pro-inflammatory response from the circulation, creating a less neurotrophic tissue environment that may contribute to neurodegeneration [14]. In an unresolved chronic inflammatory situation, microglia may not cope with this continuous stimulus and lose their homeostatic activity, becoming senescent [15]. The senescent microglia are characterized by morphological changes, impaired phagocytic activity, and huge release of inflammatory mediators [16].

Senescent cell abundance increases with obesity in peripheral tissues (adipose, hepatic, pancreatic) and the brain in experimental animals and obese humans [17], and recent studies have shown that eliminating senescent cells can alleviate obesity-induced metabolic dysfunction in several tissues [18,19]. Interestingly, cellular senescence of CNS cells is taking on an increasingly interesting role in the study of new therapeutic strategies in neurodegenerative diseases [20,21]. Thus, targeting senescent cells with natural or synthetic agents is a promising new therapeutic approach for chronic diseases [18–22]. Studies suggest that, for treating complex inflammatory disorders with multifactorial natures, a balanced modulation of more than one target can be a more efficient strategy than the single target modulation [23]. In this frame, plant species often serve as good sources for identifying these kinds of multitarget intervention. Bioactive principles present in the herbal remedy may act synergistically among them and also modulate the activity of other bioactive constituents from other plant sources [24]. In the efforts devoted to identifying innovative interventions with potential efficacy on obesity and associated complications, the aim of this study was to investigate the effectiveness of nutraceutical interventions. Dekosilhue® (DKS) is a phytocomplex composed of an association of thirteen plant extracts, used as dietary supplement to promote carbohydrate and lipid metabolism, that has recently been reported to possess anti-neuroinflammatory activity [25]. On these bases, we examine the effectiveness of DKS in attenuating the microglia senescent phenotype in an in vitro model of microglial senescence.

2. Materials and Methods

2.1. Reagents

The commercially available food supplement Dekosilhue® (DKS) (expiration date 30 September 2025; kindly provided by Gianluca Mech S.p.A., Vicenza, Italy) was used in the study. The list of constituents is reported in Table 1.

Table 1. Composition of Dekosilhue® (DKS). The content of each constituent (aqueous extract) is expressed as grams of extract in 1 L of product (g/L).

DKS Constituent	g/L
Cinnamon (*Cinnamomum zeylanicum* Blume) bark	90
Orthosiphon (*Orthosiphon stamineus* Benth) leaf	90
Green tea (*Camellia sinensis* (L.) Kuntze) leaf	90
Mate (*Ilex paraguariensis* A.St.Hill) leaf	70
Gymnema (*Gymnema sylvestre* R. Br.) leaf	70
Bean (*Phaseolus vulgaris* L.) pod	60
Pineapple (*Ananas comosus* (L.) Merr.) stem	40
Common gromwell (*Lithospermum officinale* L.) seeds	40
Horsetail (*Equisetum arvense* L.) herb	40
Curly dock (*Rumex crispus* L.) root	30
Asparagus (*Asparagus officinalis* L.) root	30
Fennel (*Foeniculum vulgare* Miller) fruit	30
Birch (*Betuta pendolo* Roth.) leaf	20

DKS was obtained using a decoction process. Briefly, the raw material (root, leaf, cortex, fruit, aerial part, stem, according to each plant) is cooked several times in water at temperatures between 97 and 99 °C. For the in vitro tests, DKS was diluted in distilled water and then diluted in media at a concentration of 100 μg/mL, as previously optimized [25]. Bacterial lipopolysaccharide (LPS) from Gram-(*Salmonella enteridis*) was purchased from Sigma-Aldrich (Milan, Italy). All chemicals and solvents used in the present study were of analytical grade or with purity above 95%.

2.2. Cell Cultures

BV2 (immortalized murine microglial cells, C57BL/6 Tema Ricerca, Genova, Italy) were kept in culture in a 75 cm^2 flask (Sarstedt, Verona, Italy) in medium containing RPMI (Sigma-Aldrich, Milan, Italy) with 10% heat-inactivated fetal bovine serum (56 °C, 30 min) (FBS, Gibco, Milan, Italy), 1% glutamine, and 1% penicillin-streptomycin solution (Merck, Milan, Italy).

A human neuroblastoma cell line (SH-SY5Y), kindly donated by Prof. Lorenzo Corsi (University of Modena and Reggio Emilia, Italy), was cultured in DMEM (Sigma-Aldrich) and F12 Ham's nutrients mixture (Sigma-Aldrich), containing 10% heat-inactivated FBS (Sigma-Aldrich, Milan, Italy), 1% L-glutamine (Sigma-Aldrich, Milan, Italy), and 1% penicillin-streptomycin solution (Sigma-Aldrich, Milan, Italy) until confluence (70–80%). The cells were grown in a humidified atmosphere with 5% CO_2 at 37 °C. EDTA-trypsin solution (Sigma-Aldrich, Milan, Italy) was used for detaching the cells from flasks, and cell counting was performed using a hemocytometer via Trypan blue staining.

2.3. Senescent Microglia Model

The senescent model was performed as previously reported [26]. BV2 cells were treated with the LPS (Merck, Darmstadt, Germany) 4 times, for 4 h/day for a total of 10 days, at a concentration of 500 ng/mL in minimal medium (RPMI with 3% FBS), as reported in Figure 1. DKS (100 μg/mL) was added for 24 h after the last LPS stimulation.

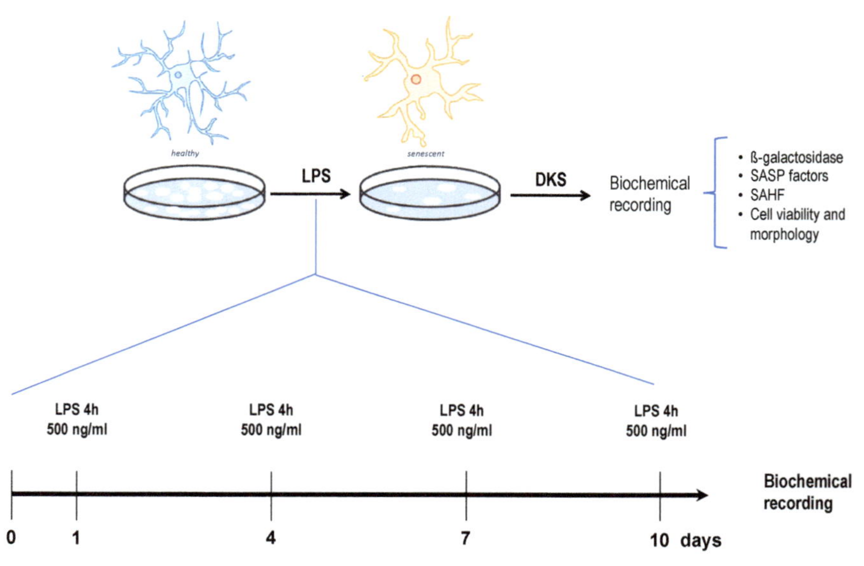

Figure 1. Schematic representation of the senescent microglia model, administration, and tests schedule.

2.4. Sulforhodamine B (SRB) Assay

The SRB test was performed to assess cell viability [27]. Briefly, cells were seeded in 96-well plates (2.0 × 10^4 cells per well) and fixed in 50% trichloroacetic acid (TCA, Merck, Darmstadt, Germany) at 4 °C for 1 h. Then, they were treated with SRB 0.4% in acetic acid

1% for 30 min at rt. Finally, Tris-HCl pH = 10 was used and absorbance at 570 nm was recorded using a multiplate reader (Biorad, Milan, Italy). Three independent experiments ($n = 3$) were carried out to evaluate the effect of each treatment. Cell viability values were normalized to the mean of the control.

2.5. Senescence-Associated Heterochromatin Foci Analysis (SAHF)

Immortalized murine microglial cells (BV2) were seeded in 24-well plates (1.0×10^5 cells per well) containing previously sterilized slides at the bottom of the wells. Following treatments, cells were fixed with 4% PFA for 30 min at 4 °C. After 3 washes with PBS, the slides were treated with a solution containing DAPI in mounting medium (90% glycerol + PBS) and images were taken (OLYMPUS BX63F fluorescence microscope connected to a PC). Three independent experiments ($n = 3$) were carried out to evaluate the effects of treatments. The DAPI intensity values were normalized to the mean of the control [26].

2.6. ß-Galactosidase Activity Assay

In a 96-well plate, 25 µL of fresh cell lysate, 80 µL of solution containing 76 µL of "buffer A" (NaH_2PO_4 100 mM, KCl 10 mM, $MgSO_4$ 1 mM) and 4 µL of ß-mercaptoethanol were added to each well. The plate was left in an oven at 37 °C for 5 min. Then, 25 µL of the chromogenic substrate, o-nitrophenyl ß-D-galactopyranoside (ONPG) 4 mg/mL in NaH_2PO_4 buffer (pH = 7.5), was added for each well, and the plate was placed back in the stove at 37 °C for 2 h until a light-yellow color was obtained. After the required time had elapsed, 45 µL of stop solution (Na_2CO_3 1 M) was added to each well and a spectrophotometer reading was taken at 405–450 nm.

2.7. Neuroprotection Model

On day 10 after LPS intermittent stimulation, DKS (100 µg/mL) was added to senescent BV2 cells. The conditioned medium was collected and centrifuged ($1000\times$ *g* for 10 min, 37 °C). The pellet was discarded, and the supernatant was stored at −80 °C. The cell viability of the SH-SY5Y cells treated with the conditioned medium from untreated or DKS-treated senescent BV2 for 48 h was assessed to evaluate the neuroprotective activity of DKS. Unstimulated BV2 medium was used as a control [28].

2.8. Western Blot (WB)

Protein samples (30 µg of protein/lane) were separated by SDS_PAGE on 10% minigels and thereafter transferred to nitrocellulose membranes for 90 min at 110 V. Membranes were blocked for 90 min in PBST (PBS with 0.1% Tween) containing 5% non-fat dry milk. After blocking membranes were incubated overnight at 4 °C with primary antibodies: anti-ß-galactosidase (Santa Cruz Biotechnology, Santa Cruz, CA, USA, Cat# sc-65670, RRID: AB_831022IBA1), anti-p-NFκB p65 (Santa Cruz Biotechnology, Santa Cruz, CA, USA, Cat# sc-136548). Blots were then rinsed three times with PBST and incubated at rt with HRP-conjugated secondary antibodies for 2 h, and then detected via Colorimetric detection Kit (Opti-4CN^{TM} Substrate Kit, BIO-RAD). Signal intensity (pixels/mm^2) was quantified using ImageJ (NIH). For each sample, the signal intensity was normalized to that of total protein stained by Ponceau S. The treatments were carried out in three independent experiments ($n = 3$), and protein expression was calculated by normalizing the values to the mean of the control.

2.9. Statistical Analysis

The data are presented as the mean ± SEM. One-way or two-way ANOVA, followed by Tukey or Bonferroni post hoc tests, was performed. $p < 0.05$ was considered statistically significant. The software GraphPad Prism (version 9.5, San Diego, CA, USA) was used.

3. Results

3.1. Effects of DKS on Microglial Senescent Cells

Repeated intermittent exposition of BV2 cells to low doses of the inflammatory agent LPS (LPS 500 ng/mL every 72 h for 4 h/day) promoted the development of a senescent phenotype. Cell viability was progressively reduced, starting from 2 days of treatment up to 10 days of intermittent stimulation, when a highly significant reduction of cell viability was observed (Figure 2A). Lower doses of LPS were ineffective. DKS (1–100 µg/mL) did not alter cell viability of unstimulated BV2 cells (Figure 2B). Thus, the experimental condition of 10 days of intermittent LPS treatment was chosen to conduct the study. For treatment effect experimentations, BV2 cells were treated overnight with DKS 100 µg/mL on day 10, after the last inflammatory stimulus, and the effects on main senescence parameters (i.e., ß-galactosidase, SAHF, cell viability, SASP) were evaluated.

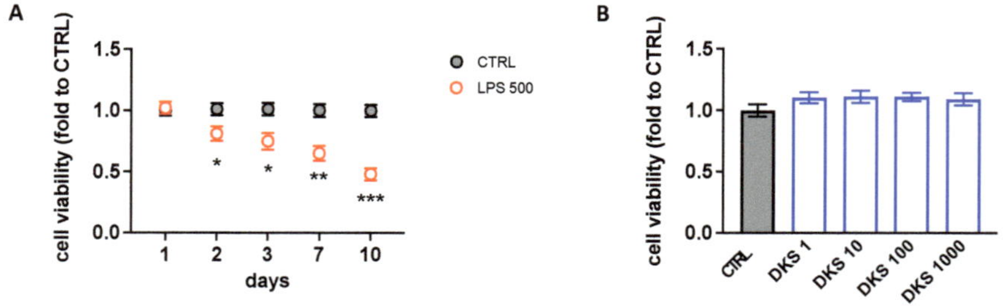

Figure 2. LPS and DKS effects on cell viability by the SRB test. (**A**) Time-course evaluation of cell viability in BV2 cells intermittently exposed to LPS (500 ng/mL). Two-way ANOVA * $p < 0.05$, ** $p < 0.01$, *** $p < 0.001$ in comparison with unstimulated control (CTRL) group. (**B**) Dose–response effect on cell viability of unstimulated BV2 cells by DSK (1, 10, 100, and 1000 µg/mL). Vertical lines represent SEM.

3.2. Effect on ß-Galactosidase Activity and Expression

ß-galactosidase (ß-gal) represents one of the main markers of cellular senescence [29]. In Figure 3A, we can observe a significant increase in ß-gal activity compared with control cells, which were not stimulated with LPS. DKS significantly reduced this activity bringing the values closer to those of the control (CT) group. This result was then confirmed by evaluating the expression of the enzyme by Western blot (WB) analysis. In Figure 3B, we can observe the increase of ß-gal protein expression induced by LPS and the reversal of this up-regulation by DKS, obtaining CTRL-like levels.

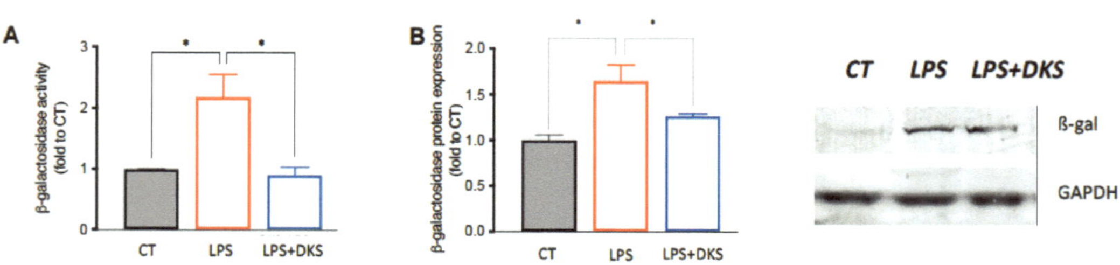

Figure 3. Effect of DKS on ß-galactosidase. (**A**) ß-galactosidase (ß-gal) activity is increased on fresh cell lysate. DKS treatment counteracted this activity. (**B**) ß-galactosidase (ß-gal) protein expression and representative blots. One-way ANOVA * $p < 0.05$.

3.3. Reduction of SAHF Formation

Senescent cells tend to form clusters of heterochromatin, resulting in a reduction in the expression of genes involved in cell replication processes, leading to cell death [30]. In senescent cells, as can be seen from the results shown in Figure 4A,B, a lower intensity in the staining of the nuclear marker DAPI and an increase in the expression of heterochromatin foci (Figure 4C) was observed compared with control cells. DKS treatment increased the intensity of DAPI staining and significantly reduced the number of heterochromatin foci compared with senescent cells.

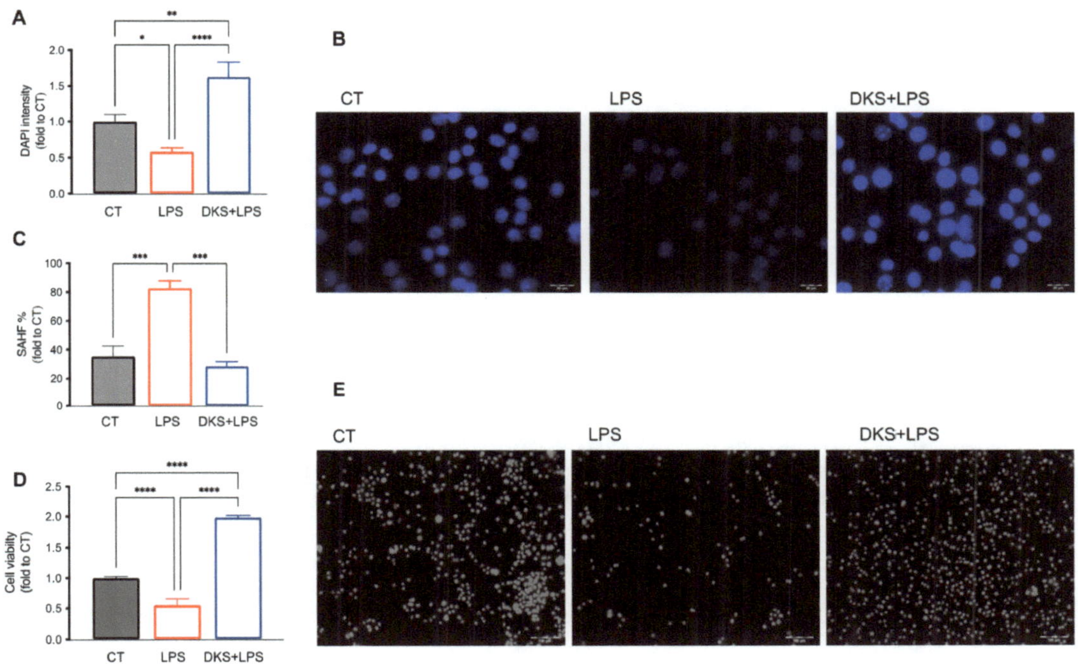

Figure 4. Effect of DKS on senescent-associated heterochromatin foci (SAHF). DAPI staining of nuclei (quantification analysis (**A**); representative image of DAPI intensity (**B**)) and SAHF formation (**C**) in BV2 senescent cells at 10 days after intermittent stimulus with LPS (500 ng/mL). DSK (100 µg/mL) reverted both LPS-induced senescence markers. (**D**) Decrease of cell viability by LPS intermittent stimulus and reversal by DSK (100 µg/mL). (**E**) Representative image of cell viability. One-way ANOVA * $p < 0.05$, ** $p < 0.01$, *** $p < 0.001$, **** $p < 0.0001$.

From Figure 4D,E, we can observe that DSK is able to significantly increase cell viability compared with senescent cells treated with LPS, suggesting a reduction in the apoptotic mechanism and an increase in cell replication.

3.4. DKS Reduced SASP Inflammatory Markers

The senescent cells take on new characteristics, including an increase in the production and secretion of cytokines, growth factors, and proteolytic enzymes, leading to the acquisition of a proinflammatory phenotype defined as "secretory phenotype associated with senescence" (SASP) [31,32]. The production of these factors appears to be controlled by complex molecular mechanisms involving the transcription factor NF-kB [33], which also represents the main driver of pro-inflammatory microglia activation. LPS-senescent microglia cells had higher levels of phosphorylated NF-kB p65 subunit protein expression, and DKS treatment counteracted this increase to values comparable to that observed for the CT group (Figure 5). These results suggested a reduction of microglial activity in the inflammatory process associated with the development of the senescent phenotype.

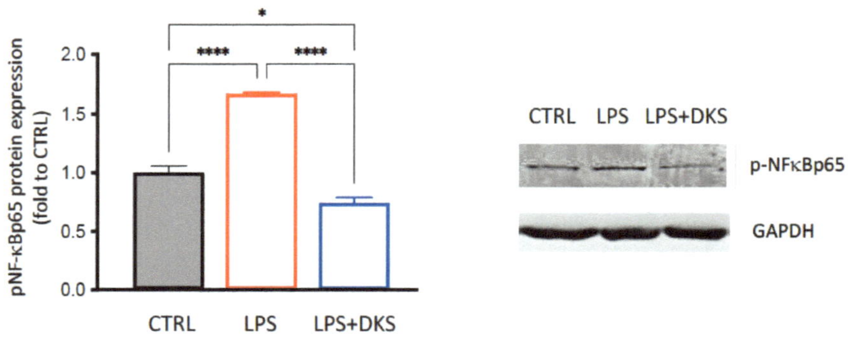

Figure 5. Effect of DKS on NF-kB pathway activation. Reversal by DKS (100 μg/mL) of increased *p*NF-kBp65 protein expression induced by LPS (500 ng/mL) intermittent stimulation. Representative blots are reported. One-way ANOVA * $p < 0.05$, **** $p < 0.0001$.

3.5. Neuroprotective Activity of DKS

SH-SY5Y cells were stimulated with medium from senescent BV2 cells, taken on day 10 from the beginning of the LPS intermittent stimulation (Figure 6A). After 48 h stimulation, neuronal cells showed a reduction of cell viability, evaluated by SRB test, compared with neurons treated with non-senescent BV2 medium (Figure 6B). SH-SY5Y cells stimulated with DKS-treated senescent BV2 medium showed a higher percentage of living cells, highlighting a neuroprotective effect of treatment on neurotoxicity induced by microglia aging (Figure 6B).

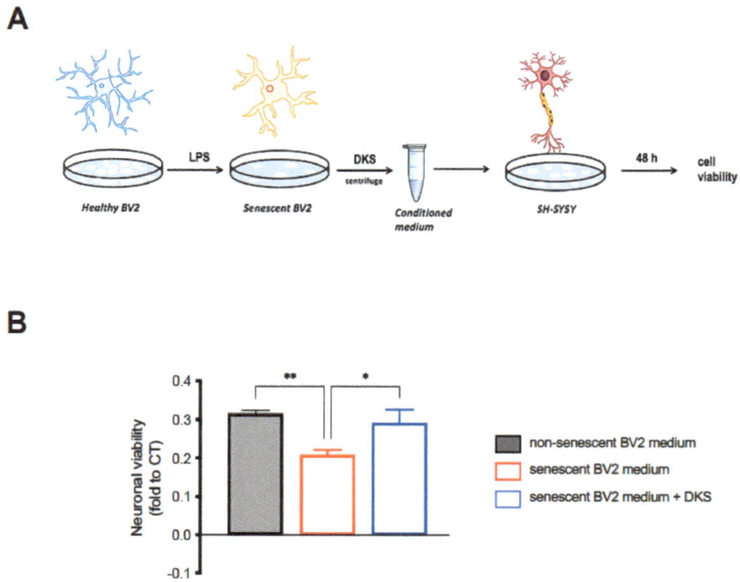

Figure 6. Neuroprotective effect of DKS. (**A**) Schematic representation of the experimental protocol. (**B**). Evaluation of neuronal viability (SRB test) in SH−SY5Y cells exposed to non−senescent medium, senescent medium or senescent medium followed by DKS (100 μg/mL) treatment. One-way ANOVA * $p < 0.05$, ** $p < 0.01$.

4. Discussion

The present study investigated the capability of the food supplement DKS to attenuate microglia senescence in an in vitro model. The results obtained showed the senotherapeutic properties of DKS and illustrated its efficacy in promoting neuroprotection against the senescent microglia-associated neurotoxicity.

Obesity is associated with profound changes in cellular function, showing a pathophysiology that recapitulates a chronic condition similar to the aging process. Indeed, obesity is a condition in which the burden of senescent cells is particularly high. An exaggerated and abnormal accumulation of senescent cells in tissues drastically increases the secretion of SASP factors, fostering the development of a persistent low-grade chronic inflammation [34]. Components of the SASP secreted by senescent adipocytes have been suggested to confer and exacerbate insulin resistance [35,36]. Several reports indicate that senescent cells accumulate in adipose tissue of obese and diabetic humans and mice [37,38] and in other organs, including the brain [17]. Thus, targeting senescent cells by means of senotherapeutics has been postulated as a new strategy to alleviate obesity and to improve obesity-associated complications [12,18].

With the aim of searching for an innovative senotherapeutic treatment potentially useful in the therapy of obesity and associated neurological complications, we investigated the antisenescence activity of DKS, a product containing polyphenol-based herbal extracts with antioxidant and anti-inflammatory properties, currently used as adjuvant in hypocaloric diet to stimulate the metabolism of lipids and carbohydrates. Indeed, natural compounds are increasingly attracting scientific interest for their beneficial effects and high tolerability. Many natural constituents, especially polyphenols from commonly used herbal preparations, are widely known to possess antioxidant and anti-inflammatory activities and their anti-obesity effect has been documented [39]. Specifically, treatments endowed with antioxidant and anti-inflammatory activity are promising candidates to mitigate cellular senescent. Of note, oxidative stress, increased in obesity [37], plays an important role in the generation of senescent cells [40] and concurs to the activation of microglia, key cells that play a prominent role not only in the immune and inflammatory response, but also in neuroprotection, food intake, and energy expenditure [41]. Indeed, activation of microglia is a key event in obesity, and in mice fed with high-fat diet, the degree of microgliosis parallels the degree of weight gain [42]. Furthermore, obesity-associated neuroinflammation may be involved in neurodegenerative alteration observed in obese subjects [43].

Exposure of BV2 immortalized murine microglia cells to LPS has been broadly used as a model of neuroinflammation. We have recently shown that, by prolonging the time of exposure and reducing the concentration of LPS, BV2 cells produced a model of microglia senescence [26]. In this model, DKS was able to attenuate ß-gal activity and expression. Increased senescence-associated ß-gal activity was the first marker used for the revelation of cellular senescence in tissues in situ [44] and it is presently recognized as one of the prominent markers of cellular senescence. DKS was also able to reduce SAHF formation, which are heterochromatin domains that induce silencing of various genes involved in the promotion of proliferation and represent a well-known marker for senescent cells [45]. Importantly, treatment of BV2 senescent cells with DKS largely reduced the expression of the phosphorylated p65 subunit of the transcription factor NF-kB, a key event in the acquisition of the SASP phenotype, restoring the basal levels of the protein. NF-kB represents one of the main regulators of proinflammatory processes, as well as driving of microglia activation and the activation of genes that could contribute to cellular senescence and SASP acquisition [46]. Thus, NF-kB represents a key target for both anti-inflammatory and anti-senescence activity. Of note, SASP activity is increased in murine senescent microglia in in vitro and in vivo models, whereas in microglia from the aged mouse brain, SASP is unaltered [16]. This evidence further correlates the microglia senescent phenotype that has acquired SASP activity with pathological conditions rather than with physiological aging.

Senotherapeutics comprise drugs that promote a selective cell death of senescent cells, termed senolytics, and drugs that suppress markers of senescence, in particular the SASP, termed senomorphics [47]. Senomorphic drugs are emerging as alternatives to target senescence-associated diseases [48]. Indeed, the SASP is involved in the secretion of cytokines, chemokines, growth factors, and reactive oxygen species that can lead to detrimental effects on neighboring tissues. Senescence then spreads, since SASP factors can convert adjacent cells to senescence. At the molecular level, senomorphics act by targeting the most important transcription factors for inflammatory mediators, i.e., NF-kB, indicating a promising senomorphic activity for DKS. Multiple commonly used drugs, including metformin [49], have been classified as senomorphics. These compounds usually improve longevity in animal models, confirming the therapeutic value of reducing the effects of senescent cells.

Neurological disturbances are associated with obesity [12]. Neuroinflammation has emerged as key risk to the development of neurological disturbances and microglia dominate as contributors to neuroinflammation [50]. Senescent glial cells are found in obese laboratory animals, and their removal attenuates anxiety-related behaviors [51]. Previous findings showed anti-neuroinflammatory properties for DKS by reducing the levels of proinflammatory mediators [25], and we thus investigated the capability of DKS to protect neurons in the presence of microglia senescent phenotype. We observed that treatment with DKS counteracted the decrease of cell viability of neuronal cells exposed to the neurotoxic insult induced by senescent BV2 medium, showing a potential neuroprotective activity. Growing evidence indicates the potential role of glucagon-like peptide-1 receptor (GLP-1R) activation in the management of obesity and obesity-related brain disorders [52]. GLP-1R activation in microglia suppress their inflammatory response and prevent neurodegeneration in obese mice, making the modulation of this pathway a promising therapeutic strategy. On these bases, we cannot exclude an involvement of GLP-1R mediated effects in the activity of DKS.

Senomorphic treatment represents a safer approach to target senescent cells, as it reduces the generation of SASP factors without causing the death of senescent cells [47]. Obesity causes a continuous production of new senescent cells, given ongoing metabolic insults, and anti-SASP strategies may require a continual treatment to maintain efficacy, given that senescent cells are not eliminated from tissue. Of course, effectiveness, tolerability, and safety must be carefully considered further for use in clinical practice. DKS contains extracts from more than ten medicinal plants, obtained through an extraction process with water at high temperatures. This phytocomplex fosters a synergistic activity between the various components, allowing the use of lower doses that limit the occurrence of potential side effects. In addition, DKS represent a multitarget approach that has been postulated to represent a more effective therapeutic strategy than single target therapies in multifactorial disorders, including obesity [53].

5. Conclusions

Obesity, an impending global threat, is not being effectively controlled by current intervention. The prevalence of obesity is rising, creating an urgent need for efficacious therapies. Growing evidence illustrates cellular senescence as a causal factor in obesity-related inflammation and metabolic derangements. Removing or blunting the effects of senescent cells can alleviate obesity-induced organ and tissue dysfunction. Our results showed the senotherapeutic activity of DKS on senescent microglia, and its neuroprotective effect toward the senescent microglia-associated neurotoxicity. Thus, DKS might hold promise as an adjunctive intervention to control inflammation-related mechanisms of obesity and obesity-related complications.

Author Contributions: Conceptualization, N.G. and V.B.; methodology, formal analysis, and data curation V.B., C.S., M.M. and N.G.; original draft preparation, N.G., C.S., M.M. and V.B.; review and editing, N.G., V.B. and L.C.; All authors have read and agreed to the published version of the manuscript.

Funding: This research received no external funding.

Institutional Review Board Statement: Not applicable.

Informed Consent Statement: Not applicable.

Data Availability Statement: The data presented in this study are available on request from the corresponding author.

Acknowledgments: Thanks to all the technical team of Gianluca Mech Srl for providing Dekosilhue.

Conflicts of Interest: The authors declare no conflict of interest.

References

1. Chooi, Y.C.; Ding, C.; Magkos, F. The epidemiology of obesity. *Metabolism* **2019**, *92*, 6–10. [CrossRef]
2. Velazquez, A.; Apovian, C.M. Updates on obesity pharmacotherapy. *Ann. N. Y. Acad. Sci.* **2018**, *1411*, 106–119. [CrossRef] [PubMed]
3. Asghar, A.; Sheikh, N. Role of immune cells in obesity induced low grade inflammation and insulin resistance. *Cell. Immunol.* **2017**, *315*, 18–26. [CrossRef] [PubMed]
4. Kawai, T.; Autieri, M.V.; Scalia, R. Adipose tissue inflammation and metabolic dysfunction in obesity. *Am. J. Physiol. Cell Physiol.* **2021**, *320*, C375–C391. [CrossRef] [PubMed]
5. Sindhu, S.; Thomas, R.; Kochumon, S.; Wilson, A.; Abu-Farha, M.; Bennakhi, A.; Al-Mulla, F.; Ahmad, R. Increased adipose tissue expression of interferon regulatory factor (IRF)-5 in obesity: Association with metabolic inflammation. *Cells* **2019**, *8*, 1418. [CrossRef]
6. Lumeng, C.N.; Saltiel, A.R. Inflammatory links between obesity and metabolic disease. *J. Clin. Investig.* **2011**, *121*, 2111–2117. [CrossRef]
7. Hotamisligil, G.S. Foundations of Immunometabolism and Implications for Metabolic Health and Disease. *Immunity* **2017**, *47*, 406–420. [CrossRef]
8. Dantzer, R. Neuroimmune interactions: From the brain to the immune system and vice versa. *Physiol. Rev.* **2018**, *98*, 477–504. [CrossRef]
9. Rivest, S. Regulation of innate immune responses in the brain. *Nat. Rev. Immunol.* **2009**, *9*, 429–439. [CrossRef]
10. Lai, K.S.P.; Liu, C.S.; Rau, A.; Lanctôt, K.L.; Köhler, C.A.; Pakosh, M.; Carvalho, A.F.; Herrmann, N. Peripheral inflammatory markers in Alzheimer's disease: A systematic review and meta-analysis of 175 studies. *J. Neurol. Neurosurg. Psychiatry* **2017**, *88*, 876–882. [CrossRef]
11. Guillemot-Legris, O.; Muccioli, G.G. Obesity-Induced Neuroinflammation: Beyond the Hypothalamus. *Trends Neurosci.* **2017**, *40*, 237–253. [CrossRef] [PubMed]
12. Fulton, S.; Décarie-Spain, L.; Fioramonti, X.; Guiard, B.; Nakajima, S. The menace of obesity to depression and anxiety prevalence. *Trends Endocrinol. Metab.* **2022**, *33*, 18–35. [CrossRef]
13. Kim, H.J.; Cho, M.H.; Shim, W.H.; Kim, J.K.; Jeon, E.-Y.; Kim, D.-H.; Yoon, S.-Y. Deficient autophagy in microglia impairs synaptic pruning and causes social behavioral defects. *Mol. Psychiatry* **2017**, *22*, 1576–1584. [CrossRef]
14. Alexaki, V.I. The impact of obesity on microglial function: Immune, metabolic and endocrine perspectives. *Cells* **2021**, *10*, 1584. [CrossRef] [PubMed]
15. Subhramanyam, C.S.; Wang, C.; Hu, Q.; Dheen, S.T. Microglia-mediated neuroinflammation in neurodegenerative diseases. *Semin. Cell Dev. Biol.* **2019**, *94*, 112–120. [CrossRef] [PubMed]
16. Stojiljkovic, M.R.; Ain, Q.; Bondeva, T.; Heller, R.; Schmeer, C.; Witte, O.W. Phenotypic and functional differences between senescent and aged murine microglia. *Neurobiol. Aging* **2019**, *74*, 56–69. [CrossRef]
17. Palmer, A.K.; Tchkonia, T.; Kirkland, J.L.; Jensen, M.D. Senescence in obesity. In *Cellular Senescence in Disease*; Academic Press: Cambridge, MA, USA, 2022; pp. 289–308.
18. Palmer, A.K.; Xu, M.; Zhu, Y.; Pirtskhalava, T.; Weivoda, M.M.; Hachfeld, C.M.; Prata, L.G.; Dijk, T.H.; Verkade, E.; Casaclang-Verzosa, G.; et al. Targeting senescent cells alleviates obesity-induced metabolic dysfunction. *Aging Cell* **2019**, *18*, e12950. [CrossRef]
19. Narasimhan, A.; Flores, R.R.; Camell, C.D.; Bernlohr, D.A.; Robbins, P.D.; Niedernhofer, L.J. Cellular Senescence in Obesity and Associated Complications: A New Therapeutic Target. *Curr. Diab. Rep.* **2022**, *22*, 537–548. [CrossRef]
20. Van Deursen, J.M. The role of senescent cells in ageing. *Nature* **2014**, *509*, 439–446. [CrossRef]
21. Kumar, A.; Bano, D.; Ehninger, D. Cellular senescence in vivo: From cells to tissues to pathologies. *Mech. Ageing Dev.* **2020**, *190*, 111308. [CrossRef]
22. Lagoumtzi, S.M.; Chondrogianni, N. Senolytics and senomorphics: Natural and synthetic therapeutics in the treatment of aging and chronic diseases. *Free. Radic. Biol. Med.* **2021**, *171*, 169–190. [CrossRef]
23. Mathew, G.; Unnikrishnan, M.K. Multi-target drugs to address multiple checkpoints in complex inflammatory pathologies: Evolutionary cues for novel "first-in-class" anti-inflammatory drug candidates: A reviewer's perspective. *Inflamm. Res.* **2015**, *64*, 747–752. [CrossRef] [PubMed]
24. Wagner, H.; Ulrich-Merzenich, G. Synergy research: Approaching a new generation of phytopharmaceuticals. *Phytomedicine* **2009**, *16*, 97–110. [CrossRef] [PubMed]

25. Borgonetti, V.; Cenci, L.; Galeotti, N. Anti-Inflammatory and Neuroprotective Effect of the Anti-Obesity Dietary Supplement Dekosilhue®in an In Vitro Model of Neuroinflammation. *Nutraceuticals* **2022**, *2*, 22–31. [CrossRef]
26. Borgonetti, V.; Galeotti, N. Rosmarinic Acid Reduces Microglia Senescence: A Novel Therapeutic Approach for the Management of Neuropathic Pain Symptoms. *Biomedicines* **2022**, *10*, 1468. [CrossRef] [PubMed]
27. Wu, F.Y.H.; Liao, W.C.; Chang, H.M. Comparison of antitumor activity of vitamins K1, K2 and K3 on human tumor cells by two (MTT and SRB) cell viability assays. *Life Sci.* **1993**, *52*, 1797–1804. [CrossRef] [PubMed]
28. Borgonetti, V.; Governa, P.; Biagi, M.; Pellati, F.; Galeotti, N. Zingiber officinale Roscoe rhizome extract alleviates neuropathic pain by inhibiting neuroinflammation in mice. *Phytomedicine* **2020**, *78*, 153307. [CrossRef]
29. Ng, P.Y.; McNeely, T.L.; Baker, D.J. Untangling senescent and damage-associated microglia in the aging and diseased brain. *FEBS J.* **2021**, *290*, 1326–1339. [CrossRef]
30. Aird, K.M.; Zhang, R. Detection of senescence-associated heterochromatin foci (SAHF). *Methods Mol. Biol.* **2013**, *965*, 185–196. [CrossRef]
31. Burton, D.G.A.; Stolzing, A. Cellular senescence: Immunosurveillance and future immunotherapy. *Ageing Res. Rev.* **2018**, *43*, 17–25. [CrossRef]
32. Hernandez-Segura, A.; de Jong, T.V.; Melov, S.; Guryev, V.; Campisi, J.; DeMaria, M. Unmasking Transcriptional Heterogeneity in Senescent Cells. *Curr. Biol.* **2017**, *27*, 2652–2660.e4. [CrossRef] [PubMed]
33. Ohanna, M.; Giuliano, S.; Bonet, C.; Imbert, V.; Hofman, V.; Zangari, J.; Bille, K.; Robert, C.; Paillerets, B.B.-D.; Hofman, P.; et al. Senescent cells develop a parp-1 and nuclear factor-κB-associated secretome (PNAS). *Genes Dev.* **2011**, *25*, 1245–1261. [CrossRef] [PubMed]
34. Franceschi, C.; Campisi, J. Chronic inflammation (Inflammaging) and its potential contribution to age-associated diseases. *J. Gerontol. Ser. A Biol. Sci. Med. Sci.* **2014**, *69*, S4–S9. [CrossRef] [PubMed]
35. Ahima, R.S. Connecting obesity, aging and diabetes. *Nat. Med.* **2009**, *15*, 996–997. [CrossRef] [PubMed]
36. Xu, M.; Palmer, A.K.; Ding, H.; Weivoda, M.M.; Pirtskhalava, T.; A White, T.; Sepe, A.; O Johnson, K.; Stout, M.B.; Giorgadze, N.; et al. Targeting senescent cells enhances adipogenesis and metabolic function in old age. *Elife* **2015**, *4*, e12997. [CrossRef] [PubMed]
37. Minamino, T.; Orimo, M.; Shimizu, I.; Kunieda, T.; Yokoyama, M.; Ito, T.; Nojima, A.; Nabetani, A.; Oike, Y.; Matsubara, H.; et al. A crucial role for adipose tissue p53 in the regulation of insulin resistance. *Nat. Med.* **2009**, *15*, 1082–1087. [CrossRef] [PubMed]
38. Schafer, M.J.; Miller, J.D.; LeBrasseur, N.K. Cellular senescence: Implications for metabolic disease. *Mol. Cell. Endocrinol.* **2017**, *455*, 93–102. [CrossRef]
39. Wang, S.; Moustaid-Moussa, N.; Chen, L.; Mo, H.; Shastri, A.; Su, R.; Bapat, P.; Kwun, I.; Shen, C.-L. Novel insights of dietary polyphenols and obesity. *J. Nutr. Biochem.* **2014**, *25*, 1–18. [CrossRef]
40. Hernandez-Segura, A.; Nehme, J.; Demaria, M. Hallmarks of Cellular Senescence. *Trends Cell Biol.* **2018**, *28*, 436–453. [CrossRef]
41. García-Cáceres, C.; Balland, E.; Prevot, V.; Luquet, S.; Woods, S.C.; Koch, M.; Horvath, T.L.; Yi, C.-X.; Chowen, J.A.; Verkhratsky, A.; et al. Role of astrocytes, microglia, and tanycytes in brain control of systemic metabolism. *Nat. Neurosci.* **2019**, *22*, 7–14. [CrossRef]
42. De Luca, S.N.; Miller, A.A.; Sominsky, L.; Spencer, S.J. Microglial regulation of satiety and cognition. *J. Neuroendocr.* **2020**, *32*, e12838. [CrossRef]
43. Woo, A.; Botta, A.; Shi, S.S.W.; Paus, T.; Pausova, Z. Obesity-Related Neuroinflammation: Magnetic Resonance and Microscopy Imaging of the Brain. *Int. J. Mol. Sci.* **2022**, *23*, 8790. [CrossRef] [PubMed]
44. Kurz, D.J.; Decary, S.; Hong, Y.; Erusalimsky, J.D. Senescence-associated ß-galactosidase reflects an increase in lysosomal mass during replicative ageing of human endothelial cells. *J. Cell Sci.* **2000**, *113*, 3613–3622. [CrossRef] [PubMed]
45. Narita, M.; Nũnez, S.; Heard, E.; Narita, M.; Lin, A.W.; Hearn, S.A.; Spector, D.L.; Hannon, G.J.; Lowe, S.W. Rb-mediated heterochromatin formation and silencing of E2F target genes during cellular senescence. *Cell* **2003**, *113*, 703–716. [CrossRef] [PubMed]
46. Salminen, A.; Kauppinen, A.; Kaarniranta, K. Emerging role of NF-kB signaling in the induction of senescence-associated secretory phenotype (SASP). *Cell Signal.* **2012**, *24*, 835–845. [CrossRef]
47. Zhang, L.; Pitcher, L.E.; Prahalad, V.; Niedernhofer, L.J.; Robbins, P.D. Targeting cellular senescence with senotherapeutics: Senolytics and senomorphics. *FEBS J.* **2022**, *290*, 1362–1383. [CrossRef]
48. Birch, J.; Gil, J. Senescence and the SASP: Many therapeutic avenues. *Genes Dev.* **2020**, *34*, 1565–1576. [CrossRef]
49. Kanigur Sultuybek, G.; Soydas, T.; Yenmis, G. NF-kB as the mediator of metformin's effect on ageing and ageing-related diseases. *Clin. Exp. Pharmacol. Physiol.* **2019**, *46*, 413–422. [CrossRef]
50. Wang, H.; He, Y.; Sun, Z.; Ren, S.; Liu, M.; Wang, G.; Yang, J. Microglia in depression: An overview of microglia in the pathogenesis and treatment of depression. *J. Neuroinflammation* **2022**, *19*, 132. [CrossRef]
51. Ogrodnik, M.; Zhu, Y.; Langhi, L.G.P.; Tchkonia, T.; Krüger, P.; Fielder, E.; Victorelli, S.; Ruswhandi, R.A.; Giorgadze, N.; Pirtskhalava, T.; et al. Obesity-Induced Cellular Senescence Drives Anxiety and Impairs Neurogenesis. *Cell Metab.* **2019**, *29*, 1061–1077.e8. [CrossRef]

52. Cui, Q.N.; Stein, L.M.; Fortin, S.M.; Hayes, M.R. The role of glia in the physiology and pharmacology of glucagon-like peptide-1: Implications for obesity, diabetes, neurodegeneration and glaucoma. *Br. J. Pharmacol.* **2022**, *179*, 715–726. [CrossRef] [PubMed]
53. Camilleri, M.; Acosta, A. Combination Therapies for Obesity. *Metab. Syndr. Relat. Disord.* **2018**, *16*, 390–394. [CrossRef] [PubMed]

nutraceuticals

Article

Melatonin Modulates Lipid Metabolism and Reduces Cardiovascular Risk in Apolipoprotein E-Deficient Mice Fed a Western Diet

Guillermo Santos-Sánchez [1,2,†], Ana Isabel Álvarez-López [1,2,†], Eduardo Ponce-España [1,2], Ana Isabel Álvarez-Ríos [3], Patricia Judith Lardone [1,2], Antonio Carrillo-Vico [1,2,*] and Ivan Cruz-Chamorro [1,2,*]

1 Instituto de Biomedicina de Sevilla, IBiS/Hospital Universitario Virgen del Rocío/CSIC/Universidad de Sevilla, 41013 Seville, Spain; gsantos-ibis@us.es (G.S.-S.); aialvarez-ibis@us.es (A.I.Á.-L.); eponce-ibis@us.es (E.P.-E.); plardone@us.es (P.J.L.)
2 Departamento de Bioquímica Médica y Biología Molecular e Inmunología, Facultad de Medicina, Universidad de Sevilla, 41009 Seville, Spain
3 Departamento de Bioquímica Clínica, Unidad de Gestión de Laboratorios, Hospital Universitario Virgen del Rocío, 41013 Seville, Spain; anai.alvarez.sspa@juntadeandalucia.es
* Correspondence: vico@us.es (A.C.-V.); icruz-ibis@us.es (I.C.-C.)
† These authors contributed equally to this work.

Abstract: Melatonin (MLT), a natural compound found in the animal and vegetable kingdom, participates in several physiological processes. MLT exerts antioxidant and anti-inflammatory activities, among others, but information about its action on lipid metabolism is still scarce. For this reason, mice deficient in apolipoprotein E (ApoE$^{-/-}$) fed a Western diet (WD) were intragastrically treated with different concentrations of MLT (2 and 9 mg/kg) for 12 weeks. The lipid parameters were quantified, and, since links between cardiovascular risk and immune function and oxidative stress have been established, we also analyzed the population of leukocytes and the oxidative stress status. Although there was no change in the weight of the mice, a significant reduction in low-density lipoprotein cholesterol (LDL-C) was observed in mice treated with the higher concentration of MLT tested in this study. Additionally, an improvement in cardiovascular risk indexes was observed. A reduction in the hepatic total cholesterol (TC) and LDL-C levels was also observed in the treated mice. Finally, a decrease in leukocytes and lymphocytes in particular, as well as an increase in the antioxidant status, were shown in MLT-treated mice. In conclusion, MLT is a promising candidate that could be considered as a possible functional ingredient capable of preventing cardiovascular risk.

Keywords: ApoE; cardiovascular diseases; hypocholesterolemia; LDL; leukocytes

Citation: Santos-Sánchez, G.; Álvarez-López, A.I.; Ponce-España, E.; Álvarez-Ríos, A.I.; Lardone, P.J.; Carrillo-Vico, A.; Cruz-Chamorro, I. Melatonin Modulates Lipid Metabolism and Reduces Cardiovascular Risk in Apolipoprotein E-Deficient Mice Fed a Western Diet. *Nutraceuticals* **2024**, *4*, 260–272. https://doi.org/10.3390/nutraceuticals4020016

Academic Editors: Ronan Lordan and Emanuel Vamanu

Received: 30 October 2023
Revised: 25 April 2024
Accepted: 6 May 2024
Published: 9 May 2024

1. Introduction

Cardiovascular diseases (CVDs), including coronary, rheumatic and congenital heart, cerebrovascular, and peripheral arterial diseases, are the leading cause of mortality worldwide (being responsible for 32% of all global deaths in 2019) and a major contributor to disability [1]. Furthermore, CVDs remain a major cause of rising health care costs, and their prevalence nearly doubled from 271 million in 1990 to 523 million in 2019 [2]. The relationship between cholesterol and CVDs has been reported by several epidemiological studies [3–5]. Low-density lipoprotein cholesterol (LDL-C) levels, also called 'bad cholesterol' as it is the main cholesterol-carrying particle in plasma, are directly associated with the subsequent risk of CVDs, due to LDL-C's ability to be deposited in the intima vessels and thus cause obstruction [6]. In contrast, increasing high-density lipoprotein cholesterol (HDL-C) levels has been accepted as a therapeutic strategy to reduce the risk of death from CVDs because HDL-C transports cholesterol from tissues to the liver, reducing the serum values [7]. However, studies have shown that when conventional lipid parameters remain apparently normal or moderately high, lipid relationships such as the Castelli risk index (CRI) I (CRI-I, considered as the ratio between total cholesterol -TC- and HDL-C) and II (CRI-II, considered as the LDL-C/HDL-C ratio), and the atherogenic index of plasma (AIP, considered as the logarithm of the ratio between triglycerides -TGs- and HDL-C) are diagnostic alternatives for the prediction of cardiovascular events [8–10] and the effectiveness of therapy [11]. Specifically, the CRI-I has been associated with the formation of coronary plaques [12,13], and the CRI-II has been shown to be a predictor of cardiovascular risk [14]. Regarding the AIP, previous studies have correlated high values of this index to the risk of cardiovascular incidence and all-cause mortality in patients with coronary heart disease [15–17].

Alternatively, an elevated count of white blood cells serves as a robust and autonomous indicator of the risk of heart-related issues in individuals of all genders, whether they have coronary heart disease or not. Furthermore, an increased quantity of leukocytes is linked to the occurrence of coronary heart disease, peripheral arterial conditions, and strokes [18]. Furthermore, alterations in lipid metabolism are closely related to alterations in the immune system and prevailing chronic systemic inflammation in people with a pathological increase in body fat [19]. Furthermore, a rise in lymphocyte count has been observed in conditions like obesity, diabetes, and cardiovascular disease. Lymphocytes are notably elevated in visceral fat, playing a significant role as a key controller of insulin resistance. [20]. For this reason, the relationship between nutritional status and immune functions has been widely studied over the years, associating malnutrition with pathological conditions [21,22]. In such studies, overweight mice fed a Western diet (WD) have shown an increase in the total number of leukocytes and lymphocyte cell numbers [23,24].

Furthermore, oxidative stress has been widely highlighted as a major risk factor in the development of the main cardiovascular diseases. This is an imbalance between the production of reactive oxygen species (ROSs) and antioxidant defenses caused by certain cardiovascular risk factors, such as diabetes, hypertension, smoking, and obesity [25]. At the cardiovascular level, oxidative stress is highly implicated in myocardial infarction, ischemia/reperfusion, and heart failure due to damage to the vascular system through lipid peroxidation, membrane damage, immune cell activation (proteases, nucleases, and protein kinases), structural remodeling, or inflammation [26].

Melatonin (MLT) is considered a pleiotropic molecule due to its participation in various physiological processes. This particular indoleamine is primarily produced by the pineal gland in accordance with a daily cycle, reaching its highest levels at nighttime and the lowest levels during the daytime in humans [27]. In recent decades, this molecule has been shown to not be exclusively synthesized in the pineal gland but also in several peripheral tissues, such as the gastrointestinal tract, immune system cells, and skin cells [28,29]. Unlike studies on its antioxidant role, studies on the effect of MLT on the regulation of lipid metabolism are scarce. Previous studies have shown that MLT can reduce serum levels of TC, LDL-C, HDL-C, and TGs. However, the conclusions obtained are variable and depend on the methodology used [30–33]; thus, more studies are required to determine the benefits of MLT for the lipid profile [34].

Furthermore, the immunomodulatory role of MLT has been extensively studied in recent years, and its efficacy in controlling inflammatory processes has been demonstrated in different diseases [29,35–37]. Previous studies have reported that MLT therapy decreases the number of leukocytes in burn-induced Wistar rats [38] and controls the count of leukocytes and lymphocytes during intense effort in adolescent athletes [39]. However, there is no evidence of the anti-inflammatory role of MLT in an animal model of cardiovascular risk.

Given that growing evidence has shown that MLT might exert a cardiovascular protective effect through the control of lipid metabolism and anti-inflammatory and antioxidant status, the aim of this study was to evaluate the effect of MLT on (i) the plasmatic and hepatic lipid profile, (ii) white blood cell population, and (iii) anti-inflammatory and oxidative stress status in an experimental model of cardiovascular disease consisting of apolipoprotein E (ApoE) knockout mice ($ApoE^{-/-}$) fed a WD.

2. Materials and Methods

2.1. Study Design

Male ApoE$^{-/-}$ mice (kindly donated by Dr. Antonio Ordoñez and Dr. Raquel del Toro), at the age of four weeks, were kept in the animal facility at the Instituto de Biomedicina de Sevilla (IBiS) under typical conditions, which included a 12 h light and 12 h dark cycle, a temperature of 22 ± 2 °C, and humidity levels below 55%. These mice were provided with unrestricted access to both water and a Western diet (Test Diet 58v8, containing 45% energy from fat, as detailed in Supplementary Table S1). When the mice turned 6 weeks old, they were randomly divided into three groups and treated intragastrically with MLT daily (Sigma Aldrich, MO, USA) 2 mg/kg (WD + MLT (2 mg/kg), *n* = 12), 9 mg/kg (WD + MLT (9 mg/kg), n = 12) or the vehicle (=ethanol; WD, *n* = 10) for 12 weeks (Figure 1). The mice were closely monitored, controlled, and observed by the researchers themselves and by the technical staff of the IBiS Animal Facility. In addition, the veterinary manager of the animal facility checked the health status of the animals every week, not recording any side effects.

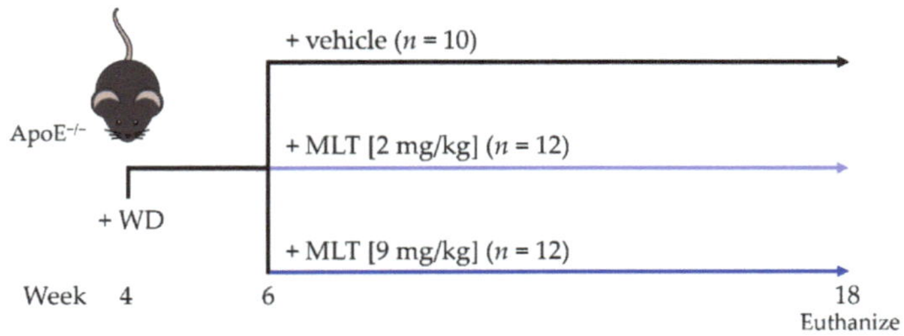

Figure 1. Experimental design of the study. ApoE$^{-/-}$, apolipoprotein E knockout mice; MLT, melatonin; WD, Western diet.

The MLT doses used in this study were chosen because they are equivalent doses of 10 mg and 50 mg MLT/day, respectively, in humans, calculated according to [40] (Figure 2). Daily food intake and individual body weight were measured weekly and recorded. At the endpoint, fasted animals were euthanized, and blood was collected in Minicollect EDTA tubes (Greiner Bio-one, Kremsmünster, Austria) by cardiac puncture. Subsequently, plasma was obtained by centrifugation (3000× *g*, 4 °C, 10 min) and stored at −20 °C until use. The animals were then perfused with phosphate-buffered saline (PBS) for 5 min using an FH100 peristaltic pump (Thermo Scientific, Vantaa, Finland), and the liver was collected and stored until use.

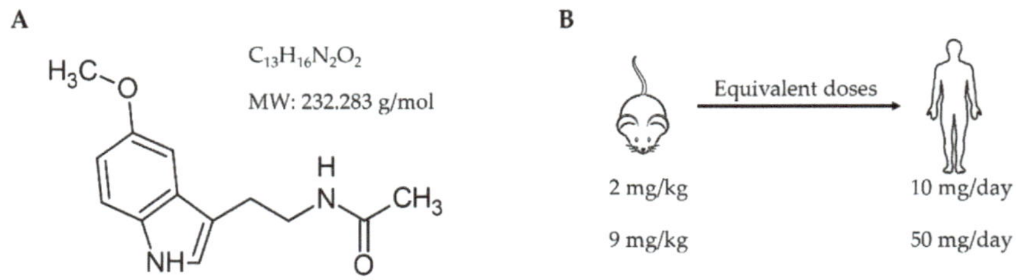

Figure 2. Chemical structure of melatonin ((**A**), CAS number: 73-31-4) and equivalent doses of it in mice and humans (**B**). MW, molecular weight.

Every experimental process adhered to the regulations established by Spanish law and conformed to the guidelines outlined in the EU Directive 2010/63/EU regarding animal experiments. Additionally, these procedures received approval from the Ethics Committee at the Virgen Macarena and Virgen del Rocío University Hospitals, with reference number 21/06/2016/105.

2.2. Plasma and Hepatic–Lipid Profile

Plasma lipid parameters (including TC, TGs, LDL-C, and HDL-C) were measured with chemiluminescence immunoassay techniques using the COBAS e601 modular analyzer (Roche Diagnostic, Basel, Switzerland). In addition, the cardiovascular disease risk indexes CRI-I, CRI-II, and AIP were calculated according to [41]. On the other hand, 100 mg liver tissues were homogenized with a TissueRuptor (Qiagen, Hilden, Germany), and the hepatic TC, HDL-C, LDL-C, and TG concentrations were measured in the supernatants produced by Cobas Integra 400 (Roche Diagnostics, Indianapolis, IN, USA) at the 'Estación Biológica de Doñana' (EBD-CSIC, Seville, Spain).

2.3. White Blood Cell (WBC) Count

White blood cells were quantified in blood samples using the SYSMEX XE 5000 Hematology Analyzer (Sysmex Europe GmbH, Norderstedt, Germany) fluorescence flow cytometer.

2.4. Plasma ELISA

To confirm that MLT could act as an anti-inflammatory molecule, TNF was quantified by a commercial enzyme-linked immunosorbent assay (BD OptEIA™ Mouse TNF (Mono/Mono) ELISA Set, BD Biosciences, San Jose, CA, USA).

Briefly, the plasma of mice was incubated overnight with anti-mouse TNF antibody in a precoated 96-well plate. A biotinylated anti-mouse TNF antibody and streptavidin–HRP conjugate enzyme were used to detect the TNF cytokine. The addition of tetramethylbenzidine (TMB; Sigma-Aldrich, Saint Louis, MO, USA) led to the development of a color that was read at 450 nm with a CLARIOstar Plus microplate reader (BMG Labtech, Ortenberg, Germany) once the reaction was stopped by HCl.

2.5. Plasma Antioxidant Capacity

To test the role of MLT in the antioxidant capacity, the Trolox equivalent antioxidant capacity (TEAC) assay was performed. Briefly, 140 µL of 2,2-azino-bis-(3-ethylbenzothiazoline-6-sulfonic acid) (ABTS, Sigma-Aldrich) radical solution was mixed with 10 µL of plasma. After 30 min of incubation at 30 °C, the ABTS radical content was quantified using a CLARIOstar Plus microplate reader (BMG Labtech) at 730 nm. Then, the values were extrapolated by a Trolox (Sigma-Aldrich) standard curve.

2.6. Statistical Analysis

The data are presented as the mean value accompanied by the standard error of the mean (SEM). Statistical analysis involved the use of non-parametric Mann–Whitney U tests or two-way ANOVA, followed by post hoc corrections, and statistical significance was established at p-values equal to or less than 0.05. The data underwent analysis using GraphPad Prism v.8 (GraphPad Software, San Diego, CA, USA).

3. Results

3.1. Melatonin Does Not Alter the Body Weight of Mice

As shown in Figure 3A, no significant differences in body weight (p = 0.815) were observed between the three experimental groups throughout the experiment. No differences were shown in basal or final body weight (Figure 3B,C) between the control group (basal: 21.29 ± 0.48 g; final: 26.32 ± 0.45 g) and the groups given 2 mg/kg MLT (basal: 20.38 ± 0.84 g, p = 0.594; final: 26.16 ± 0.82 g, p = 0.987) or 9 mg/kg MLT (basal: 20.37 ± 0.82 g, p = 0.571; final: 26.57 ± 0.87 g, p = 0.959). Additionally, the mice did not show significant differences in body weight gain when treated with MLT at 2 mg/kg (5.78 ± 0.39 g, p = 0.288) and 9 mg/kg (6.23 ± 0.39 g, p = 0.118) compared to the control group (5.03 ± 0.49 g) (Figure 3D).

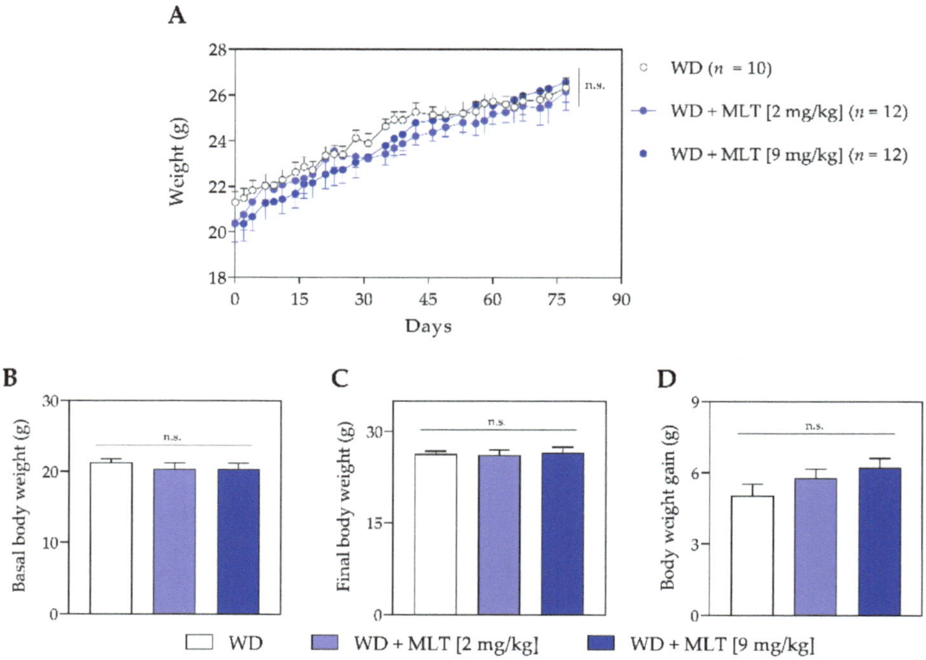

Figure 3. Body weight monitored over time (**A**), basal body weight (**B**), final weight (**C**), and body weight gain (**D**) in the in vivo experiments. Data were represented as mean ± SEM. n.s., not significant. MLT, melatonin; WD, Western diet.

3.2. MLT Improves the Plasmatic Lipid Profile and Reduces the Risk of Cardiovascular Disease

To investigate the lipid-lowering effect of MLT, the plasma lipid profile and the main cardiovascular risk indexes were analyzed. As shown in Table 1, a significant reduction was observed in LDL-C values (−24.6%, $p = 0.008$) in mice treated with 9 mg/kg MLT, compared to the control group. No significant differences were observed in the values of TC, TGs, or HDL-C between the MLT groups compared to the WD group ($p > 0.05$).

Table 1. Plasma lipid profile.

Biochemical Parameter	WD (mg/dL)	WD + MLT (2 mg/kg) (% of Control)	p-Value	WD + MLT (9 mg/kg) (% of Control)	p-Value
TC	506.10 ± 12.34	106.20 ± 8.81	0.548	87.12 ± 6.59	0.151
TG	99.38 ± 0.72	110.80 ± 14.89	0.706	105.20 ± 9.37	0.683
LDL-C	373.20 ± 36.81	112.20 ± 12.10	0.643	75.43 ± 4.39	0.008
HDL-C	113.70 ± 33.59	103.40 ± 5.41	0.548	111.00 ± 11.16	0.691

Results are expressed as a percentage of the control group and represent the mean ± SEM of each group. TC, total cholesterol; TGs, triglycerides; LDL-C, low-density lipoprotein cholesterol; HDL-C, high-density lipoprotein cholesterol; MLT, melatonin.

In addition, cardiovascular risk was evaluated in each group by calculating the indexes CRI I, CRI II, and AIP. Although no differences in CRI I or II were observed at the lowest concentration (2 mg/kg) (104.50 ± 12.27% and 110.80 ± 15.81%, respectively, $p > 0.05$), they were reduced at the highest concentration of MLT (9 mg/kg) (CRI I: 80.35 ± 7.04%, $p = 0.016$ and CRI II: 74.93 ± 6.27%, $p = 0.016$) (Figure 4). In addition, mice treated with MLT showed a reduction in the AIP of 51.74% and 42.31% at both concentrations of MLT tested ($p < 0.05$), respectively (Figure 4).

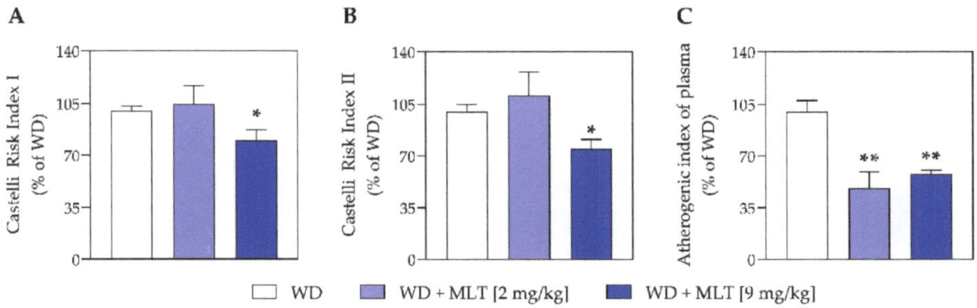

Figure 4. Evaluation of cardiovascular disease risk through Castelli risk index I (TC/HDL-C) (**A**) and II (LDL-C/HDL-C) (**B**) and atherogenic index of plasma (Log(TG/HDL-C)) (**C**). Results are expressed as a percentage of the control group and represent the mean ± SEM of each group. * $p \leq 0.05$, ** $p \leq 0.01$ vs. WD group; HDL-C, high-density lipoprotein cholesterol; LDL-C, low-density lipoprotein cholesterol; MLT, melatonin; TC, total cholesterol; TGs, triglycerides; WD, Western diet.

3.3. MLT Treatment Decreases Hepatic Lipids

As shown in Figure 5, the levels of TC (Figure 5A) and LDL-C (Figure 5C) were reduced by 21.57 and 14.17%, respectively, after treatment with 9 mg/kg MLT, without significant differences in 2 mg/kg MLT-treated mice. On the other hand, TG (Figure 5B) and HDL-C (Figure 5D) levels remained unchanged for both groups treated with MLT.

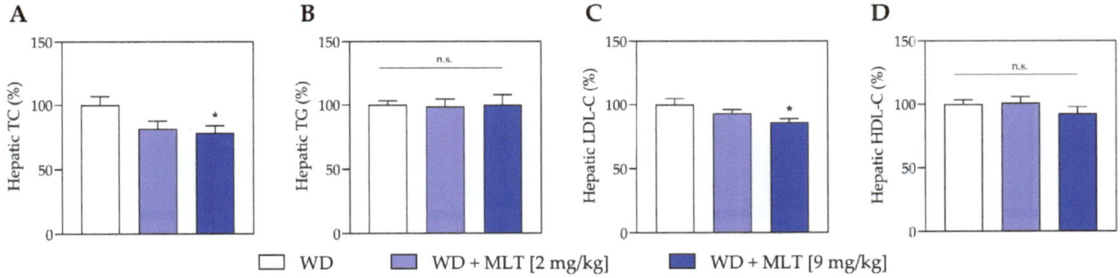

Figure 5. Hepatic TC (**A**), TG (**B**), LDL-C (**C**), and HDL-C (**D**) content in the three experimental groups. Results are expressed as a percentage of the control group and represent the mean ± SEM of each group. * $p \leq 0.05$ vs. WD. n.s., not significant. HDL-C, high-density lipoprotein cholesterol; LDL-C, low-density lipoprotein cholesterol; MLT, melatonin; TC, total cholesterol; TGs, triglycerides; WD, Western diet.

3.4. MLT Reduces Lymphocytosis

To determine the immune status after MLT treatment, the total number of leukocytes, as well as their subpopulations (lymphocytes, monocytes, and granulocytes) was quantified in the plasma sample.

As shown in Table 2, although there were no significant differences in the number of leukocytes in the mice treated with 2 mg/kg MLT compared to the control group ($p > 0.05$), the daily ingestion of 9 mg/kg MLT significantly reduced the number of leukocytes (−50.36%, $p = 0.011$). In particular, MLT treatment reduced the lymphocyte populations by 56% ($p = 0.030$), while the monocyte and granulocyte subpopulations were not altered by the MLT treatment.

Table 2. Hemogram.

Cells (×10^3/μL)	WD	MLT (2 mg/kg)	MLT (9 mg/kg)
Leukocytes	1.49 ± 0.21	1.14 ± 0.16	0.74 ± 0.15 *
Lymphocytes	0.99 ± 0.18	0.76 ± 0.14	0.43 ± 0.13 *
Monocytes	0.069 ± 0.024	0.067 ± 0.021	0.062 ± 0.032
Granulocytes	0.37 ± 0.06	0.31 ± 0.03	0.33 ± 0.04

Results are expressed as the number of cells ± SEM of each experimental group. *, $p \leq 0.05$ with respect to the control group. MLT, melatonin; WD, Western diet.

3.5. MLT Reduces the Number of TNF Pro-Inflammatory Cytokines and Improves the Antioxidant Capacity

The immunomodulatory effect of MLT was corroborated by quantifying the pro-inflammatory cytokine TNF in the plasma of mice. As shown in Figure 6A, MLT treatment reduced the TNF concentration in a dose-dependent manner, with a statistical difference for the treatment with MLT 9 mg/kg being observed (a reduction of 52.6%, $p = 0.044$, with respect to the WD group). The antioxidant capacity of MLT was measured by the ABTS radical scavenging assay in plasma. Figure 6B shows a significant increase in the TEAC values in plasma of mice treated with MLT 9mg/kg. Specifically, the treatment increased the TEAC values by 58.5 ± 29.70% compared to the control group (WD). No differences were observed for 2 mg/kg MLT-treated mice.

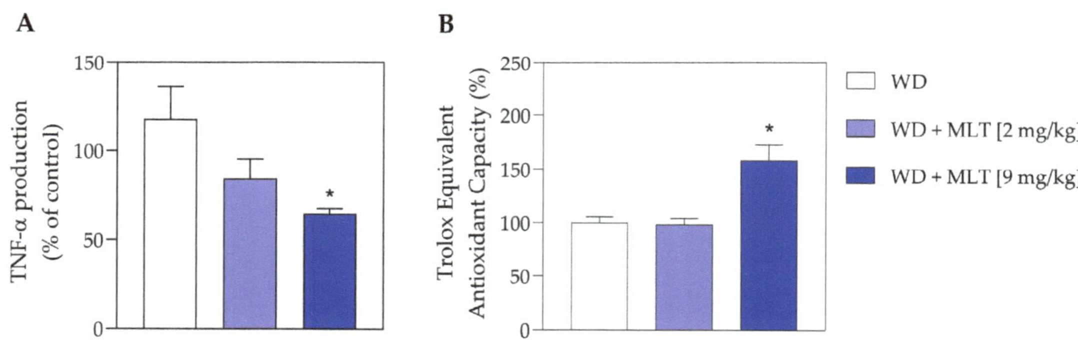

Figure 6. Effect of MLT treatment on plasma anti-inflammatory and antioxidant status evaluated by an enzyme-linked immunosorbent assay (ELISA) (**A**) and the Trolox equivalent antioxidant capacity (TEAC) assay (**B**). Results are expressed as a percentage of the control group and represent the mean ± SEM of each group. * $p \leq 0.05$ with respect to the Western diet (WD) group. MLT, melatonin.

4. Discussion

This study highlights the cardioprotective effect of MLT therapy in an ApoE$^{-/-}$ mouse model characterized by a greater susceptibility to cardiovascular accidents. In fact, ApoE$^{-/-}$ mice display poor lipoprotein clearance with subsequent accumulation of cholesterol ester-enriched particles in the blood, which promotes the development of atherosclerotic plaques [42].

The present study shows, for the first time, that treatment with MLT for 12 weeks reduces the LDL-C concentration and cardiovascular risk indexes in WD-fed ApoE$^{-/-}$ mice. Additionally, MLT treatment decreases the plasma levels of lymphocytes and improves the plasma antioxidant status. The effects of MLT were not related to the body weight gain of the mice, which remained unchanged for the three experimental groups.

High plasma concentrations of TC, TG, and LDL-C, as well as low plasma concentrations of HDL-C, among other things, are risk factors for the onset and progression of CVDs [25,41]. Similarly, a previous study showed that an increase of 1.0 mmol/L in LDL-C was associated with an increased absolute risk of myocardial infarction in individuals aged 70–100 years [43]. On the contrary, an increase of 1 mg/dL in HDL-C reduces the risk of coronary heart disease by 2% in men and 3% in women [44]. It is remarkable that in the present study, 12 weeks of MLT treatment at 9 mg/kg reduced plasma LDL-C by −24.6%, while HDL-C levels were not altered. Furthermore, a strong association has been shown between CVDs, metabolic dysfunction-associated fatty liver disease (MAFLD), and the accumulation of liver fat (steatosis) [45]. In this sense, the present study shows for the first time that treatment with 9 mg/kg MLT reduced the liver total cholesterol and LDL-C in WD-fed ApoE$^{-/-}$ mice, which is in agreement with previous studies performed in rats [46,47]. Although these previous studies showed the effect of MLT treatment on serum cholesterol and LDL-C concentration [48–50], none of these were performed in a specific model of hypercholesterolemia and cardiovascular disease. In fact, these previous studies were focused on the effect of MLT on a specific increase in LDL-C caused by nicotine administration [51], cigarette smoke [52], diabetes induction [53], or a diet modification [31,54].

In addition, MLT reduces the CRI I, CRI II, and AIP, which are used as optimal indicators of cardiovascular risk [41]. Bhardwaj et al. have reported that CRIs can contribute significantly to the estimation of the risk of coronary artery disease, especially when the absolute values of the plasma lipid parameters do not change markedly [55]. Furthermore, Quispe and colleagues proposed that the CRI I should be considered for additional risk assessment in the primary prevention population, specifically in high-CV-risk individuals, such as patients with diabetes [56]. Regarding the AIP ratio, this is a novel indicator of dyslipidemia, and patients with coronary artery disease (CAD) have significantly higher values compared to healthy controls [57]. Furthermore, the AIP has been described as an independent predictor of CAD [57]. Although it was previously shown that daily treatment of rats fed a high-cholesterol diet with intraperitoneal MLT at 12.5 mg/kg decreased the CRI II [31], there are no previous studies related to the effect of MLT on the CRI I and AIP. Thus, to our knowledge, this is the first study to describe the potential of this molecule to reduce the CRI I and II, and AIP in a mouse model with hypercholesterolemia and cardiovascular risk.

Taking into account the present results, together with previous studies, we suggest that MLT could be a good candidate to prevent the development of CVDs and treat pathologies that cause an imbalance in the lipid profile, such as metabolic syndrome and obesity, among others.

In addition, the role of MLT in the modulation of the immune system has been extensively studied over the last years in different contexts, such as in autoimmune diseases [58], infections [59], and even pathologies associated with metabolic syndrome, including neuroinflammation [60]. Mice fed with WD have an increased number of blood leukocytes and lymphocytes [23]. According to these data, elevated cholesterol levels are widely known to predispose individuals to a pro-inflammatory state through a systemic increase in leukocytes and, to a greater extent, the lymphocytes and soluble mediators secreted by these cells [61]. Furthermore, a previous study has shown that body fat affects the number of circulating leukocytes and lymphocytes in children [24]. In numerous studies, the capacity of MLT to modulate the immune response in different diseases has been described, as well as the association of this molecule with low blood levels of leukocytes and/or lymphocytes [29,62]. Winklewski et al. showed that MLT treatment significantly decreases the number of leukocytes and lymphocytes in ethanol-intoxicated mice [63], and other authors have shown that MLT treatment reduces the number of lymphocytes in animals with zymosan-induced peritonitis [64]. Furthermore, recent studies show the ability of MLT to not only control the number of immune cells such as leukocytes and/or lymphocytes in various pathological situations, but also to modify its pro/anti-inflammatory profile, favoring resolution of the disease [35,58,65]. However, there is no evidence of the effect of MLT on blood leukocyte and/or lymphocyte levels in mice fed a high-fat diet. This study is the first to report the effect of MLT on the levels of lymphocytes and leukocytes in ApoE$^{-/-}$ mice fed with a WD. Furthermore, since a decrease in the population of lymphocytes in the blood, as well as in the plasma levels of TNF, was observed in this study, we could say that MLT contributes to the decrease in the subpopulation of Th1 lymphocytes, being characterized by the production of TNF and involved in inflammatory processes [66]. Also, M1 pro-inflammatory macrophages are responsible for the production of TNF. Our results indicate a slight decrease in monocytes, which could contribute to the decrease in M1-phenotype macrophages and therefore to the decrease in TNF. This is in agreement with previous results which demonstrated that MLT promotes the polarization of macrophages to an M2-type anti-inflammatory profile [67]. In this way, MLT has been shown to reduce TNF levels in the blood in women with polycystic ovary syndrome, patients with COVID-19, and anemic patients with chronic kidney disease, among other populations [68–70].

Finally, oxidative stress is widely known for its role in the generation and development of cardiovascular diseases. ApoE$^{-/-}$ mice have been shown to have a high basal pro-oxidative status compared to C57BL/6 mice due to the antioxidant role of ApoE [71]. In the present work, oxidative stress was also increased through the intake of a WD. Furthermore, it has been widely demonstrated that the consumption of fat-rich diets increases oxidative stress, and in turn, increases the risk of developing cardiovascular disease [72]. In the present investigation, the intake of 9 mg/kg MLT daily for 12 weeks was capable of alleviating the oxidative effects caused by the consumption of the WD, with no improvements being observed when the daily concentration consumed was 2 mg/kg MLT. Due to the close relationship between oxidative stress and cardiovascular diseases, it is concluded that the consumption of MLT, in addition to leading to the previously mentioned effects, would reduce the risk of cardiovascular disease, through the reduction in oxidative stress. The fact that MLT controls the number of systemic lymphocytes as well as the lipid parameters in WD-fed ApoE$^{-/-}$ mice, a murine model of hypercholesterolemia, atherosclerosis, and metabolic syndrome, indicates that this molecule could be used to restore the lipid and anti-inflammatory imbalance generated in the development of cardiovascular events.

In addition, it is important to note that chronobiotic MLT is not the same as MLT, which is administered in high doses to treat different diseases. Both have different treatment guidelines to obtain the greatest effectiveness without harming the patient's health. Specifically, therapy with high doses of MLT has been tested in patients with Charcot–Marie–Tooth neuropathy (70 mg/day for 6 months) [73], in multiple sclerosis (25 mg/day for 6 months) [74], and neoplastic cachexia (20 mg/day for 3 months) [75], among other diseases, and in all of them, there was evidence of its effectiveness as an anti-inflammatory or antioxidant without showing side effects that could compromise the patients' health. Furthermore, a series of articles reinforce the hypothesis that increasing inflammatory responses leads to the suppression of nocturnal MLT production and that MLT administration to control inflammatory processes could at the same time compensate for this loss of nocturnal MLT [76–79]. In line with this, it would also be interesting to delve deeper into the possibilities offered by chronotherapy in MLT treatment in the future, the objective of which is to understand the impact that biological rhythms have on the response to a given therapy to optimize its action, maximize the benefits, and minimize possible adverse effects.

As a limitation of our study, the effect of MLT globally on the number of leukocytes and lymphocytes was analyzed, but not the effect that MLT could have on the different immune subpopulations. In fact, different subsets of white blood cells play different roles, and some are even opposite [80,81]. However, there are previous studies in which MLT therapy in an inflammatory context decreases the number of CD4 cells and macrophages, and more specifically the potentially pathogenic Th1 cells (characterized by the production of TNF), while promoting the regulatory responses mediated by Tregs [35,58]. Regarding macrophage subpopulations, MLT favors polarization from M1 (characterized by the production of TNF) towards the M2 profile, promoting an anti-inflammatory environment [67]. Our results support the hypothesis that MLT could reduce these pro-inflammatory subpopulations (Th1 cells and M1 macrophages) due to a decrease in TNF levels.

Given all the above and that MLT is a pleiotropic molecule with a wide range of functions, we suggest that the consumption of MLT could also control other key risk factors, such as oxidative stress, involved in these pathologies.

5. Conclusions

MLT could be considered a functional ingredient to prevent or treat the development of CVDs derived from high cholesterol intake, since in addition to controlling plasma and liver TC and LDL-C levels and reducing CV risk indexes, it decreases systemic inflammation and oxidative stress due to excessive fat consumption. However, more research is needed to decipher the molecular mechanisms by which MLT exerts these actions and to develop clinical trials to determine the effect of MLT in patients with cardiovascular risk.

Supplementary Materials: The following supporting information can be downloaded at: https://www.mdpi.com/article/10.3390/nutraceuticals4020016/s1, Table S1: Western diet composition.

Author Contributions: The following are the authors' contributions: Conceptualization: A.C.-V., I.C.-C. and P.J.L.; Methodology: G.S.-S., I.C.-C., A.I.Á.-R., A.I.Á.-L. and E.P.-E.; Resources: A.C.-V.; Formal analysis: G.S.-S., I.C.-C., A.I.Á.-L. and E.P.-E.; Drafting of the manuscript: G.S.-S., A.I.Á.-L. and A.C.-V.; Supervision: A.C.-V., I.C.-C. and P.J.L.; Funding acquisition: A.C.-V. and P.J.L. All authors have read and agreed to the published version of the manuscript.

Funding: This research was funded by the Andalusian Government Ministry of Health PC-0111-2016-0111, PEMP-0085-2020 (co-financed with FEDER funds, Resolution of 7 July 2021 of the General Secretary for Research, Development and Innovation in Health, which called for grants to finance research, development, and innovation in biomedicine and health sciences in Andalusia by 2021) and the PAIDI Program from the Andalusian Government (CTS160). G.S.-S. was supported by FPU grants from the Spanish Ministerio de Educación, Cultura y Deporte, (FPU16/02339). I.C.-C. was supported by a postdoctoral fellowship from the Andalusian Government Ministry of Economy, Knowledge, Business, and University (DOC_00587/2020). A.I.Á.-L. was funded by the Andalusian Government Ministry of Health (PI-0136-2019 and PEMP-0085-2020). E.P.-E. was supported by the VI Program of Inner Initiative for Research and Transfer of University of Seville (VI PPIT-US).

Institutional Review Board Statement: All experimental procedures were conducted under the Spanish legislation and under the EU Directive 2010/63/EU for animal experiments and was approved by the Virgen Macarena and Virgen del Rocío University Hospitals Ethical Committee (reference 21/06/2016/105).

Informed Consent Statement: Not applicable.

Data Availability Statement: Data are included within the article and Supplementary Materials.

Acknowledgments: We thank all the staff from the IBiS Animal Facility for their valuable assistance.

Conflicts of Interest: The authors declare no conflicts of interest.

References

1. WHO Organization. Cardiovascular Diseases (CVDs). Available online: https://www.who.int/news-room/fact-sheets/detail/cardiovascular-diseases-(cvds) (accessed on 20 February 2024).
2. Roth, G.A.; Mensah, G.A.; Johnson, C.O.; Addolorato, G.; Ammirati, E.; Baddour, L.M.; Barengo, N.C.; Beaton, A.Z.; Benjamin, E.J.; Benziger, C.P. Global burden of cardiovascular diseases and risk factors, 1990–2019: Update from the GBD 2019 study. *J. Am. Coll. Cardiol.* **2020**, *76*, 2982–3021. [CrossRef]
3. Anderson, K.M.; Castelli, W.P.; Levy, D. Cholesterol and mortality: 30 years of follow-up from the Framingham study. *JAMA* **1987**, *257*, 2176–2180. [CrossRef]
4. Berger, S.; Raman, G.; Vishwanathan, R.; Jacques, P.F.; Johnson, E.J. Dietary cholesterol and cardiovascular disease: A systematic review and meta-analysis. *Am. J. Clin. Nutr.* **2015**, *102*, 276–294. [CrossRef]
5. Targher, G.; Bonapace, S.; Byrne, C. Does high LDL-cholesterol cause cardiovascular disease? *Expert Rev. Clin. Pharmacol.* **2019**, *12*, 91. [CrossRef] [PubMed]
6. Trinick, T.; Duly, E. *Hyperlipidemia: Overview*; Caballero, B., Ed.; Academic Press: Waltham, MA, USA, 2013.
7. Bardagjy, A.S.; Steinberg, F.M. Relationship between HDL functional characteristics and cardiovascular health and potential impact of dietary patterns: A narrative review. *Nutrients* **2019**, *11*, 1231. [CrossRef] [PubMed]
8. Criqui, M.H.; Golomb, B.A. Epidemiologic aspects of lipid abnormalities. *Am. J. Med.* **1998**, *105*, 48s–57s. [CrossRef] [PubMed]

9. Akpınar, O.; Bozkurt, A.; Acartürk, E.; Seydaoğlu, G. A new index (CHOLINDEX) in detecting coronary artery disease risk. *Anadolu Kardiyol. Derg.* **2013**, *13*, 315–319. [CrossRef]
10. Edwards, M.K.; Blaha, M.J.; Loprinzi, P.D. Atherogenic Index of Plasma and Triglyceride/High-Density Lipoprotein Cholesterol Ratio Predict Mortality Risk Better Than Individual Cholesterol Risk Factors, Among an Older Adult Population. *Mayo Clin. Proc.* **2017**, *92*, 680–681. [CrossRef]
11. Dobiásová, M.; Frohlich, J.; Sedová, M.; Cheung, M.C.; Brown, B.G. Cholesterol esterification and atherogenic index of plasma correlate with lipoprotein size and findings on coronary angiography. *J. Lipid Res.* **2011**, *52*, 566–571. [CrossRef]
12. Cai, G.; Shi, G.; Xue, S.; Lu, W. The atherogenic index of plasma is a strong and independent predictor for coronary artery disease in the Chinese Han population. *Medicine* **2017**, *96*, e8058. [CrossRef]
13. Olamoyegun, M.A.; Oluyombo, R.; Asaolu, S.O. Evaluation of dyslipidemia, lipid ratios, and atherogenic index as cardiovascular risk factors among semi-urban dwellers in Nigeria. *Ann. Afr. Med.* **2016**, *15*, 194–199. [CrossRef] [PubMed]
14. Millán, J.; Pintó, X.; Muñoz, A.; Zúñiga, M.; Rubiés-Prat, J.; Pallardo, L.F.; Masana, L.; Mangas, A.; Hernández-Mijares, A.; González-Santos, P.; et al. Lipoprotein ratios: Physiological significance and clinical usefulness in cardiovascular prevention. *Vasc. Health Risk Manag.* **2009**, *5*, 757–765. [PubMed]
15. Hadaegh, F.; Khalili, D.; Ghasemi, A.; Tohidi, M.; Sheikholeslami, F.; Azizi, F. Triglyceride/HDL-cholesterol ratio is an independent predictor for coronary heart disease in a population of Iranian men. *Nutr. Metab. Cardiovasc. Dis.* **2009**, *19*, 401–408. [CrossRef] [PubMed]
16. Bittner, V.; Johnson, B.D.; Zineh, I.; Rogers, W.J.; Vido, D.; Marroquin, O.C.; Bairey-Merz, C.N.; Sopko, G. The TG/HDL cholesterol ratio predicts all cause mortality in women with suspected myocardial ischemia a report from the Women's ischemia syndrome evaluation (WISE). *Am. Heart J.* **2009**, *157*, 548. [CrossRef]
17. Bertoluci, M.C.; Quadros, A.S.; Sarmento-Leite, R.; Schaan, B.D. Insulin resistance and triglyceride/HDLc index are associated with coronary artery disease. *Diabetol. Metab. Syndr.* **2010**, *2*, 11. [CrossRef] [PubMed]
18. Madjid, M.; Fatemi, O. Components of the complete blood count as risk predictors for coronary heart disease: In-depth review and update. *Tex. Heart Inst. J.* **2013**, *40*, 17–29. [PubMed]
19. Hubler, M.J.; Kennedy, A.J. Role of lipids in the metabolism and activation of immune cells. *J. Nutr. Biochem.* **2016**, *34*, 1–7. [CrossRef] [PubMed]
20. Winer, S.; Chan, Y.; Paltser, G.; Truong, D.; Tsui, H.; Bahrami, J.; Dorfman, R.; Wang, Y.; Zielenski, J.; Mastronardi, F.; et al. Normalization of obesity-associated insulin resistance through immunotherapy. *Nat. Med.* **2009**, *15*, 921–929. [CrossRef]
21. Keusch, G.T. The history of nutrition: Malnutrition, infection and immunity. *J. Nutr.* **2003**, *133*, 336S–340S. [CrossRef]
22. Mustafa, A.; Ward, A.; Treasure, J.; Peakman, M. T lymphocyte subpopulations in anorexia nervosa and refeeding. *Clin. Immunol. Immunopathol.* **1997**, *82*, 282–289. [CrossRef]
23. Maysami, S.; Haley, M.J.; Gorenkova, N.; Krishnan, S.; McColl, B.W.; Lawrence, C.B. Prolonged diet-induced obesity in mice modifies the inflammatory response and leads to worse outcome after stroke. *J. Neuroinflammation* **2015**, *12*, 140. [CrossRef] [PubMed]
24. Zaldivar, F.; McMurray, R.; Nemet, D.; Galassetti, P.; Mills, P.; Cooper, D. Body fat and circulating leukocytes in children. *Int. J. Obes.* **2006**, *30*, 906–911. [CrossRef] [PubMed]
25. Dubois-Deruy, E.; Peugnet, V.; Turkieh, A.; Pinet, F. Oxidative stress in cardiovascular diseases. *Antioxidants* **2020**, *9*, 864. [CrossRef] [PubMed]
26. Mourino-Alvarez, L.; Sastre-Oliva, T.; Corbacho-Alonso, N.; Barderas, M.G. Oxidative Stress in Cardiovascular Diseases. In *Importance of Oxidative Stress and Antioxidant System in Health and Disease*; IntechOpen: London, UK, 2022.
27. Zawilska, J.B.; Skene, D.J.; Arendt, J. Physiology and pharmacology of melatonin in relation to biological rhythms. *Pharmacol. Rep.* **2009**, *61*, 383–410. [CrossRef] [PubMed]
28. Chen, C.-Q.; Fichna, J.; Bashashati, M.; Li, Y.-Y.; Storr, M. Distribution, function and physiological role of melatonin in the lower gut. *World J. Gastroenterol. WJG* **2011**, *17*, 3888. [CrossRef] [PubMed]
29. Carrillo-Vico, A.; Lardone, P.J.; Álvarez-Sánchez, N.; Rodríguez-Rodríguez, A.; Guerrero, J.M. Melatonin: Buffering the immune system. *Int. J. Mol. Sci.* **2013**, *14*, 8638–8683. [CrossRef] [PubMed]
30. Hussain, S.A.R. Effect of melatonin on cholesterol absorption in rats. *J. Pineal Res.* **2007**, *42*, 267–271. [CrossRef] [PubMed]
31. Chan, T.; Tang, P. Effect of melatonin on the maintenance of cholesterol homeostasis in the rat. *Endocr. Res.* **1995**, *21*, 681–696. [CrossRef]
32. Elias, A.; Nelson, B.; Obianime, W.A. Beneficial effects of melatonin and alpha lipoic acid on lopinavir/ritonavir-induced alterations in lipid and glucose levels in male albino rats. *J. Med. Sci.* **2016**, *85*, 46–53.
33. Allagui, M.S.; Hachani, R.; Saidi, S.; Feriani, A.; Murat, J.C.; Kacem, K. Pleiotropic protective roles of melatonin against aluminium-induced toxicity in rats. *General. Physiol. Biophys.* **2015**, *34*, 415–424.
34. Mohammadi-Sartang, M.; Ghorbani, M.; Mazloom, Z. Effects of melatonin supplementation on blood lipid concentrations: A systematic review and meta-analysis of randomized controlled trials. *Clin. Nutr.* **2018**, *37*, 1943–1954. [CrossRef] [PubMed]
35. Álvarez-Sánchez, N.; Cruz-Chamorro, I.; Díaz-Sánchez, M.; Sarmiento-Soto, H.; Medrano-Campillo, P.; Martínez-López, A.; Lardone, P.J.; Guerrero, J.M.; Carrillo-Vico, A. Melatonin reduces inflammatory response in peripheral T helper lymphocytes from relapsing-remitting multiple sclerosis patients. *J. Pineal Res.* **2017**, *63*, e12442. [CrossRef] [PubMed]

36. Carrillo-Vico, A.; Lardone, P.J.; Naji, L.; Fernández-Santos, J.M.; Martín-Lacave, I.; Guerrero, J.M.; Calvo, J.R. Beneficial pleiotropic actions of melatonin in an experimental model of septic shock in mice: Regulation of pro-/anti-inflammatory cytokine network, protection against oxidative damage and anti-apoptotic effects. *J. Pineal Res.* **2005**, *39*, 400–408. [CrossRef] [PubMed]
37. Nabavi, S.M.; Nabavi, S.F.; Sureda, A.; Xiao, J.; Dehpour, A.R.; Shirooie, S.; Silva, A.S.; Baldi, A.; Khan, H.; Daglia, M. Anti-inflammatory effects of Melatonin: A mechanistic review. *Crit. Rev. Food Sci. Nutr.* **2019**, *59*, S4–S16. [CrossRef] [PubMed]
38. Kurniawan, M.F.; Utami, S.B.; Fulyani, F.; Kresnoadi, E.; Wicaksono, S.A. Melatonin prevented the elevation of leukocyte count and the decreased of hematocrit levels in burn-induced Wistar Rats. *Bali Med. J.* **2021**, *10*, 668–672. [CrossRef]
39. Cheikh, M.; Makhlouf, K.; Ghattassi, K.; Graja, A.; Ferchichi, S.; Kallel, C.; Houda, M.; Souissi, N.; Hammouda, O. Melatonin ingestion after exhaustive late-evening exercise attenuate muscle damage, oxidative stress, and inflammation during intense short term effort in the following day in teenage athletes. *Chronobiol. Int.* **2020**, *37*, 236–247. [CrossRef] [PubMed]
40. Reagan-Shaw, S.; Nihal, M.; Ahmad, N. Dose translation from animal to human studies revisited. *FASEB J.* **2008**, *22*, 659–661. [CrossRef] [PubMed]
41. Kalelioglu, T.; Genc, A.; Karamustafalioglu, N.; Emul, M. Assessment of cardiovascular risk via atherogenic indices in patients with bipolar disorder manic episode and alterations with treatment. *Diabetes Metab. Syndr. Clin. Res. Rev.* **2017**, *11*, S473–S475. [CrossRef] [PubMed]
42. Lo Sasso, G.; Schlage, W.K.; Boué, S.; Veljkovic, E.; Peitsch, M.C.; Hoeng, J. The Apoe$^{-/-}$ mouse model: A suitable model to study cardiovascular and respiratory diseases in the context of cigarette smoke exposure and harm reduction. *J. Transl. Med.* **2016**, *14*, 146. [CrossRef]
43. Mortensen, M.B.; Nordestgaard, B.G. Elevated LDL cholesterol and increased risk of myocardial infarction and atherosclerotic cardiovascular disease in individuals aged 70–100 years: A contemporary primary prevention cohort. *Lancet* **2020**, *396*, 1644–1652. [CrossRef]
44. Gordon, D.J.; Probstfield, J.L.; Garrison, R.J.; Neaton, J.D.; Castelli, W.P.; Knoke, J.D.; Jacobs, D.R., Jr.; Bangdiwala, S.; Tyroler, H.A. High-density lipoprotein cholesterol and cardiovascular disease. Four prospective American studies. *Circulation* **1989**, *79*, 8–15. [CrossRef] [PubMed]
45. Roca-Fernandez, A.; Banerjee, R.; Thomaides-Brears, H.; Telford, A.; Sanyal, A.; Neubauer, S.; Nichols, T.E.; Raman, B.; McCracken, C.; Petersen, S.E. Liver disease is a significant risk factor for cardiovascular outcomes-a UK Biobank study. *J. Hepatol.* **2023**, *79*, 1085–1095. [CrossRef] [PubMed]
46. Pan, M.; Song, Y.L.; Xu, J.M.; Gan, H.Z. Melatonin ameliorates nonalcoholic fatty liver induced by high-fat diet in rats. *J. Pineal Res.* **2006**, *41*, 79–84. [CrossRef] [PubMed]
47. Sun, H.; Huang, F.-f.; Qu, S. Melatonin: A potential intervention for hepatic steatosis. *Lipids Health Dis.* **2015**, *14*, 75. [CrossRef] [PubMed]
48. Shabani, A.; Foroozanfard, F.; Kavossian, E.; Aghadavod, E.; Ostadmohammadi, V.; Reiter, R.J.; Eftekhar, T.; Asemi, Z. Effects of melatonin administration on mental health parameters, metabolic and genetic profiles in women with polycystic ovary syndrome: A randomized, double-blind, placebo-controlled trial. *J. Affect. Disord.* **2019**, *250*, 51–56. [CrossRef] [PubMed]
49. Cichoz-Lach, H.; Celinski, K.; Konturek, P.; Konturek, S.; Slomka, M. The effects of L-tryptophan and melatonin on selected biochemical parameters in patients with steatohepatitis. *J. Physiol. Pharmacol.* **2010**, *61*, 577. [PubMed]
50. Agil, A.; Navarro-Alarcón, M.; Ruiz, R.; Abuhamadah, S.; El-Mir, M.Y.; Vázquez, G.F. Beneficial effects of melatonin on obesity and lipid profile in young Zucker diabetic fatty rats. *J. Pineal Res.* **2011**, *50*, 207–212. [CrossRef] [PubMed]
51. Hendawy, A.K.; El-Toukhey, N.E.S.; AbdEl-Rahman, S.S.; Ahmed, H.H. Ameliorating effect of melatonin against nicotine induced lung and heart toxicity in rats (retracted). *Environ. Sci. Pollut. Res.* **2021**, *28*, 35628–35641. [CrossRef] [PubMed]
52. Ma, B.; Chen, Y.; Wang, X.; Zhang, R.; Niu, S.; Ni, L.; Di, X.; Han, Q.; Liu, C. Cigarette smoke exposure impairs lipid metabolism by decreasing low-density lipoprotein receptor expression in hepatocytes. *Lipids Health Dis.* **2020**, *19*, 88. [CrossRef]
53. Ebaid, H.; Bashandy, S.A.; Abdel-Mageed, A.M.; Al-Tamimi, J.; Hassan, I.; Alhazza, I.M. Folic acid and melatonin mitigate diabetic nephropathy in rats via inhibition of oxidative stress. *Nutr. Metab.* **2020**, *17*, 6. [CrossRef]
54. Wang, L.; McFadden, J.W.; Yang, G.; Zhu, H.; Lian, H.; Fu, T.; Sun, Y.; Gao, T.; Li, M. Effect of melatonin on visceral fat deposition, lipid metabolism and hepatic lipo-metabolic gene expression in male rats. *J. Anim. Physiol. Anim. Nutr.* **2021**, *105*, 787–796. [CrossRef]
55. Bhardwaj, S.; Bhattacharjee, J.; Bhatnagar, M.; Tyagi, S.; Delhi, N. Atherogenic index of plasma, castelli risk index and atherogenic coefficient-new parameters in assessing cardiovascular risk. *Int. J. Pharm. Biol. Sci.* **2013**, *3*, 359–364.
56. Quispe, R.; Elshazly, M.B.; Zhao, D.; Toth, P.P.; Puri, R.; Virani, S.S.; Blumenthal, R.S.; Martin, S.S.; Jones, S.R.; Michos, E.D. Total cholesterol/HDL-cholesterol ratio discordance with LDL-cholesterol and non-HDL-cholesterol and incidence of atherosclerotic cardiovascular disease in primary prevention: The ARIC study. *Eur. J. Prev. Cardiol.* **2020**, *27*, 1597–1605. [CrossRef] [PubMed]
57. Wu, T.T.; Gao, Y.; Zheng, Y.Y.; Ma, Y.T.; Xie, X. Atherogenic index of plasma (AIP): A novel predictive indicator for the coronary artery disease in postmenopausal women. *Lipids Health Dis.* **2018**, *17*, 197. [CrossRef] [PubMed]
58. Álvarez-Sánchez, N.; Cruz-Chamorro, I.; López-González, A.; Utrilla, J.C.; Fernández-Santos, J.M.; Martínez-López, A.; Lardone, P.J.; Guerrero, J.M.; Carrillo-Vico, A. Melatonin controls experimental autoimmune encephalomyelitis by altering the T effector/regulatory balance. *Brain Behav. Immun.* **2015**, *50*, 101–114. [CrossRef]
59. Begum, R.; Mamun-Or-Rashid, A.N.M.; Lucy, T.T.; Pramanik, M.K.; Sil, B.K.; Mukerjee, N.; Tagde, P.; Yagi, M.; Yonei, Y. Potential Therapeutic Approach of Melatonin against Omicron and Some Other Variants of SARS-CoV-2. *Molecules* **2022**, *27*, 6934. [CrossRef]

60. Mansouri, S.; Salari, A.A.; Abedi, A.; Mohammadi, P.; Amani, M. Melatonin treatment improves cognitive deficits by altering inflammatory and neurotrophic factors in the hippocampus of obese mice. *Physiol. Behav.* **2022**, *254*, 113919. [CrossRef] [PubMed]
61. Ellulu, M.S.; Patimah, I.; Khaza'ai, H.; Rahmat, A.; Abed, Y. Obesity and inflammation: The linking mechanism and the complications. *Arch. Med. Sci.* **2017**, *13*, 851–863. [CrossRef] [PubMed]
62. Obayashi, K.; Saeki, K.; Kurumatani, N. Higher melatonin secretion is associated with lower leukocyte and platelet counts in the general elderly population: The HEIJO-KYO cohort. *J. Pineal Res.* **2015**, *58*, 227–233. [CrossRef]
63. Kurhaluk, N.; Sliuta, A.; Kyriienko, S.; Winklewski, P.J. Melatonin restores white blood cell count, diminishes glycated haemoglobin level and prevents liver, kidney and muscle oxidative stress in mice exposed to acute ethanol intoxication. *Alcohol Alcohol.* **2017**, *52*, 521–528. [CrossRef]
64. Kepka, M.; Szwejser, E.; Pijanowski, L.; Verburg-van Kemenade, B.L.; Chadzinska, M. A role for melatonin in maintaining the pro-and anti-inflammatory balance by influencing leukocyte migration and apoptosis in carp. *Dev. Comp. Immunol.* **2015**, *53*, 179–190. [CrossRef]
65. Medrano-Campillo, P.; Sarmiento-Soto, H.; Álvarez-Sánchez, N.; Álvarez-Ríos, A.I.; Guerrero, J.M.; Rodríguez-Prieto, I.; Castillo-Palma, M.J.; Lardone, P.J.; Carrillo-Vico, A. Evaluation of the immunomodulatory effect of melatonin on the T-cell response in peripheral blood from systemic lupus erythematosus patients. *J. Pineal Res.* **2015**, *58*, 219–226. [CrossRef]
66. Moss, R.B.; Moll, T.; El-Kalay, M.; Kohne, C.; Soo Hoo, W.; Encinas, J.; Carlo, D.J. Th1/Th2 cells in inflammatory disease states: Therapeutic implications. *Expert. Opin. Biol. Ther.* **2004**, *4*, 1887–1896. [CrossRef]
67. Nardo, L.; Rezzani, R.; Facchetti, L.; Favero, G.; Franco, C.; Abdelhafez, Y.G.; Badawi, R.D.; Guindani, M.; Seo, Y.; Pampaloni, M. Beneficial effects of melatonin on apolipoprotein-E knockout mice by morphological and 18F-FDG PET/CT assessments. *Int. J. Mol. Sci.* **2020**, *21*, 2920. [CrossRef] [PubMed]
68. Mousavi, R.; Alizadeh, M.; Asghari Jafarabadi, M.; Heidari, L.; Nikbakht, R.; Babaahmadi Rezaei, H.; Karandish, M. Effects of melatonin and/or magnesium supplementation on biomarkers of inflammation and oxidative stress in women with polycystic ovary syndrome: A randomized, double-blind, placebo-controlled trial. *Biol. Trace Elem. Res.* **2022**, *200*, 1010–1019. [CrossRef] [PubMed]
69. Esmaeili Gouvarchin Ghaleh, H.; Hosseini, A.; Aghamollaei, H.; Fasihi-Ramandi, M.; Alishiri, G.; Saeedi-Boroujeni, A.; Hassanpour, K.; Mahmoudian-Sani, M.-R.; Farnoosh, G. NLRP3 inflammasome activation and oxidative stress status in the mild and moderate SARS-CoV-2 infected patients: Impact of melatonin as a medicinal supplement. *Z. Naturforsch. C* **2022**, *77*, 37–42. [CrossRef]
70. Hameed, E.N.; Tukmagi, H.A.; Haydar, F.; Allami, H.C. Melatonin improves erythropoietin hyporesponsiveness via suppression of inflammation. *Rev. Recent. Clin. Trials* **2019**, *14*, 203–208. [CrossRef]
71. Pereira, S.S.; Teixeira, L.G.; Aguilar, E.C.; Matoso, R.O.; Soares, F.L.; Ferreira, A.V.; Alvarez-Leite, J.I. Differences in adipose tissue inflammation and oxidative status in C57BL/6 and ApoE−/− mice fed high fat diet. *Anim. Sci. J.* **2012**, *83*, 549–555. [CrossRef] [PubMed]
72. Jiang, S.; Liu, H.; Li, C. Dietary regulation of oxidative stress in chronic metabolic diseases. *Foods* **2021**, *10*, 1854. [CrossRef]
73. Chahbouni, M.; López, M.D.S.; Molina-Carballo, A.; De Haro, T.; Muñoz-Hoyos, A.; Fernández-Ortiz, M.; Guerra-Librero, A.; Acuña-Castroviejo, D. Melatonin treatment reduces oxidative damage and normalizes plasma pro-inflammatory cytokines in patients suffering from Charcot-Marie-Tooth neuropathy: A pilot study in three children. *Molecules* **2017**, *22*, 1728. [CrossRef]
74. Sánchez-López, A.L.; Ortiz, G.G.; Pacheco-Moises, F.P.; Mireles-Ramírez, M.A.; Bitzer-Quintero, O.K.; Delgado-Lara, D.L.; Ramírez-Jirano, L.J.; Velázquez-Brizuela, I.E. Efficacy of melatonin on serum pro-inflammatory cytokines and oxidative stress markers in relapsing remitting multiple sclerosis. *Arch. Med. Res.* **2018**, *49*, 391–398. [CrossRef] [PubMed]
75. Lissoni, P.; Paolorossi, F.; Tancini, G.; Barni, S.; Ardizzoia, A.; Brivio, F.; Zubelewicz, B.; Chatikhine, V. Is there a role for melatonin in the treatment of neoplastic cachexia? *Eur. J. Cancer* **1996**, *32*, 1340–1343. [CrossRef]
76. Fernandes, P.A.; Cecon, E.; Markus, R.P.; Ferreira, Z.S. Effect of TNF-α on the melatonin synthetic pathway in the rat pineal gland: Basis for a 'feedback'of the immune response on circadian timing. *J. Pineal Res.* **2006**, *41*, 344–350. [CrossRef] [PubMed]
77. Pontes, G.N.; Cardoso, E.C.; Carneiro-Sampaio, M.M.; Markus, R.P. Injury switches melatonin production source from endocrine (pineal) to paracrine (phagocytes)–melatonin in human colostrum and colostrum phagocytes. *J. Pineal Res.* **2006**, *41*, 136–141. [CrossRef] [PubMed]
78. Pontes, G.N.; Cardoso, E.C.; Carneiro-Sampaio, M.M.; Markus, R.P. Pineal melatonin and the innate immune response: The TNF-α increase after cesarean section suppresses nocturnal melatonin production. *J. Pineal Res.* **2007**, *43*, 365–371. [CrossRef] [PubMed]
79. Tamura, E.K.; Fernandes, P.A.; Marçola, M.; Cruz-Machado, S.d.S.; Markus, R.P. Long-lasting priming of endothelial cells by plasma melatonin levels. *PLoS ONE* **2010**, *5*, e13958. [CrossRef] [PubMed]
80. Zielinski, C.E. T helper cell subsets: Diversification of the field. *Eur. J. Immunol.* **2023**, *53*, 2250218. [CrossRef]
81. Hou, P.; Fang, J.; Liu, Z.; Shi, Y.; Agostini, M.; Bernassola, F.; Bove, P.; Candi, E.; Rovella, V.; Sica, G. Macrophage polarization and metabolism in atherosclerosis. *Cell Death Dis.* **2023**, *14*, 691. [CrossRef]

nutraceuticals

Article

The Acute Effects of a Commercially Available Caffeinated and Caffeine-Free Thermogenic Dietary Supplement on Resting Energy Expenditure, Hunger, and Hemodynamic Responses

Kworweinski Lafontant [1,2], Jacob Broeckel [1,3], Kara Phillips [1,4], Yasamian Alsayed [1], Wayne A. Ayers-Creech [1], Yuto Ohigashi [1], John Solis [1], Cassidy Bale [1], Arielle Parks [1], Scott Dankel [5] and Bill I. Campbell [1,*]

1 Performance & Physique Enhancement Laboratory, Exercise Science Program, University of South Florida, Tampa, FL 33629, USA; kworweinski.lafontant@ucf.edu (K.L.); broeckelj@tamu.edu (J.B.); kara.phillips@unco.edu (K.P.); yalsayed@usf.edu (Y.A.); ayerscreechw@usf.edu (W.A.A.-C.); yutoohigashi@usf.edu (Y.O.); jesolis1@usf.edu (J.S.); cassidybale@usf.edu (C.B.); arielleparks@usf.edu (A.P.)
2 Institute of Exercise Physiology and Rehabilitation Science, University of Central Florida, Orlando, FL 32816, USA
3 Exercise & Sport Nutrition Laboratory, Human Clinical Research Facility, Department of Health & Kinesiology, Texas A&M University, College Station, TX 77843, USA
4 Department of Kinesiology, Nutrition, and Dietetics, College of Natural and Health Sciences, University of Northern Colorado, Greeley, CO 80639, USA
5 Health and Exercise Science Department, School of Nursing & Health Professions, Rowan University, Glassboro, NJ 08028, USA; dankel47@rowan.edu
* Correspondence: bcampbell@usf.edu; Tel.: +1-813-974-4766

Abstract: There has been a rise in popularity of "stimulant-free" or caffeine-free fat loss supplements, but it is not well understood whether those fat loss supplements are effective at enhancing thermogenesis without caffeine's influence. The purpose of this study was to examine the effects of a caffeinated and non-caffeinated commercially available fat loss supplement on resting energy expenditure (REE), hunger, and hemodynamic variables in healthy adults. Twenty-five healthy male and female participants completed three separate laboratory visits after overnight fasts. Baseline assessments of REE, subjective hunger, heart rate (HR), and blood pressure (BP) were followed by ingestion of a caffeinated (Phoenix, Legion®; CAF), non-caffeinated (Phoenix Caffeine-Free, Legion®; NCAF), or placebo (PL) fat loss supplement. REE, hunger, HR, and BP assessments were repeated at 60-, 120-, and 180-min post-ingestion. CAF, but not NCAF, significantly elevated REE greater than PL at all time points ($p < 0.05$). NCAF significantly reduced hunger compared to CAF and PL at the 120-min time point ($p = 0.006$). CAF significantly increased diastolic BP 60-min post-ingestion and significantly increased systolic BP 120- and 180-min post-ingestion compared to NCAF and PL. Further research is warranted with respect to investigating non-caffeinated ingredients and their effects on REE.

Keywords: stimulants; non-stimulant; fat loss; fat burner; metabolism

Citation: Lafontant, K.; Broeckel, J.; Phillips, K.; Alsayed, Y.; Ayers-Creech, W.A.; Ohigashi, Y.; Solis, J.; Bale, C.; Parks, A.; Dankel, S.; et al. The Acute Effects of a Commercially Available Caffeinated and Caffeine-Free Thermogenic Dietary Supplement on Resting Energy Expenditure, Hunger, and Hemodynamic Responses. *Nutraceuticals* **2024**, *4*, 82–93. https://doi.org/10.3390/nutraceuticals4010006

Academic Editors: Ivan Cruz-Chamorro and Guillermo Santos Sánchez

Received: 18 November 2023
Revised: 26 January 2024
Accepted: 8 February 2024
Published: 21 February 2024

1. Introduction

The ability to effectively induce body composition changes is the goal of many diet and exercise intervention programs for those with excess adiposity. For some individuals with overweight and obesity, interventions have extended beyond diet and exercise intervention programs to include the use of dietary supplements (i.e., thermogenic fat burners) [1]. The use of thermogenic supplements (i.e., fat burners) extends beyond this population to fitness-minded individuals and physique athletes that desire extremely low levels of fat mass [2,3]. Recent investigations have identified that commercially available thermogenic supplements and ingredients typically contained in such products positively induce acute changes in energy intake and energy expenditure, and when taken chronically may elicit positive changes in body composition [4,5].

One of the more common ingredients found in thermogenic supplements is caffeine. With decades of research on caffeine, it has a strong scientific backing to support its application to aid in fat loss. Numerous clinical trials have shown acute caffeine ingestion to significantly increase metabolic rate, plasma free fatty acid concentration and fat oxidation in normal weight and obese individuals [6,7]. These observed beneficial adaptations observed are derived through its release of catecholamines which act on adrenergic receptors, catalyzing a critical step in the process of fat oxidation [8], which is reflected in a lower respiratory exchange ratio (RER). For these reasons, caffeine has landed itself as one of the most common ingredients found within commercially available thermogenic supplements. However, additional effects from the stimulatory nature of catecholamine release associated with caffeine consumption include increased heart rate (HR), blood pressure (BP), and feelings of anxiety and restlessness [9]. Due to these drawbacks, dietary supplement manufacturers have formulated "stimulant-free" supplements to meet the demands of individuals who desire a thermogenic effect without additional stimulation of the central nervous system. Since caffeine is typically omitted from stimulant-free supplements, the term "caffeine-free" is often used interchangeably.

A wide variety of ingredients in caffeine-free products have been utilized by supplement manufacturers such as extracts from tree bark, herbs, and various plant compounds. Commercially available supplements commonly use multiple ingredients in conjunction to provide individual unique benefits or work synergistically with one another to aid in weight loss. In the present study, the ingredients under investigation include *Caralluma fimbrata, Coleus forskohlii* (Forskolin), *Mucuna pruriens* (L-DOPA), *Griffonia simplicifolia* (5-Hydroxytryptophan; 5-HTP), *Kaempferia parviflora* (Black Ginger), *Aframomum melegueta* (Grains of Paradise), *Laminaria jaoponica aresch*, and caffeine.

Caralluma fimbrata, a flowering plant native to India, has been consumed in times of famine as it is hypothesized to have appetite suppressant effects [10]. While the results on its effectiveness regarding fat loss are mixed, *Caralluma fimbrata* supplementation has been shown to be effective in reducing both waist circumference and hip-to-waist ratio [10]. Herbs such as Forskolin have been shown in previous studies to specifically increase cAMP production in animal models [11,12]; cAMP is a regulatory step in the process of fat breakdown (lipolysis) [13]. *Mucuna pruriens*, a legume from regions of Africa and southwest Asia, contains naturally high levels of L-DOPA. While the primary mechanisms of action attributed to L-DOPA are cognitive, they may provide application to individuals seeking fat loss. Previous investigations have shown supplementation of L-DOPA to be effective in increasing circulating levels of dopamine and serotonin in healthy male participants [14]. Increases in these hormones may aid in mental clarity, alertness and reduction in feelings of anxiety commonly associated with prolonged caloric restriction and energy deficits [15].

The supplements in the present study (Phoenix and Phoenix Caffeine-Free, Legion®, Clearwater, FL, USA) also contain *Griffonia simplicifolia*, standardized to 98% 5-HTP. A byproduct of the naturally occurring amino acid L-Tryptophan, in clinical trials 5-HTP has demonstrated an ability to reduce overall calorie consumption, decrease carbohydrate intake, and increase early feelings of satiety in obese individuals consuming a hypocaloric diet [16,17]. *Kaempferia parviflora*, known as Thai Ginseng or Black Ginger, has been shown in animal models to decrease body weight via a reduction in voluntary feed intake over an 8-week period [18]. Additionally, supplementation of this extract may increase activation of brown adipose tissue and subsequent total body energy expenditure [19]. There is evidence to suggest that chronic supplementation of Grains of Paradise may favorably increase metabolic rate [20]. However, the single investigation that observed these effects included exposure to cold therapy and brown fat as prerequisites for inclusion, which may have impacted the ability to generalize the results [20]. Finally, *Laminaria jaoponica aresch* is an extract of sea kelp, native to southeast Asia. Occurring naturally within this brown alga is fucoxanthin, a carotenoid suggested to promote numerous health benefits including the inhibition of fat cell proliferation and improvements in glucose control in diabetic populations [21]. In a previous randomized, placebo-controlled, clinical trial, significant reductions in body weight, waist circumference and body fat were observed with chronic supplementation of fucoxanthin [22].

While it is understood that the aforementioned ingredients may have a positive influence on fat loss and resting energy expenditure (REE) individually, the effect of their combination is less clear. Therefore, the primary purpose of this study was to investigate the effects of a commercially available multi-ingredient caffeinated and caffeine-free supplement on resting energy expenditure in metabolically healthy men and women. A secondary aim was to determine whether the multi-ingredient thermogenic supplement had an acute effect on the suppression of hunger. Tertiary objectives for this study were to determine the effects of the thermogenic dietary supplement on resting HR and BP.

2. Materials and Methods

2.1. Participants

Thirty men and women between the ages of 18 and 39 years volunteered to participate in this study. A total of 5 participants dropped out/did not complete all required lab visits due to personal reasons unrelated to the study, leaving only 25 participants (women, n = 18; men, n = 7) included in the data analysis. Characteristics of analyzed participants are presented in Table 1. Participants varied in training and supplementation history; thus, a within-subjects protocol was used. The research protocol was approved by the University of South Florida Institutional Review Board (IRB ID: STUDY001825), and all ethical guidelines set forth in the Declaration of Helsinki were adhered to. All participants provided written informed consent prior to participation in this study. Participants were required to be between the ages of 18 and 50 years and free from metabolic, cardiovascular, and pulmonary diseases. Additionally, participants were excluded if they were pregnant or nursing, actively under treatment for any medical conditions, or had any sensitivities/allergies to the dietary supplements' ingredients. Participants were instructed to maintain their typical dietary habits throughout the duration of the study.

Table 1. Participant characteristics.

Characteristic	All Participants (N = 25)	Female Participants (n = 18)	Male Participants (n = 7)
Age (years)	23.0 ± 4.5	22.9 ± 5.0	23.3 ± 3.1
Weight (kg)	68.9 ± 15.9	65.1 ± 17.1	78.6 ± 4.9
Height (cm)	164 ± 9.9	159.0 ± 7.1	176 ± 5.0
BMI (kg/m^2)	25.6 ± 4.7	25.6 ± 5.5	25.5 ± 2.0

Data are presented as mean ± standard deviation.

2.2. Experimental Design

This study utilized a randomized, triple-blind, placebo-controlled cross-over design. Each participant was scheduled to visit the lab on 3 separate days, with each visit being randomly assigned to a supplement condition: caffeinated (CAF, Phoenix, Legion®); non-caffeinated (NCAF, Phoenix Caffeine-Free, Legion®); or placebo (PL; inert ingredients). The ingredients contained in CAF and NCAF are presented in Table 2. All visits were consistently held in the same laboratory and were initiated between 6:15 a.m. and 6:30 a.m. The lab was climate-controlled; the mean temperature, barometric pressure, and humidity were 22.9 °C, 760.1 mmHg, and 46.7%, and these were stable across each of the visits with coefficient of variance values of 1.9%, 0.3%, and 3.1%, respectively. Participants reported to the lab following overnight fast (≥8 h) and abstinence from exercise for at least 24 h.

Table 2. List of active ingredients for each supplement condition.

Phoenix (CAF)		Pheonix Caffeine-Free (NCAF)	
Ingredient	**Dose**	**Ingredient**	**Dose**
Vitamin B6 (Pyridoxine HCl)	1.7 mg	Vitamin B6 (Pyridoxine HCl)	1.7 mg
Vitamin B12 (Methylcobalamin)	2.4 mcg	Vitamin B12 (Methylcobalamin)	2.4 mcg
Iodine (Potassium Iodide)	240 mcg	Iodine (Potassium Iodide)	240 mcg
Caralluma fimbriata [Aerial Parts] Extract 20:1	1000 mg	*Caralluma fimbriata* [Aerial Parts] Extract 20:1	1000 mg
Coleus forskohlii [Root] Extract (Standardized to contain 20% Forskolin)	250 mg	*Coleus forskohlii* [Root] Extract (Standardized to contain 20% Forskolin)	250 mg
Mucuna pruriens [Seed] Extract (Standardized to 98% L-DOPA)	153 mg	*Mucuna pruriens* [Seed] Extract (Standardized to 98% L-DOPA)	153 mg
Griffonia simplicifolia [Seed] Extract (Standardized to 98% 5-HTP)	153 mg	*Griffonia simplicifolia* [Seed] Extract (Standardized to 98% 5-HTP)	153 mg
Black Ginger (*Kaempferia parviflora*) [Root] Extract (Standardized to 2.5% Dimethoxyflavone)	100 mg	Black Ginger (*Kaempferia parviflora*) [Root] Extract (Standardized to 2.5% Dimethoxyflavone)	100 mg
Grains of Paradise (*Aframomum melegueta*) [Seed] (Standardized to 12.5% 6-Paradol)	30 mg	Grains of Paradise (*Aframomum melegueta*) [Seed] (Standardized to 12.5% 6-Paradol)	30 mg
Laminaria japonica aresch [Whole Plant] Extract (Standardized to 50% Fucoxanthin)	16 mcg	*Laminaria japonica aresch* [Whole Plant] Extract (Standardized to 50% Fucoxanthin)	16 mcg
Caffeine Anhydrous	200 mg	-	-

Doses are provided for a single serving of each supplement, which is delivered in three vegetarian capsules.

2.3. Testing Sessions

Participants were encouraged to use the restroom and void their bladders at the start of each lab visit. Following that, body weight and height were assessed with a physician beam scale (Health-O-Meter™, Model 402KL, McCook, IL, USA). Next, participants rested in a reclined position for 15 min prior to REE assessments. Indirect calorimetry was used to assess REE with a ParvoMedics TrueOne® 2400 Canopy System (ParvoMedics, Inc., Salt Lake City, UT, USA). The device was calibrated immediately preceding the baseline REE assessment and recalibrated between assessments as time allowed. During each REE assessment, participants were instructed to relax, lie motionless, breathe normally, and refrain from speaking or falling asleep for the entirety of the 20 min assessment. The initial 5 min of each assessment were discarded, and the final 15 min of data collected were used to calculate REE [23]. After the REE assessment, participants sat with their legs uncrossed and had their resting HR and BP recorded using a previously validated automated, oscillometric BP measurement device (Omron 5 series Model BP742, Lake Forest, IL, USA) [24]. HR and BP were assessed in triplicate with each measurement separated by 2 min, and the average of the three measurements was recorded.

Figure 1 presents the sequence of events during each testing session. REE was assessed at baseline, 60-, 120-, and 180-min post-ingestion. After the baseline assessment, participants ingested 3 capsules of either CAF, NCAF, or PL, with water under the supervision of research personnel. All capsules were identical in shape, smell, size, color, and taste. Immediately after each REE assessment, the participants had their resting HR and BP assessed along with a hunger VAS scale. The hunger VAS scale was presented in Likert-style, with 7 possible responses on a continuous line ranging from "very hungry" to "full".

2.4. Statistical Analysis

All statistics were computed using SPSS version 28.0 (IBM Corp., Armonk, NY, USA). The dependent variables of REE, RER, HR, systolic blood pressure (SBP) and diastolic blood pressure (DBP) were analyzed using a 3 (within factor of condition: CAF, NCAF, PL) $\times$ 4 (within factor of time: baseline, 60-, 120-, 180-min) repeated measures analysis of variance (RMANOVA). When an interaction was present, changes from baseline were compared at each time point using a RMANOVA. When no interactions were present, the main effects of each condition were analyzed. The total energy expenditure over the 180-min period was also calculated by dividing the individual REE at each time point by 24 and then implementing the trapezoidal method to obtain the area under the curve for the entire 180-min window. An RMANOVA across total energy expenditure during this window was then computed to compare differences in total energy expenditure. Hunger scores were analyzed using a non-parametric Friedman test. Statistical significance was set at $p \leq 0.05$.

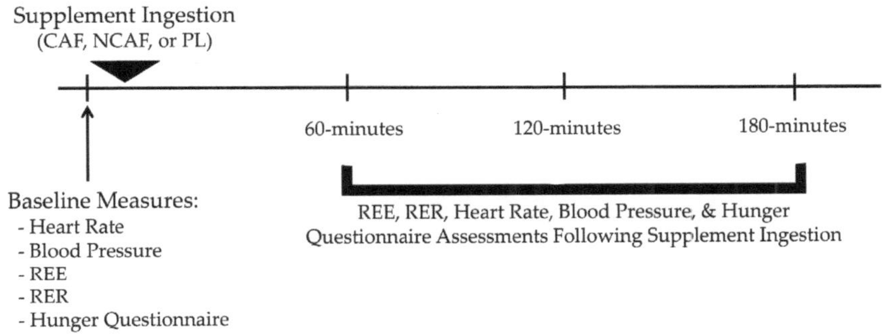

Figure 1. An overview of testing sessions.

3. Results

3.1. Resting Energy Expenditure (REE)

There was a significant interaction ($p < 0.001$) for REE (Figure 2). There was a significant difference for the change in REE the 60-min time point ($p = 0.019$) with a larger increase in REE resulting from CAF compared to NCAF ($p = 0.025$) and PL ($p = 0.022$). At the 120-min time point, there was a greater increase in REE in CAF compared to NCAF ($p < 0.001$) and PL ($p < 0.001$). At the 180-min time point, the changes in REE were statistically different ($p < 0.001$) with CAF increasing to a greater extent than NCAF ($p < 0.001$) and PL ($p = 0.014$). At the 180-min time point, PL also increased to a greater extent than NCAF ($p = 0.043$). When examining total energy expenditure during the 180-min window (Figure 3), there was a statistically significant difference ($p = 0.003$) with CAF resulting in a greater energy expenditure than PL ($p < 0.001$). No significant differences were observed between CAF and NCAF ($p = 0.070$) nor NCAF and PL ($p = 0.095$).

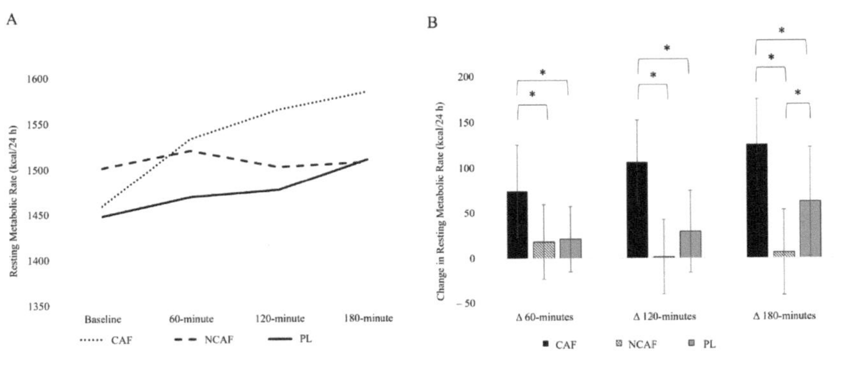

Figure 2. (**A**) REE values over the 180-min study period; (**B**) change from baseline in Resting Energy Expenditure; error bars represent 95% confidence intervals; * indicates statistical significance at the $p \leq 0.05$ level.

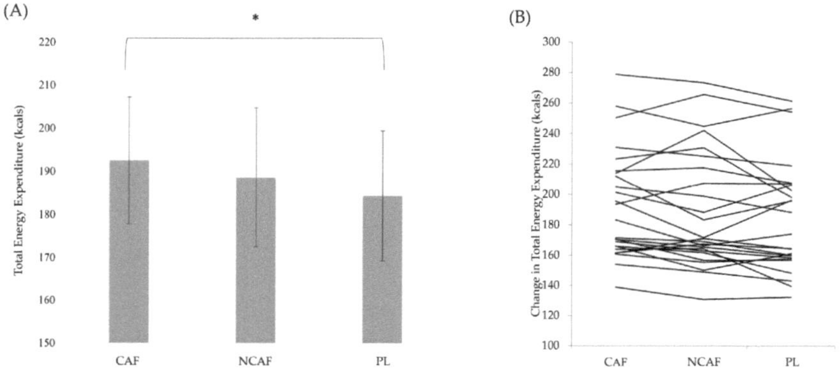

Figure 3. (**A**) Total energy expenditure over the 180-min study period; (**B**) individual responses for changes in total energy expenditure, with each line indicating one individuals REE values across each of the trials; * indicates statistical significance at the $p \leq 0.05$ level.

3.2. Respiratory Exchange Ratio (RER)

There was a significant interaction for RER ($p = 0.009$), but no post hoc analyses were statistically significant. There was no difference in the change in RER at 60-min ($p = 0.093$), 120-min ($p = 0.873$), or 180-min ($p = 0.735$) time points. Table 3 outlines the changes in RER and other secondary outcomes between time points.

Table 3. Respiratory exchange ratio, hemodynamic variables, and hunger scores.

	Baseline	Δ 60-min	Δ 120-min	Δ 180-min
Respiratory Exchange Ratio				
CAF	0.83 (0.06)	0.029 (0.007, 0.052)	0.006 (−0.014, 0.026)	0.008 (−0.014, 0.030)
NCAF	0.84 (0.04)	0.007 (−0.011, 0.026)	0.010 (0.011, 0.031)	0.019 (0.000, 0.038)
PL	0.85 (0.05)	0.008 (−0.008, 0.024)	0.005 (−0.027, 0.037)	0.009 (−0.021, 0.039)
Heart Rate (beats/min)				
CAF	65 (9)	−1.3 (−3.9, 1.3)	−0.5 (−3.0, 1.9)	−1.2 (−3.2, 0.8)
NCAF	68 (12)	0.3 (−2.6, 3.2)	−0.8 (−3.6, 2.0)	−3.3 (−6.5, −0.1) [b]
PL	65 (11)	−0.3 (−3.0, 2.5)	−0.8 (−3.5, 1.9)	0.7 (−1.8, 3.2)
Systolic Blood Pressure (mmHg)				
CAF	119 (14)	3.2 (0.5, 6.0)	5.5 (2.4, 8.5) [a,b]	6.3 (2.6, 10.1) [a,b]
NCAF	119 (13)	0.7 (−2.2, 3.6)	0.4 (−2.3, 3.2)	1.3 (−1.4, 4.0)
PL	119 (12)	0.0 (−2.9, 2.9)	1.3 (−1.4, 4.1)	1.5 (−1.5, 4.5)
Diastolic Blood Pressure (mmHg)				
CAF	76 (10)	6.1 (3.7, 8.5) [a,b]	4.0 (1.9, 6.2) [a]	4.3 (1.2, 7.5)
NCAF	75 (9)	1.5 (−0.9, 3.9)	1.1 (−0.8, 3.0)	2.9 (1.2, 4.6)
PL	76 (8)	0.7 (−1.3, 2.8)	1.2 (−1.3, 3.6)	1.3 (−0.5, 3.1)
Hunger Score				
CAF	3 (3, 4)	0 (−1, 0)	0 (−1, 0.25) [a]	−1 (−2, 0)
NCAF	4 (3, 4)	0 (−1, 0)	−1 (−2, −1)	−1 (−2, −1)
PL	4 (3, 4)	0 (−1, 0)	−1 (−1, 0) [a]	−1 (−2, 0)

Baseline data are presented as mean (standard deviation). Changes from baseline are presented at each of the 60-, 120-, and 180-min time points along with the 95% confidence intervals to demonstrate the variability of the intervention. Hunger scores are presented as medians (25th, 75th percentiles). Subscripts next to each value indicate a significant difference ($p \leq 0.05$) from another condition at each specific time point as follows: [a] = different from NCAF, [b] = different from PL.

3.3. Heart Rate and Blood Pressure

There was a significant interaction for HR ($p = 0.029$). There were no differences in the change in HR at 60-min ($p = 0.6540$) or 120-min ($p = 0.975$) time points. There was a difference at the 180-min time point ($p = 0.016$), with HR decreasing to a greater extent in

NCAF than in PL (p = 0.007), but no differences were present between CAF and NCAF (p = 0.160) nor CAF and PL (p = 0.112).

There was a significant interaction for SBP (p = 0.028). While there were no differences in the change in SBP at the 60-min (p = 0.186) time point, there was at both the 120-min (p = 0.007) and 180-min (p = 0.016) time points. At both 120-min and 180-min time points, SBP increased to a greater extent in CAF compared to NCAF (120-min, p = 0.006; 180-min, p = 0.013) and PL (120-min, p = 0.035; 180-min, p = 0.025). No differences were present between NCAF and PL at either 120-min (p = 0.465) or 180-min (p = 0.896) time points.

There was a significant interaction for DBP (p = 0.013). There was a significant difference in the increase in DBP at the 60-min time point (p = 0.002) with greater increases present in CAF compared to NCAF (p = 0.015) and PL (p = 0.001). No differences were present between NCAF and PL (p = 0.597). There were no differences in the change in DBP at 120-min (p = 0.052) or 180-min (p = 0.138) time points.

3.4. Hunger Scores

There was no difference in the change in hunger at 60-min (p = 0.949) or 180-min (p = 0.302) time points. At the 120-min time point, the change in hunger did differ (p = 0.006) with hunger scores decreasing to a greater extent in NCAF relative to both CAF (p = 0.018) and PL (p = 0.014). No differences were present between CAF and PL (p = 0.648).

4. Discussion

The primary results of this study were that the CAF treatment was able to elevate REE when compared to NCAF and PL at all time points, and that the NCAF was able to suppress perceptions of hunger two hours post-ingestion compared to CAF and PL treatments. Presumably, with an increase in energy expenditure and a decrease in energy intake (resulting from a reduction in hunger), ingestion of the multi-ingredient thermogenic supplements used in this study would promote fat loss outcomes over longer periods of supplementation. Relative to REE, the CAF treatment significantly elevated REE by 5.1%, 7.2%, and 8.6% at 60-min, 120-min, and 180-min post-ingestion respectively. This is consistent with previous research with respect to both effect and magnitude [6,25]. Previous work from our laboratory using similar caffeine-containing multi-ingredient thermogenic supplements reported nearly identical increases in REE three hours post-ingestion, with increases of approximately 9% compared to pre-supplementation baseline levels [26–28]. NCAF did not impact energy expenditure differently than the PL condition at any time point, despite limited evidence for ingredients found in both CAF and NCAF supplements to increase energy expenditure [19,20,29]. A potential explanation for this observed phenomenon may be an interaction effect, or lack thereof. CAF and NCAF contained identical ingredients and dosages with the sole exception of 200 mg caffeine anhydrous present in CAF but not NCAF. This single-ingredient deletion, coupled with the strength of randomized, placebo-controlled, triple-blinded, cross-over study design, allows for NCAF's lack of effect on REE to be reasonably attributed to the absence of caffeine.

Previous literature on the ingredients of the present thermogenic supplements, as outlined earlier, suggest that each ingredient was included to either suppress appetite, increase energy expenditure, or upregulate fat metabolism. Caffeine is well understood to achieve all three of those desired outcomes on its own [7,8,25,30] and thus could still potentially bring about the observed effects of CAF without any of the other ingredients. For the ingredients in the NCAF treatment, consideration of their efficacy without the presence of caffeine, as well as their dosing, is warranted.

Forskolin is understood to upregulate lipolysis by increasing activation of cAMP. Interestingly, Forskolin's ability to increase cAMP activation is maximized when in the presence of both adenosine deaminase and a methylxanthine [11]. This is due to adenosine deaminase's ability to degrade adenosine, an inhibitor of cAMP, as well as methylxanthines' ability to block adenosine from binding to adenylate cyclase and to inhibit the degradation of cAMP [11]. Caffeine is a methylxanthine [31], meaning Forskolin's ability to upregulate lipolysis is less robust when caffeine is not present. Black Ginger has evidence supporting its efficacy in increasing brown adipose tissue activation when containing a concentration of 4.07% dimethoxyflavone, which is the primary polymethoxyflavonoid in Black Ginger extract and is responsible for inhibiting cAMP phosphodiesterase, an enzyme that degrades cAMP [19]. CAF and NCAF contain a concentration of 2.5% dimethyoxyflavone but, given NCAF's lack of effect on REE, it is likely that caffeine is able to inhibit cAMP phosphodiesterase in CAF regardless of Black Ginger [11].

Grains of Paradise was another ingredient contained in both CAF and NCAF treatments. Previous research in healthy women reported that chronic (4 weeks) Grains of Paradise supplementation at a dosage of 30 mg/day, which was the same amount contained in CAF and NCAF treatments, significantly increased whole-body energy expenditure [29]. In the present study, only an acute dosage was assessed for its effectiveness of increasing resting energy expenditure. Considering NCAF's lack of an REE response, it is plausible that repeated daily dosages of Grains of Paradise across several weeks are needed to induce an increase in resting energy expenditure. Mucuna pruriens has some evidence supporting its effect on increasing levels of epinephrine and norepinephrine in humans with 5 g of crushed seed powder ingested daily [14]. L-DOPA is the active ingredient in Mucuna pruriens, and although both CAF and NCAF contain 153 mg of Mucuna pruriens standardized at 98% L-DOPA, it is not possible to determine whether the dose contained in the present supplements is optimal compared to the doses in previous research due to the unknown concentration of L-DOPA in 5 g of crushed seed powder. Nonetheless, caffeine is understood to increase the release of epinephrine and norepinephrine itself, which may help further explain the increase in REE observed in CAF but not in NCAF, especially in the possible event of the L-DOPA dose in CAF and NCAF being suboptimal.

RER was not different between any conditions, despite CAF's impact on REE. Caffeine has been previously shown to support weight loss and fat loss [32], suggesting caffeine favors increasing lipid metabolism. A recent meta-analysis by Conger and colleagues [33] suggests caffeine's effect on lipid metabolism is more robustly measured via blood biomarkers rather than via whole-body gas exchange, which may explain a lack of observed difference in RER from the CAF group to NCAF or PL.

There were minimal differences in hunger scores between the groups, with the only difference observed being a decrease in hunger at the 120-min time point for the NCAF group. *Caralluma fimbriata* found in both CAF and NCAF supplements has limited evidence to suggest appetite suppression [10]; however, the inclusion of this ingredient in both supplements does not explain the sole observed difference during the experiment. There is evidence that caffeine exerts modest reductions in hunger, but the data are not consistent [30,34,35], suggesting high interindividual variability. *Griffonia simplicifolia* is understood to decrease energy intake, predominantly through a reduction in carbohydrate consumption, at dosages of 750 mg or 900 mg of 5-HTP. Standardized to 98% 5-HTP, CAF and NCAF both provide 153 mg of *Griffonia simplicifolia*, which may be suboptimal and provide some explanation of CAF and NCAF's ability to impact hunger throughout each of the three-hour post-supplementation time frame. Furthermore, it may be possible that the interaction of caffeine with *Caralluma fimbriata*, *Griffonia simplicifolia*, and the other included ingredients may have produced a null effect on hunger that was not observed in NCAF due to its exclusion of caffeine.

Additional outcomes for this study included BP and HR responses between supplement conditions. The CAF condition had no effect on HR at any time point. This finding is consistent with prior work in our laboratory reporting that caffeine-containing thermogenic supplements had no effect on resting heart rate three hours post-ingestion [26–28]. The only observed difference in HR was a significant decrease in the NCAF group at the 180-min time point when compared to PL. Considering this significant decrease in tandem with the lack of observed increase in REE, there is plausibility in the combined effect of NCAF's ingredients being somewhat depressive if not inert. More research on the interactive mechanistic pathways of each ingredient is needed to better elucidate the cause behind the observed significant decrease in HR at the 180-min time point. Focusing on BP, CAF induced a significant increase in systolic blood pressure at 120- and 180-min time points compared to NCAF and PL. This observation was also consistent with previous work from our laboratory relative to consumption of a caffeine-containing thermogenic supplement and systolic blood pressure [26,28]. This is likely due to caffeine's known effect on cardiac contractility; it can be speculated that although not measured, stroke volume may have increased, leading to an increase in BP despite the absence of significant change in HR. While caffeine is well understood to acutely increase BP, caffeine ingestion has been shown to have positive effects on overall long-term cardiometabolic health and other diseases [36].

Despite the inherent strengths of the triple-blind placebo-controlled design, the present study is not without limitations. Inclusion of a caffeine-only condition would have better elucidated the impact caffeine has on desired outcome variables when combined, alone, and absent from the other active ingredients in CAF and NCAF. Future studies investigating caffeinated and non-caffeinated thermogenic supplements would benefit from including a caffeine-only condition. Another potential limitation of the design was not accounting for training status nor caffeine habituation differences among the subjects. A recent meta-analysis by Carvalho and colleagues [37] concluded that caffeine habituation did not influence the ergogenic potential of caffeine; however, more research is needed on the effects of caffeine habituation on metabolic response. Furthermore, the cross-over nature of the present study design suggests that any existing effect of caffeine habituation would have been equally present between all three supplement conditions. Additionally, the present study did not compare the results of the male participants to that of the female participants in a separate analysis, as that was not the purpose of this study. Some limited previous literature has shown sex differences in caffeine metabolism [38], making this an interesting area for future research. One delimitation of the present study is the lack of body weight adjusted dosing. Given that the purpose of this study was to examine the effects of a commercially available thermogenic supplement, it is important to recognize that the general population has abundant access to these types of pre-blended commercially available products. Therefore, the use of this supplement with pre-determined dosing provides greater generalizability of results for the average consumer. Finally, the time points chosen for data collection could be considered both a limitation and delimitation. The present study captured the results at 60-, 120-, and 180-min post-ingestion of the supplements. Potential differences between conditions could have remained unobserved between time points; however, the use of intermittent rather than continuous testing allowed for additional data (HR, BP, hunger scores) to be collected and therefore resulted in a more robust overall design.

5. Conclusions

The primary results of this study were that the CAF treatment was able to elevate REE when compared to NCAF and PL at all time points, and NCAF was able to suppress perceptions of hunger two hours post-ingestion compared to CAF and PL treatments. Additionally, the caffeinated supplement did not influence subjective hunger or heart rate responses but did increase acute systolic blood pressure responses. The non-caffeinated supplement exhibited a modest effect via hunger suppression and may find utility as a fat loss aid through that mechanism. However, further research is warranted for the ingredients of both thermogenic supplements, specifically focused on the impact of caffeine, varying ingredient dosages, and ingredient interaction effects on resting energy expenditure, fat metabolism, and appetite suppression in humans. Future research examining adaptations to chronic supplementation with Phoenix and Phoenix Caffeine-Free would also be beneficial for further elucidating the overall efficacy of these commercially available supplements.

Author Contributions: Conceptualization, B.I.C.; methodology, B.I.C.; formal analysis, S.D.; investigation, B.I.C., K.L., J.B., K.P., Y.A., W.A.A.-C., Y.O., J.S., C.B. and A.P.; resources, B.I.C., K.L., J.B. and K.P.; data curation, B.I.C., K.L., J.B., K.P. and J.S.; writing—original draft preparation, B.I.C., K.L., J.B. and K.P.; writing—review and editing, B.I.C., K.L., J.B., K.P., Y.A., W.A.A.-C., Y.O., J.S., C.B., A.P. and S.D.; visualization, B.I.C., K.L. and S.D.; supervision, B.I.C., K.L., J.B. and K.P.; project administration, B.I.C.; funding acquisition, B.I.C. All authors have read and agreed to the published version of the manuscript.

Funding: This research was funded by The Florida High Tech Corridor and Legion Athletics, Inc., grant number 1776108900.

Institutional Review Board Statement: The study was conducted in accordance with the Declaration of Helsinki and approved by the Institutional Review Board of the University of South Florida (ID: STUDY001825, approved 25 February 2022).

Informed Consent Statement: Informed consent was obtained from all subjects involved in the study.

Data Availability Statement: The data presented in this study are available on request from the corresponding author, BIC (ethical reasons).

Acknowledgments: The authors would like to acknowledge Karina Noboa for her assistance with piloting this study, as well as Alexis Belcher, Malena Sellen, Zachary Warhul, Rashed Daher, Eric Velazquez, Savannah Ericksen, Gretchen Shelton, Indira Alur, Cassandra Resler, and Andrew Heath for their assistance with data collection.

Conflicts of Interest: BIC has received grants and contracts to conduct research on dietary supplements; has served as a paid consultant for industry; has received honoraria for speaking at conferences and writing lay articles about sports nutrition ingredients and topics; is a member of the International Protein Board that disseminates knowledge on protein and protein products; has served as an expert witness on behalf of the plaintiff and defense in cases involving dietary supplements; and receives compensation for writing and providing educational services related to exercise and nutrition-related topics. No other authors have conflicts of interest to report. All researchers involved in this study independently collected, analyzed, and interpreted the results from this study, and have no financial interests concerning the outcome of the study. The funders had no role in the design of the study; in the collection, analyses, or interpretation of data; in the writing of the manuscript; or in the decision to publish the results.

References

1. Clark, J.E.; Welch, S. Comparing effectiveness of fat burners and thermogenic supplements to diet and exercise for weight loss and cardiometabolic health: Systematic review and meta-analysis. *Nutr. Health* **2021**, *27*, 445–459. [CrossRef]
2. Chappell, A.J.; Simper, T.; Barker, M.E. Nutritional strategies of high level natural bodybuilders during competition preparation. *J. Int. Soc. Sports Nutr.* **2018**, *15*, 4. [CrossRef]
3. Rukstela, A.; Lafontant, K.; Helms, E.; Escalante, G.; Phillips, K.; Campbell, B.I. Bodybuilding Coaching Strategies Meet Evidence-Based Recommendations: A Qualitative Approach. *J. Funct. Morphol. Kinesiol.* **2023**, *8*, 84. [CrossRef]
4. Nagao, T.; Komine, Y.; Soga, S.; Meguro, S.; Hase, T.; Tanaka, Y.; Tokimitsu, I. Ingestion of a tea rich in catechins leads to a reduction in body fat and malondialdehyde-modified LDL in men. *Am. J. Clin. Nutr.* **2005**, *81*, 122–129. [CrossRef]
5. Nagao, T.; Hase, T.; Tokimitsu, I. A green tea extract high in catechins reduces body fat and cardiovascular risks in humans. *Obesity (Silver Spring)* **2007**, *15*, 1473–1483. [CrossRef]
6. Acheson, K.J.; Zahorska-Markiewicz, B.; Pittet, P.; Anantharaman, K.; Jequier, E. Caffeine and coffee: Their influence on metabolic rate and substrate utilization in normal weight and obese individuals. *Am. J. Clin. Nutr.* **1980**, *33*, 989–997. [CrossRef] [PubMed]
7. Dulloo, A.G.; Geissler, C.A.; Horton, T.; Collins, A.; Miller, D.S. Normal caffeine consumption: Influence on thermogenesis and daily energy expenditure in lean and postobese human volunteers. *Am. J. Clin. Nutr.* **1989**, *49*, 44–50. [CrossRef] [PubMed]
8. Acheson, K.J.; Gremaud, G.; Meirim, I.; Montigon, F.; Krebs, Y.; Fay, L.B.; Gay, L.J.; Schneiter, P.; Schindler, C.; Tappy, L. Metabolic effects of caffeine in humans: Lipid oxidation or futile cycling? *Am. J. Clin. Nutr.* **2004**, *79*, 40–46. [CrossRef] [PubMed]
9. Temple, J.L.; Bernard, C.; Lipshultz, S.E.; Czachor, J.D.; Westphal, J.A.; Mestre, M.A. The safety of ingested caffeine: A comprehensive review. *Front. Psychiatry* **2017**, *8*, 80. [CrossRef] [PubMed]
10. Kuriyan, R.; Raj, T.; Srinivas, S.K.; Vaz, M.; Rajendran, R.; Kurpad, A.V. Effect of Caralluma fimbriata extract on appetite, food intake and anthropometry in adult Indian men and women. *Appetite* **2007**, *48*, 338–344. [CrossRef] [PubMed]
11. Litosch, I.; Hudson, T.H.; Mills, I.; Li, S.Y.; Fain, J.N. Forskolin as an activator of cyclic AMP accumulation and lipolysis in rat adipocytes. *Mol. Pharmacol.* **1982**, *22*, 109–115.

12. Okuda, H.; Morimoto, C.; Tsujita, T. Relationship between cyclic AMP production and lipolysis induced by forskolin in rat fat cells. *J. Lipid Res.* **1992**, *33*, 225–231. [CrossRef] [PubMed]
13. Lass, A.; Zimmermann, R.; Oberer, M.; Zechner, R. Lipolysis—A highly regulated multi-enzyme complex mediates the catabolism of cellular fat stores. *Prog. Lipid Res.* **2011**, *50*, 14–27. [CrossRef]
14. Shukla, K.K.; Mahdi, A.A.; Ahmad, M.K.; Shankhwar, S.N.; Rajender, S.; Jaiswar, S.P. Mucuna pruriens improves male fertility by its action on the hypothalamus-pituitary-gonadal axis. *Fertil. Steril.* **2009**, *92*, 1934–1940. [CrossRef]
15. Perez-Cornago, A.; Ramirez, M.J.; Zulet, M.A.; Martinez, J.A. Effect of dietary restriction on peripheral monoamines and anxiety symptoms in obese subjects with metabolic syndrome. *Psychoneuroendocrinology* **2014**, *47*, 98–106. [CrossRef] [PubMed]
16. Cangiano, C.; Ceci, F.; Cascino, A.; Del Ben, M.; Laviano, A.; Muscaritoli, M.; Antonucci, F.; Rossi-Fanelli, F. Eating behavior and adherence to dietary prescriptions in obese adult subjects treated with 5-hydroxytryptophan. *Am. J. Clin. Nutr.* **1992**, *56*, 863–867. [CrossRef]
17. Cangiano, C.; Laviano, A.; Del Ben, M.; Preziosa, I.; Angelico, F.; Cascino, A.; Rossi-Fanelli, F. Effects of oral 5-hydroxy-tryptophan on energy intake and macronutrient selection in non-insulin dependent diabetic patients. *Int. J. Obes. Relat. Metab. Disord.* **1998**, *22*, 648–654. [CrossRef]
18. Yoshino, S.; Kim, M.; Awa, R.; Kuwahara, H.; Kano, Y.; Kawada, T. Kaempferia parviflora extract increases energy consumption through activation of BAT in mice. *Food Sci. Nutr.* **2014**, *2*, 634–637. [CrossRef] [PubMed]
19. Matsushita, M.; Yoneshiro, T.; Aita, S.; Kamiya, T.; Kusaba, N.; Yamaguchi, K.; Takagaki, K.; Kameya, T.; Sugie, H.; Saito, M. Kaempferia parviflora extract increases whole-body energy expenditure in humans: Roles of brown adipose tissue. *J. Nutr. Sci. Vitaminol.* **2015**, *61*, 79–83. [CrossRef]
20. Sugita, J.; Yoneshiro, T.; Hatano, T.; Aita, S.; Ikemoto, T.; Uchiwa, H.; Iwanaga, T.; Kameya, T.; Kawai, Y.; Saito, M. Grains of paradise (*Aframomum melegueta*) extract activates brown adipose tissue and increases whole-body energy expenditure in men. *Br. J. Nutr.* **2013**, *110*, 733–738. [CrossRef]
21. Maeda, H.; Hosokawa, M.; Sashima, T.; Murakami-Funayama, K.; Miyashita, K. Anti-obesity and anti-diabetic effects of fucoxanthin on diet-induced obesity conditions in a murine model. *Mol. Med. Rep.* **2009**, *2*, 897–902. [CrossRef]
22. Abidov, M.; Ramazanov, Z.; Seifulla, R.; Grachev, S. The effects of Xanthigen in the weight management of obese premenopausal women with non-alcoholic fatty liver disease and normal liver fat. *Diabetes Obes. Metab.* **2010**, *12*, 72–81. [CrossRef]
23. Horner, N.K.; Lampe, J.W.; Patterson, R.E.; Neuhouser, M.L.; Beresford, S.A.; Prentice, R.L. Indirect calorimetry protocol development for measuring resting metabolic rate as a component of total energy expenditure in free-living postmenopausal women. *J. Nutr.* **2001**, *131*, 2215–2218. [CrossRef]
24. Coleman, A.; Freeman, P.; Steel, S.; Shennan, A. Validation of the Omron MX3 Plus oscillometric blood pressure monitoring device according to the European Society of Hypertension international protocol. *Blood Press. Monit.* **2005**, *10*, 165–168. [CrossRef]
25. Poehlman, E.T.; Despres, J.P.; Bessette, H.; Fontaine, E.; Tremblay, A.; Bouchard, C. Influence of caffeine on the resting metabolic rate of exercise-trained and inactive subjects. *Med. Sci. Sports Exerc.* **1985**, *17*, 689–694. [CrossRef]
26. Campbell, B.I.; Colquhoun, R.J.; Zito, G.; Martinez, N.; Kendall, K.; Buchanan, L.; Lehn, M.; Johnson, M.; St Louis, C.; Smith, Y.; et al. The effects of a fat loss supplement on resting metabolic rate and hemodynamic variables in resistance trained males: A randomized, double-blind, placebo-controlled, cross-over trial. *J. Int. Soc. Sports Nutr.* **2016**, *13*, 14. [CrossRef]
27. Campbell, B.I.; Perry, R.; Horsley, J.; Aguilar, D.; Shimshock, T.; Fox, C.; Vargas, A.; Colenso-Semple, L. A commercially available thermogenic dietary supplement increases resting metabolic rate in physically active males: A randomized, double-blind, placebo-controlled investigation. *J. Diet Suppl.* **2020**, *17*, 150–160. [CrossRef]
28. Campbell, B.I.; Zito, G.; Colquhoun, R.; Martinez, N.; Kendall, K.; Buchanan, L.; Lehn, M.; Johnson, M.; St Louis, C.; Smith, Y.; et al. The effects of a single-dose thermogenic supplement on resting metabolic rate and hemodynamic variables in healthy females—A randomized, double-blind, placebo-controlled, cross-over trial. *J. Int. Soc. Sports Nutr.* **2016**, *13*, 13. [CrossRef] [PubMed]
29. Sugita, J.; Yoneshiro, T.; Sugishima, Y.; Ikemoto, T.; Uchiwa, H.; Suzuki, I.; Saito, M. Daily ingestion of grains of paradise (*Aframomum melegueta*) extract increases whole-body energy expenditure and decreases visceral fat in humans. *J. Nutr. Sci. Vitaminol.* **2014**, *60*, 22–27. [CrossRef] [PubMed]
30. Schubert, M.M.; Grant, G.; Horner, K.; King, N.; Leveritt, M.; Sabapathy, S.; Desbrow, B. Coffee for morning hunger pangs. An examination of coffee and caffeine on appetite, gastric emptying, and energy intake. *Appetite* **2014**, *83*, 317–326. [CrossRef] [PubMed]
31. Saboury, A.A.; Divsalar, A.; Ataie, G.; Amanlou, M.; Moosavi-Movahedi, A.A.; Hakimelahi, G.H. Inhibition study of adenosine deaminase by caffeine using spectroscopy and isothermal titration calorimetry. *Acta Biochim. Pol.* **2003**, *50*, 849–855. [CrossRef]
32. Tabrizi, R.; Saneei, P.; Lankarani, K.B.; Akbari, M.; Kolahdooz, F.; Esmaillzadeh, A.; Nadi-Ravandi, S.; Mazoochi, M.; Asemi, Z. The effects of caffeine intake on weight loss: A systematic review and dos-response meta-analysis of randomized controlled trials. *Crit. Rev. Food Sci. Nutr.* **2019**, *59*, 2688–2696. [CrossRef]
33. Conger, S.A.; Tuthill, L.M.; Millard-Stafford, M.L. Does caffeine increase fat metabolism? a systematic review and meta-analysis. *Int. J. Sport Nutr. Exerc. Metab.* **2023**, *33*, 112–120. [CrossRef]
34. Grant, C.L.; Coates, A.M.; Dorrian, J.; Paech, G.M.; Pajcin, M.; Della Vedova, C.; Johnson, K.; Kamimori, G.H.; Fidock, J.; Aidman, E.; et al. The impact of caffeine consumption during 50 hr of extended wakefulness on glucose metabolism, self-reported hunger and mood state. *J. Sleep Res.* **2018**, *27*, e12681. [CrossRef] [PubMed]

35. Schubert, M.M.; Irwin, C.; Seay, R.F.; Clarke, H.E.; Allegro, D.; Desbrow, B. Caffeine, coffee, and appetite control: A review. *Int. J. Food Sci. Nutr.* **2017**, *68*, 901–912. [CrossRef] [PubMed]
36. Grosso, G.; Godos, J.; Galvano, F.; Giovannucci, E.L. Coffee, caffeine, and health outcomes: An umbrella review. *Annu. Rev. Nutr.* **2017**, *37*, 131–156. [CrossRef]
37. Carvalho, A.; Marticorena, F.M.; Grecco, B.H.; Barreto, G.; Saunders, B. Can I have my coffee and drink it? A systematic review and meta-analysis to determine whether habitual caffeine consumption affects the ergogenic effect of caffeine. *Sports Med.* **2022**, *52*, 2209–2220. [CrossRef]
38. Lassen, M.L.; Byrne, C.; Sheykhzade, M.; Wissenberg, M.; Hurry, P.K.; Schmedes, A.V.; Kjaer, A.; Hasbak, P. Sex differences and caffeine impact in adenosine-induced hyperemia. *J. Nucl. Med.* **2022**, *63*, 431–437. [CrossRef] [PubMed]

nutraceuticals

Review

Seaweed as a Safe Nutraceutical Food: How to Increase Human Welfare?

João Cotas *, Joana O. Tavares, Rita Silva and Leonel Pereira

Marine Resources, Conservation and Technology, Marine Algae Lab, CFE—Centre for Functional Ecology—Science for People & Planet, Department of Life Sciences, University of Coimbra, 3000-456 Coimbra, Portugal; joana02tavares@gmail.com (J.O.T.); leonel.pereira@uc.pt (L.P.)
* Correspondence: jcotas@uc.pt

Abstract: Seaweeds have been utilized for millennia in Asian countries, although they have only more recently become popular in Western society. They began to be used in ancient times because of their long-term properties and, nowadays, seaweeds are being targeted as a potential tool to combat climate change. There are not many laws governing seaweeds because they have just lately been utilized as food. However, guidelines are being developed to regulate their manufacture and use. Because of seaweed's tendency to accumulate components, whether helpful or poisonous, limited doses of certain substances have been established to prevent consumer overdosage. Aside from chemical safety, microbiological safety is important for people, and preventing any pathogen from spreading and infecting seaweeds is critical. As a result, systems and ways to safeguard consumers must be developed. Because various seaweed species have varied compositions, certain seaweeds may be safer nutraceuticals than others. To ensure the safety of seaweed-based food items, the HACCP (Hazard Analysis Critical Control Point) system needs to be used. The majority of seaweeds consumed come from aquaculture; however, others come from wild harvesting. To ensure the success of the cultures, the waters must be tested for chemicals and biological risks, as well as for the pH, salinity, and temperature. Seaweeds have enormous promise in many industries, but in the food industry, they are beginning to play a major role, and seizing the chances to produce innovative, safe, and sustainable food sources is strongly advised. This critical review investigates the real potential of seaweed as a human food source and as a nutraceutical solution. This review also focuses on the usage of seaweed as a food product and the procedures required to prepare it. In addition, it compiles information on the applicable legislation and regulations, and it addresses the lengthy road that has to be traveled to increase human well-being by employing a new food source in a controlled manner while simultaneously reducing the human population's health problems.

Keywords: human welfare; new methods; food safety; food security; bioavailability

Citation: Cotas, J.; Tavares, J.O.; Silva, R.; Pereira, L. Seaweed as a Safe Nutraceutical Food: How to Increase Human Welfare? *Nutraceuticals* **2024**, *4*, 323–362. https://doi.org/10.3390/nutraceuticals4030020

Academic Editors: Ivan Cruz-Chamorro and Guillermo Santos Sánchez

Received: 23 May 2024
Revised: 13 June 2024
Accepted: 21 June 2024
Published: 29 June 2024

1. Introduction

The world's population is expected to reach ten billion in the next three decades. Food production must be increased by 70% [1]. In addition, meat output will double in the same time period. As a result, the hunt for alternative food sources is critical [2]. Agriculture and intensive farming have led to arable land saturation and restricted availability of fresh water [3,4]. Thus, aquatic organisms can be key for food safety and security due to being cultivated in aquatic systems, with usage of terrestrial land, where agriculture, farms, and food industries near coastal areas can be used to transform and work the seaweeds into the food industry [5].

Seaweeds are one of the most promising sustainable food types. Their capacity to absorb CO_2 from the water and atmosphere is an excellent way to combat climate change [6]. Furthermore, they have the ability for rapid development in the water, which simplifies manufacturing methods and decreases production costs [4]. Despite the fact that seaweeds have been used for millennia in nations such as China, Japan, and Korea, they were only recently introduced as a food source in Western civilizations [5]. After WWII, it was discovered that there was inadequate protein consumption due to exponential population expansion. To combat this nutrient deficiency during the war, seaweeds played an important role, as they are high in various macronutrients and micronutrients, including vitamins, minerals and proteins [7,8]. Also, nowadays, seaweed could play this important role again to mitigate malnutrition and food insufficiency. The nutritional potential of seaweeds is directly tied to their biochemical profiles and bioactive qualities, which are known to vary greatly between species [9]. Isolated polysaccharides (e.g., alginate and fucoidan), proteins (phycobiliproteins), polyphenols (e.g., phlorotannins), carotenoids (e.g., fucoxanthin), and n-3 long-chain polyunsaturated fatty acids (e.g., eicosapentaenoic acid) are among the highlights [10]. The biochemical composition and functional properties of algae feedstocks have been extensively studied over the last few decades, aided greatly by the advancement and development of new techniques that enable high-resolution profiling of proteins, lipids, polysaccharides, pigments, and others [11–13]. Overall, seaweeds are a sustainable supply of natural high-value bioactive chemicals with the potential to produce novel human nutrition products.

They have also been employed as a new food due to their polymers, such as carrageenan, agar, and alginate, as well as their gelling, emulsifying, and thickening capabilities [14,15].

The FAO and WHO published a report that evaluated the current food safety information regarding seaweed produced from both wild stocks and aquaculture and proposed more discussion as well as international advice [16]. The report noted that morbidities and deaths associated with eating seaweed are uncommon, but it warned that the limited evidence raises worries that some risks may be present in seaweed [17]. These risks include chemical hazards such as heavy metals (principally inorganic arsenic and cadmium), persistent organic pollutants (e.g., dioxins and polychlorinated biphenyls), radionuclides and pesticide residues; microbiological hazards (e.g., *Salmonella* spp., *Bacillus* spp., and norovirus); physical hazards (e.g., metal pieces, glass splinters, crustacean shells, micro- and nanoplastics); and allergens [17]. Thus, there is a need for technology and methods to ensure seaweed's potential for human welfare and reduce the inherent risks to human health.

This critical review attempts to explore seaweed as food for humans. It focuses on the application of seaweed as a food product and the methods needed to make it safe to eat. In addition, it gathers information on the relevant legislation and regulations and discusses the long road that needs to be traveled to improve human well-being by using a new food source, in a controlled way, and also to decrease the health problems of the human population.

2. Seaweeds as a Possible Nutraceutical Food Source

Seaweeds are a basic ingredient in South Asiatic cuisine; however, there has been a considerable increase in seaweed consumption in European countries over the last decade, from traditional sushi to afternoon snacks and novel foods (processed or not) [18]. The current move in Western diets has shifted to more plant-based and sustainable food sources, and it is expected to amplify in the next years. Thus, it is expected that by 2050, 0.1% of our seas would be dedicated to growing seaweed as a food source, producing 15 times more seaweed than is currently produced to meet the rising world demand [19]. Despite the recent spectacular rise of seaweed as a food product, outside of Asia, there are presently no food standards governing the safety and quality of seaweed, posing a potential harm to consumer health [20].

Seaweed is often regarded as a 'superfood' in the Western world, connected with healthy lifestyles and sustainability. Its significant nutritional value is derived from macro- and micronutrients such as vitamin B_{12}, dietary fibers, and omega-3 fatty acids, among others, and it is a rich source of various bioactive substances that have health advantages (e.g., polyphenols, sulphated polysaccharides, pigments) [20]. However, for seaweed to be a potential safe nutraceutical food source, there is a need to overcome several steps from biomass production until the market (Figure 1), although there is need for a more robust system.

In general, compared to green and brown algae, red algae contains a high amount of proteins, reaching 47% (*Pyropia tenera*) of the dry matter [21]. On the other hand, the lipids in these seaweeds present relatively lower contents. Also, there are not significant differences between red and brown algae, as they both revealed low fat and high fiber content. Red algae contains soluble fibers such as sulphated galactans (agars and carrageenans), xylans and floridean starch [22–24]. Brown seaweeds demonstrate lower protein content when compared with the other seaweeds but have more mineral and phenolic compound content [23].

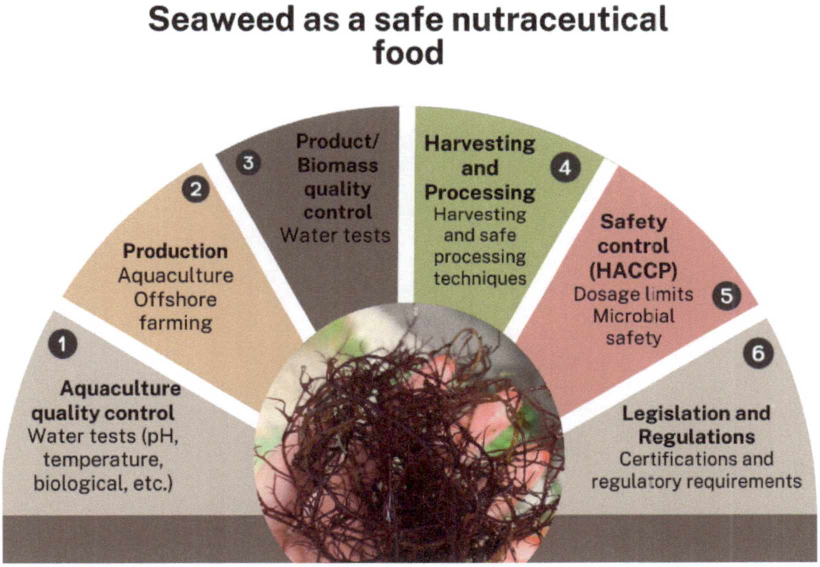

Figure 1. The real and normal steps of seaweed biomass as a human food source.

2.1. A Good Source of Nutrients?

Seaweeds are presented as a sustainable source of protein and dietary fibers from the sea, decreasing the pressure on wild-capture fisheries and being an eco-friendly alternative to meat protein [25]. Seaweed mariculture seems like a perfectly fitting solution to the secure food and feed demand, all the while avoid placing additional pressure on arable/available land and freshwater resources (leaving aside possible land-based cultivation of certain seaweed species) [26]. Studies generally agree that seaweeds contain high-quality proteins with essential amino acids (lysine, methionine, to name a couple, depending on the strain) and are a rich source of other bioactives, including taurine, lipids, carotenoids, and pigments [27–29].

In reality, seaweeds are a low-calorie meal, making them appealing for inclusion in the human daily diet (Table 1). In addition, the incorporation of vitamins, fibers, and proteins provides a high nutritional value as well as a variety of health advantages [18]. Polysaccharides from seaweeds (Figure 2), for example, have a positive effect on the digestive system but are calorie-free, unlike fibers. Because of their positive properties, these biological molecules might be employed to create innovative and functional meals, as well as implemented in pharmacological and medical applications [30–33].

Table 1. Nutrient composition of some edible seaweed (% dry weight) (adapted from Pereira [32,33]; Guiry and Guiry [24]).

Species	Common Name	Proteins	Ashes	Dietary Fibers	Carbohydrates	Lipids
Alaria esculenta (P)	Winged Kelp	9–20	-	42.86	46–51	1–2
Caulerpa lentillifera (C)	Green Caviar	10–13	24–37	33	38–59	0.86–1.11
C. racemosa (C)	Sea Grapes	17.8–18.4	7–19	64.9	33–41	9.8
Chondrus crispus (R)	Irish Moss	11–21	21.08	10–34	55–68	1.0–3.0
Codium fragile (C)	Dead Man's Fingers	8–11	21–39	5.1	39–67	0.5–1.5
Eisenia bicyclis (P)	Arame	7.5	9.72	10–12	60.6	0.1
Fucus spiralis (P)	Spiral Wrack	10.77	-	63.88	-	-
F. vesiculosus (P)	Bladder Wrack	3–14	14–30	45–59	46.8	1.9
Gracilaria chilensis (R)	Penco	13.7	18.9	-	66.1	1.3
Himanthalia elongata (P)	Sea Spaghetti	5–15	30–36	33–37	44–61	0.5–1.1
Laminaria digitata (P)	Oarweed	8–15	37.59	37.3	48	1.0
Palmaria palmata (R)	Dulse	8–35	15–30	28.57	46–56	0.7–3
Porphyra umbilicalis (R)	Purple laver	29–39	12	29–35	43	0.3
Pyropia tenera (R)	Nori	33–47	20.5	12–35	44.3	0.7
Pyropia yezoensis (R)	Nori Seaweed	31–44	7.8	48.6	44.4	2.1
Saccharina japonica (P)	Sweet Kelp	7.5	26.63	10–36	51.9	1.0
S. latissima (P)	Sugar Kelp	6–26	34.78	30	52–61	0.5–1.1
Sargassum fusiforme (P)	Hizikia	11.6	19.77	17–62	30.6	1.4
Ulva compressa (C)	Tape Weed	21–27	18.6	33–45	48.2	0.3
U. lactuca (C)	Sea Lettuce	10–25	12.9	29–38	36–43	0.6–1.6
U. australis (C)	Lacy Sea Lettuce	20–26	-	-	47.0	-
U. rigida (C)	Glasán	18–19	28.6	38–41	43–56	0.9–2.0
U. reticulata (C)	Ribbon Sea Lettuce	17–20	-	65.7	50–58	1.7–2.3
Undaria pinnatifida (P)	Wakame	12–23	26–39	16–46	45–51	1.5–4.5

C—Chlorophyta (green algae); R—Rhodophyta (red algae); P—Phaeophyceae (brown algae).

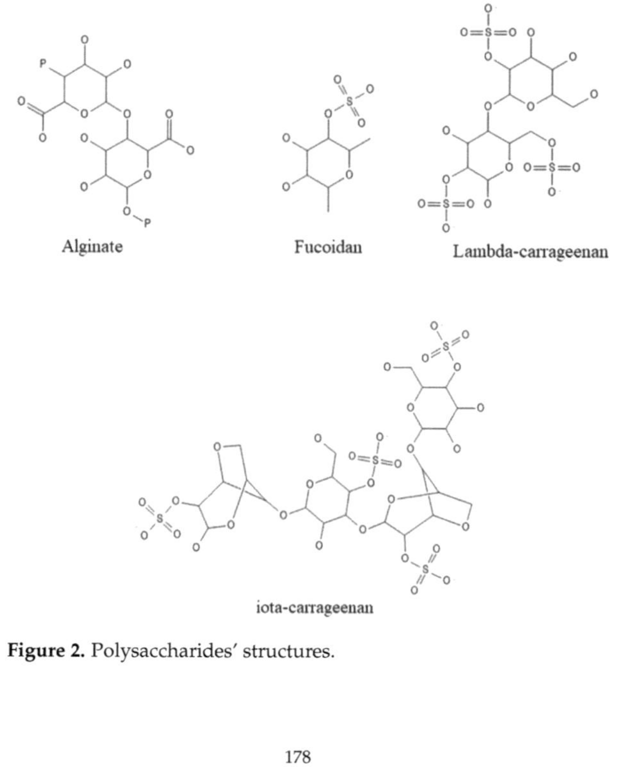

Figure 2. Polysaccharides' structures.

Seaweed bioactive substances have fueled nutraceutical interest in these functional meals. Polysaccharides (for instance, alginate, fucoidan, ulvan, agar, and carrageenan), proteins (for example, amino acids, phycobiliproteins (for example, phycoerythrin) (Figure 3), carotenoids (beta-carotene and fucoxanthin (Figure 3)), phenolic compounds (such as phlorotannins), vitamins (particularly vitamins A, B, C, D, E, and K), essential minerals (such as calcium, iron, iodine, magnesium, and potassium) and polyunsaturated fatty acids (namely ω-3 fatty acids) constitute the interesting group of compounds [34].

These seaweed-derived compounds have been studied in the treatment of human diseases and pathologies such as hyperglycemia, diabetes, metabolic disorders, cancer, pathogenic diseases, aging, obesity, bone-related diseases, and neurodegenerative and cardiovascular diseases [4,8,10,35–38].

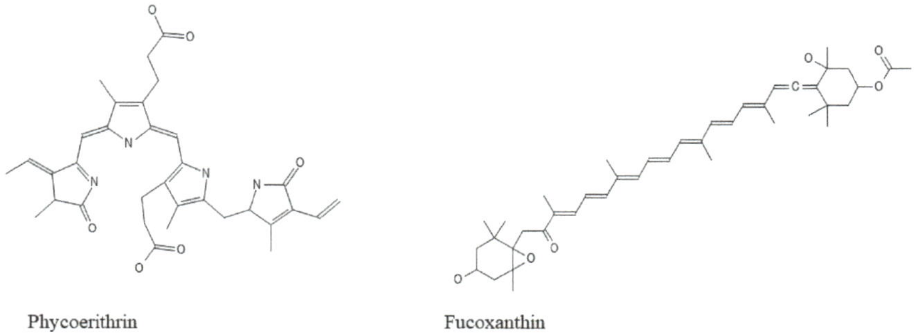

Figure 3. Pigments' structures.

There are about 600 edible seaweed species recognized globally, with *Porphyra/Pyropia* sp. (Rhodophyta), *Undaria pinnatifida* (Figure 4), and *Saccharina/Laminaria* sp. (Phaeophyceae) being the main three used in Asian meals, known as the food product names "nori", "wakame", and "kombu" [39]. Although their contents vary greatly according to the species, location, growing/production circumstances, and harvest season, they are all capable of delivering essential macro- and micronutrients. Macronutrients are proteins, lipids, and carbohydrates that must be consumed in bigger amounts on a regular basis to provide energy. Micronutrients are vitamins, minerals, and trace elements that, although required in minute amounts, are essential for maintaining key processes [4].

Figure 4. *Undaria pinnatifida* young specimen.

Seaweeds are complete protein sources because they include all nine essential amino acids (EAAs) required for protein synthesis, tissue repair, and nutrition absorption: histidine, isoleucine, leucine, lysine, methionine, phenylalanine, threonine, tryptophan, and valine [40,41]. We are unable to synthesize EAAs, so we must consume them through our food to ensure optimal organ function. Seaweeds have very similar patterns in terms of the non-essential amino acids (NEAAs), with aspartic and glutamic acid accounting for a significant portion (20–32%) of the total amino acids [42,43]. In fact, significant amounts of these two amino acids are responsible for the seaweeds' characteristic taste and "umami" flavor [39,43].

Seaweeds are made up of hydrocarbons (e.g., squalene), sterols (e.g., cholesterol), and mostly fatty acids (FAs), the amount of which varies depending on the species, environmental circumstances, and life cycle phases [39,44,45]. Myristic acid (C14:0) and palmitic acid (C16:0), the respective monounsaturated versions, palmitoleic acid (C16:1) and oleic acid (C18:1), and linoleic acid (C18:3), the most prevalent PUFA, were the most abundant FAs discovered in the three groups of examined seaweeds [45–47]. Seaweed structural polysaccharides are similar to terrestrial plant polysaccharides and primarily consist of celluloses, hemicelluloses, xylans, and mannans, whereas storage polysaccharides, such as carrageenan, alginate, and agar, are more specific to seaweed species and are the most commercially exploited components of seaweeds [15,48]. These textural and stabilizing storage polysaccharides are commonly extracted by the hydrocolloid industry and employed in culinary applications [49]. Vitamins, like micronutrients, are required for optimal physiological function in minute quantities and their solubility determines their classification. Vitamins A, D, E, and K are examples of fat-soluble vitamins. These necessitate appropriate dietary fat intake because lipids are required for absorption, transport, and cellular uptake. Water-soluble vitamins include vitamin C, vitamin B_1 (thiamine), vitamin B_2 (riboflavin), vitamin B_6, vitamin B_{12}, niacin, pantothenic acid, biotin, and folate [4].

The mineral profile of seaweeds varies according to the species, geographical location of harvest, wave exposure, seasonal, annual, environmental and physiological factors, type of processing, and method of mineralization. However, the mineral content, especially I, Fe, Mg, Ca, P, Na, and K, is abundant in seaweeds [8,50].

Due to their bioactive compounds and enrichment in other nutritional compounds, they are considered an excellent nutraceutical raw source for human welfare [51]. There is also interest in nutraceutical preparations for health benefits, and macroalgae offer promise as dietary supplements, with ingestion connected with positive health impacts due to the components they contain. Nutraceutical products are non-specific biological medications intended to improve general health, alleviate signs and symptoms, and prevent cancer. The term "nutraceutical" is a combination of the words "nutrient" (a nutritious dietary component) and "pharmaceutical" (a therapeutic drug) [52]. Seaweed bioactive compounds have the potential to play an important therapeutic role in illness prevention in humans. However, for this concept to be applied in the fulfilment of the concept, there is a need to certify that seaweeds promote health benefits with reduced secondary effects [52,53].

2.2. *A Danger to Human Welfare?*

Seaweeds absorb minerals and important trace elements from their immediate surroundings due to their unique cell wall structure with excellent biosorption characteristics. This great accumulation capability, however, may cause seaweed to accumulate potentially toxic materials found in its surroundings [54]. As a result, hazardous metals, including cadmium, lead, mercury, inorganic arsenic, and iodine, to mention a few, are frequently present at quantities many orders of magnitude greater than in water [55]. Unregulated distribution of seaweed species containing these components in high quantities may have negative health consequences for the unwary consumer. Certain seaweed species can have extraordinarily high quantities of iodine, resulting in excessive iodine levels when even modest amounts are ingested [56]. Although a necessary element of thyroid hormone production, excessive iodine intake can cause thyroid gland malfunction. In contrast, mercury, lead, cadmium, and inorganic arsenic have no biological purpose in the human body and can be toxic in even minute doses [57]. Furthermore, lead is categorized as potentially carcinogenic and neurotoxic, while inorganic arsenic is recognized as a "class I carcinogen", and their presence in seaweed puts customers' health at risk [20,58]. It is also important to note that various seaweed species have varying biosorption preferences for these hazardous substances [59]. Furthermore, extrinsic factors such as the geographic location, aquatic habitat, season, sample or processing methods can all influence the elements' composition, even within the same seaweed species. Several food regulators are worried about consumer health and safety as a result of seaweed intake due to its possible toxicological profile [60].

3. Seaweed as Food: Regulations

The usage of seaweeds, as well as their production and marketing, have changed dramatically in recent decades, particularly in Europe, where their use is not always customary [61,62]. However, this relatively young European sector is beset by a lack of rules and norms governing the production and use of seaweeds [63]. This absence of controls is especially concerning since seaweeds collect and retain substances that, when consumed, can be toxic to people. Biosorption of heavy metals, such as mercury or arsenic, as well as the mineral iodine, in particular, need regulatory attention and the establishment of appropriate intake recommendations [63,64]. Only a few laws for seaweeds and seaweed products exist in Europe to ensure safe consumption. These regulations cover aspects of biology (including the authorization of seaweed species) as well as the composition of imported and locally produced products. The EU has also set out to conduct risk evaluations to define appropriate levels of ingestion and, lastly, to inform laws. Adaptations and extensions of the present legislation and guidelines are expected based on these risk assessments as well as the developing nature of the industry in Europe [63,65].

Outside of Europe, regulatory systems governing seaweeds and seaweed-based foods are in various stages of development. The market for edible seaweeds and seaweed products in the United States, like in Europe, is currently increasing [66]. In the United States, a lack of national legislation has recently resulted in the provision of an extensive set of guidelines and regulations for the safe consumption of seaweeds and seaweed-derived products by the Connecticut Sea Grant in collaboration with the Connecticut Department of Agriculture Bureau of Aquaculture [67]. By bridging the gap in national legislation and locally different state rules, these latest recommendations constitute an important step toward the industry's efficient and safe expansion, which is required to satisfy the rising demand [66].

In Asia, where seaweeds have long been used and consumed, regulatory frameworks are more developed than in places of the world where the seaweed business is a relatively new sector. More information on the particular rules in China and Japan, as well as the other regions mentioned above, may be found in Campbell et al. [68].

The different types of seaweed do not impact the regulations, as they are considered similar between them, although seaweed compounds approved for the food industry (carrageenans, alginates and agar) are separated and very controlled in terms of the origin and extraction and purification methods. The major critical point is the mineral content of seaweeds, which includes all types of seaweeds; however, brown seaweeds are more prone to accumulate heavy metals [4].

How the Regulations Can Make the Seaweed Secure, Reducing the Risks to Human Health

The primary goal of law and regulation is to create and translate policies concerning the environment and/or health into rights and responsibilities, as well as to establish means to guarantee that such obligations are met [69].

Legislation is an important aspect of every business, especially if it impacts all the parties involved in the project, such as the primary producers and end-product makers. The legislation's goal is to safeguard both consumers and producers while also controlling the market for each commodity [63].

The accumulation level of pollutants is considerable due to the peculiarities of seaweeds' cell wall structure and their growth environment [70]. To ensure that humans are not endangered, legislation was enacted to establish the maximum amounts of harmful components in macroalgae, to regulate the amount of certain compounds ingested and to prevent food poisoning [63].

Since macroalgae has only recently gained popularity as a food product in the EU (European Union), the quantity of the laws on seaweeds in the EU is limited. The European Commission in 2006 established limit amounts of harmful chemicals for various meals in Commission Regulation (EC) No. 1881/2006 [65,71]. However, the EU has not set any limits for the amount of cadmium or inorganic arsenic found in various seaweed products [71].

As previously stated, macroalgae food items are not heavily regulated; nevertheless, macroalgae-derived additives such as carrageenan, agar, alginate, and others are controlled under (EC) No. 1333/2008 [65,72]. Because of their emulsifying, thickening, and gelling qualities, they have been employed. These items are designated by the numbers E401–E407a, and creating these categories will aid in the control of these substances as well as in their identification when needed [63].

Despite these negative and positive impacts of the human food source and food regulations, there are various aspects that can be checked to observe and make seaweed a future secure food source [26].

4. Seaweed as a Future Secure Food

Seaweeds are rich in nutrients and contains high concentrations of proteins, amino acids and minerals [73]. However, because of their tendency to retain substances such as heavy metals and other hazardous chemicals, their consumption must be continuously regulated [74]. Anthropogenic activities began to expand significantly around two decades ago, increasing the number of harmful substances released into the atmosphere and seas. All of these actions have an effect on algae, either directly or indirectly [73,75].

Biological, chemical, and physical dangers are the three types of hazards evaluated in relation to food safety [76]. As previously stated, seaweeds are excellent storage for toxic compounds [70], which are the most dangerous to seaweeds [63]. Chemical intoxication can cause a variety of health issues, and excessive use of certain substances can harm the neurological, circulatory, enzymatic, endocrine, and immunological systems [77,78]. Monitoring and limiting human consumption of seaweeds is critical to this.

4.1. Seaweed Aquaculture

Even though aquaculture produces most of the seaweed consumed globally (almost 97%), involving key instruments for seaweed food safety and nutritional, water composition analysis is uncommon in offshore and inshore aquaculture (Figure 5) [79].

Figure 5. *Ulva* sp. inshore cultivation in Mondego river estuary, Figueira da Foz, Portugal.

Thus, the nutritional profile of seaweed can have a large range of values due to the intrinsic differences across species and specimen, but also to changes in the nutrient concentrations in water, water temperature, and salinity [80,81].

Thus, these variations can make seaweed aquaculture a risky method for ensuring food safety (due to the possibility of heavy metal accumulation in seaweed); nowadays, to overcome this, chemical analysis and nutritional evaluation per cultivated seaweed batch are required [4], but it is expensive and one of the major issues in offshore aquaculture. Seaweed aquaculture may need novel techniques to regulate or manage the nutritional content of seaweeds, such as the use of data simulation to anticipate nutritional values and ensure the best nutritional values [38]. In comparison to wild specimens and offshore production, nutritional values in inshore aquaculture can be more precisely monitored and standardized since water and abiotic elements can be readily regulated [38,82].

Now, farming is being enhanced by incorporating technology that can be further explored to provide an overall insight and optimize the number and quality of seaweeds farmed, lowering the cultivation expenses, low seaweed quality, and nutrient pollution [83].

The next step in seaweed aquaculture involves the informatization of the cultivation system, allowing for more controlled cultivation systems and predictions of seaweeds' nutritional value using a specific kinetic model to obtain similar values for different batches of seaweed cultivated in the same area, thereby improving seaweeds' safety after cultivation by eliminating concerns such as water nutrient fluctuations and potential contaminations [81]. Sensors that provide real-time data regarding critical information in the offshore system, such as the nutrients, pH, temperature, and salinity, can also be incorporated. In this scenario, inshore culture is the optimal growing method for improving seaweed food safety; nevertheless, this safety comes at a cost to seaweed output [52,53,83]. Thus, in the future, cost-effective cultivation technologies (both offshore and onshore) that improve food safety while lowering costs will be required [83,84].

Green seaweeds (Chlorophyta), of which there are 2200 species, may grow to a maximum height of 1 m. They are mostly farmed in inshore systems or harvested wild. Red seaweeds (Rhodophyta), which number 6100 species, are effective photosynthesizers in deeper seas. They vary in length and are comparable to green seaweeds. They are grown for direct consumption and the polysaccharide extraction business. They are typically grown in nearshore systems (off-bottom line, floating raft, and basket methods). They are mostly farmed in longlines in offshore circumstances; after *Porphyra* sp., the giant brown seaweeds (Kelps–Phaeophyceae) are the most produced seaweeds in the world [23,83].

4.2. Seaweed Storage

Seaweed storage requirements are determined by variables such as the time required between harvest and processing, climatic conditions, the quality of the collected material, expenses, energy requirements, and environmental consequences. Temporary seaweed storage might occur at the seaweed farm, aboard ships, in the harbor, near the food processing business, or near stores. Pre-processing (for example, drying) is frequently necessary before seaweed may be preserved. There has been very little study on these crucial areas [85].

As seaweeds have a short shelf life, methods to reduce the possible biological dangers are critical. Dehydrating the seaweeds is one method. By drying the product, the water content decreases, which means there is less accessible water [86,87], reducing the possibility for infections to proliferate and reproduce and ensuring food safety [88]. Drying processes extend the shelf life of seaweeds but can have a detrimental influence on their chemical composition and their bioactive qualities [89].

4.3. Seaweed Commercial Products

For large-scale seaweed processing, drying technologies are unfeasible. Sun-drying needs wide areas and is weather-dependent; however, it is low-cost and already employed in feed production. Oven-drying (or convective air-drying) is an energy-intensive and costly industrial process. Other processes, such as lyophilization, are mostly utilized in businesses where the target components are small molecules rather than algae [31,90–92].

Another safe method to ingest seaweeds is in fermented foods. This is a food produced mostly of fermented and salted cabbages, but it has lately been tested with seaweeds. The acidity caused by the low pH and high salt concentration prevents undesired microbes from developing, therefore preserving the product [88]. When compared to terrestrial biomass, seaweeds have distinct carbohydrates, which provide a difficulty, particularly owing to the presence of mannitol and laminarin. Based on this understanding, current terrestrial biomass methods cannot be directly adapted to macroalgae biomass, and the selection of suitable microbes is critical for the success of seaweed fermentation [31].

Opportunities for marketing seaweeds as fresh vegetables are growing as a result of the increasing demand in foods that are minimally processed [4]. and call for more research on how the storage temperature affects the quality of fresh seaweeds and how to monitor the seaweed quality during refrigerated storage [85]. Because fresh seaweeds are highly perishable and begin to deteriorate quickly after harvest, rapid analysis techniques and more analysis are needed. Thus, the application of Modified Atmosphere Packaging has been demonstrated to be a viable strategy for conserving minimally processed seaweed, outperforming the efficacy of vacuum packing [93]. It should be highlighted that the seaweed species can display diverse behaviors depending on the treatments used, emphasizing the necessity for rapid investigation of this new preservation strategy [93].

Furthermore, the presence of hazardous chemicals, such as toxins or heavy metals, in seaweeds must be carefully monitored [94]. In this situation, it was discovered that solar-drying, boiling, and seaweed dehydration, as well as other methods of processing (such as washing or cooking), lowered the toxins and the concentrations of other volatile chemicals (such as iodine or arsenic) in seaweeds [95–98]. Thus, it is critical to have scientific data about the heat treatment process as well as the chemical characterization of the resulting product in order to obtain the best method of maintaining the important properties of seaweed; it is also critical to find new methods of heat treatment that do not destroy the nutritional value of seaweeds.

The safety of seaweeds and the quality starts not in biochemicals but in the certifications and guidelines for the food industry due to be a crucial point to guarantee the security of the seaweed to the food industry, without exposing it to external risks between cultivation and the final product.

4.4. *Guidelines for Food Safety in Industry*

Ensuring food safety in the food industry entails following norms and regulations to avoid foodborne diseases and protect consumers. The handling and packing of seaweed by the food industry chain, for example, can be a hotspot for cross-contamination with viruses, bacteria, fungus, protozoa, organic compounds such as prions, natural poisons, and persistent organic pollutants [13,99]. If preservation measures are inadequate, for example, after harvesting the seaweed, contaminating organisms may proliferate. Guidelines for preventing these contaminations, such as Hazard Analysis Critical Control Points (HACCPs) and Good Manufacturing Practices (GMP), are, nevertheless, thoroughly defined in food security standards. Food safety can also be enhanced by the application of ISO2200. After-harvest biomass management might avoid product deterioration while ensuring a low level of pollutants. The major critical point of danger in food safety is mostly due to man-handling the seaweed from the harvest until the seaweed consumption.

4.4.1. Good Manufacturing Practices (GMPs)

The food business is one of the world's most significant industries. As the world's population grows, so does the need for food goods. However, as demand grows, so does the responsibility for ensuring the safety and quality of food items. Implementing Good Manufacturing Practices (GMPs) is a critical step in ensuring the safety and quality of seaweed [100]. GMPs define the operating conditions and regulations required to ensure cleanliness across the food chain and in its manufacturing. GMPs are a set of standards that outline management and handling activities with the goal of guaranteeing safe food production conditions. They are also important in the design and operation of facilities, as well as in the creation of food-related processes and products. The Codex Alimentarius designed GMPs with the primary goal of protecting customers. It comprises numerous essential operating criteria and processes that food firms must follow [101].

GMPs are norms and processes that are meant to ensure the safety and quality of food items. These principles and practices apply to all areas of food production, including food product manufacturing, processing, packaging, and storage. The primary goal of GMPs is to decrease the risks involved with food manufacturing while also ensuring that food products are safe and of high quality [100].

GMP compliance is critical for the food business for the following reasons:

1. Ensure food safety: The fundamental goal of GMPs is to ensure food product safety. GMPs ensure that food items are free of contamination, adulteration, and other potentially dangerous chemicals [100].
2. Meeting regulatory requirements: GMP implementation is a legal obligation in several countries. Companies that do not follow GMPs face legal action, penalties, and closure [100].
3. Improving brand reputation: Companies that use GMPs are seen as responsible and dependable. This boosts the company's reputation and consumer trust [100].
4. Improving efficiency: Using GMPs may help businesses improve efficiency by decreasing waste, minimizing downtime, and increasing overall output [100].
5. Improving product quality: Good Manufacturing Practices (GMPs) guarantee that food items are of high quality and fulfill customer expectations. This can boost consumer happiness and loyalty [100].

4.4.2. HACCPs

Seaweeds contain a variety of important and hazardous components, and humans have no control over their existence in the wild. For seaweeds to be edible in terms of the mineral concentration, precautions must be taken to prevent overdosage, such as adhering to the DRI (Daily Recommended Intake) guidelines for seaweeds. These numbers indicate that a daily intake of one product will not impair one's health [54].

To reduce the danger of mineral overdosage, analysis can be performed to define and quantify those components in specific goods. There is, however, the HACCP (Hazard Analysis Critical Control Point) system, which is a methodology for establishing food safety processes and procedures [102]. The primary purpose of implementing this system is to prevent, for example, harm to humans resulting from food hazards, whether physical, biological, or chemical hazards, or to reduce them to an acceptable level [103]. Implementing this approach in the seaweed area is yet another way to safeguard consumers from mineral overdosage. Analyzing the essential control points entails identifying particular processes in the seaweed-processing process that are most likely to cause harm and measuring the three major risks. For physical risks, it is necessary to check to see whether there is anything that should not be there, such as plastics. Analyses must be performed to detect the presence of bacteria, viruses, or parasites in biological risks. Chemical analysis is required for chemical dangers, which include minerals and heavy metals, to ensure that the limits have not been exceeded. All of the procedures mentioned above must be paired with an investigation of the water in which the seaweeds were formed, because the environment in which they grow is also important for their composition [102,103].

By analyzing the risks and hazards in food, HACCP implementation protects both consumers and the industries that use it. Once these businesses are HACCP-certified, they can assure people that the food they produce is safe for customers [103].

4.4.3. ISO 22000

ISO 22000 is an international standard that provides rules for food safety management systems. The standard is intended to assist organizations in implementing efficient food safety management systems and ensuring that food items are safe to consume. The standard is based on the concepts of Hazard Analysis Critical Control Points (HACCPs) and integrates GMP features [104].

While GMP requirements ensure safe and sanitary food manufacturing, ISO 22000 takes a broader approach to food safety management by covering the whole food supply chain, from raw material acquisition to final product delivery [101,105].

Companies must take a systematic approach to food safety management, which involves identifying possible risks, establishing control measures, and assessing the efficacy of those measures. GMP implementation is a critical step in guaranteeing the safety and quality of food products. GMPs establish norms and procedures for all areas of food production, including the manufacture, processing, packing, and storage of food items. ISO 22000 establishes a framework for GMP implementation in the food business [100].

It is critical to regulate storage and manufacturing environments using the essential features of ISO 22000: The standard requires the implementation of an FSMS (Food Safety Management System) that addresses food safety hazards at every stage of the food chain (from production until the consumer's home), basic conditions and activities necessary for maintaining a hygienic environment throughout the food chain, a systematic approach to identifying, evaluating, and controlling food safety hazards, an increased focus on upper management's involvement in establishing and upholding the FSMS, and continuous attempts to improve. ISO 22000 adheres to the high-level structure (HLS) typical of other ISO management system standards, which includes understanding the internal and external factors that can impact the FSMS, the role of top management and quality control experts in establishing food safety policies and the objectives for that specific industry, risk management, objectives, and planning to achieve food safety in that targeted company. There is a need for support (mostly, general awareness, communication and scientific evaluations). In addition, frequent performance evaluations are required to monitor, measure, analyze, and evaluate performance, as well as internal and external audits and management reviews to enhance performance and rectify nonconformances [105,106].

Through all the steps that need to be addressed, there is a key element related to human welfare: "Seaweed can be a real keystone to the human food chain? And how we can enhance the food safety?".

5. Seaweed Food's Real Potential: How Can Be Checked?

There is a requirement in the food sector to ensure product safety and quality; however, seaweeds live in areas where it is more difficult to ensure their safety and quality using just typical procedures for plants and vegetables [4]. As a result, the safety of seaweeds' consumption and understanding of their quality is currently a hot topic in order to ensure a "new" food source that can substitute the plants' part in the human diet, thereby reducing the pressure on the terrestrial ecosystem [10,107]. Still, the seaweed safety check is similar to terrestrial vegetables, being only necessary for the nutritional composition and bacterial analysis (Figure 6).

To evaluate the bioavailability of seaweed nutrients and bioactive substances in simple raw material, in food or in meals, it is necessary to understand their content, structure, interactions with other dietary components, and destiny in the human body after ingestion. Thus, to observe all the data above and to fully understand the complex chemical reaction, there is a need to develop a system that can support the food agencies regarding the ingredients and their real potential [8,108].

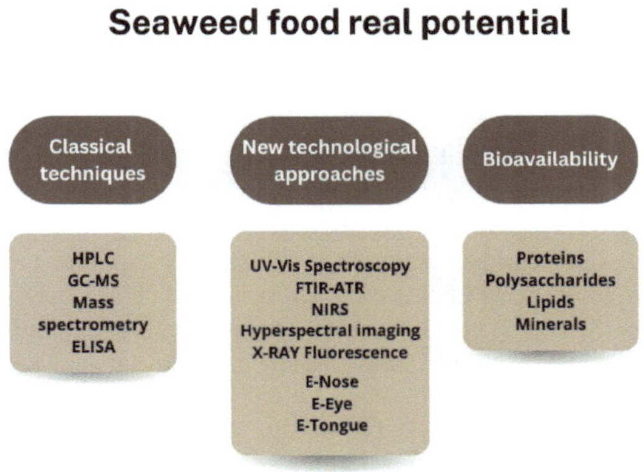

Figure 6. Standard methodologies for determining the nutritional composition of seaweeds and possible new technological steps toward food safety in seaweed consumption.

5.1. Chemical and Biochemical Techniques: New Approaches

The methodologies and approaches used for seaweed nutritional composition as direct food are based on the Association of Official Analytical Chemists (AOAC) international standards, similar to other food sources, which are based on standard procedures [109]. As a result, the nutritional content analysis of seaweed is the same as that of other foods, such as vegetables. However, seaweeds include substances other than necessary nutrients, such as phenolic compounds, pigments, fatty acids, minerals, and perhaps heavy metals [10]. As a result, seaweed can absorb potentially dangerous metals or pollutants [10]. Other types of study are needed to provide additional assurance that seaweed is an excellent and safe dietary source [110]. Food safety concerns have become critical in promoting public health safety and the financial viability of global food companies throughout the global food supply chain [110]. Despite the newest technology, detection methods, legislation, and consumer education on food safety and quality, there is still an increase in foodborne disease outbreaks throughout the world, despite the latest food standards and customer expectations [110].

Different lab-scale techniques like microbial analysis, microscopic examination, gas chromatography-mass spectrometry, liquid chromatography, differential scanning calorimetry, and nuclear magnetic resonance are used, but most of these techniques are costly, time-consuming and required technical experts [111]. Biological techniques, such as culture-dependent microbiological techniques to quantify viable bacteria, nucleic acid detection technology (e.g., multiplex PCR), biochemical detection techniques, and immunological detection techniques are examples of common traditional microbial detection techniques, which can be prone to errors and also take some time for analysis when talking about fresh or perishable food [112].

5.1.1. Classical Techniques

By utilizing sensing platforms, the great majority of food contaminations, such as heavy metals, pathogens, mycotoxins, pesticides, veterinary medications, herbicides, and unlawful additions, may be examined [113]. HPLC (high-performance liquid chromatography), GC-MS (gas chromatography-mass spectrometer), MS (mass spectrometry), and ELISA (enzyme-linked immunosorbent assay) are common conventional detection methods in food science that can provide high sensitivity and selectivity in terms of sensing approaches for the detection of various types of food contaminations [114,115]. Traditional sensing platforms, on the other hand, need professional operators to deploy samples and expensive devices, which is time-consuming and costly [116]. Furthermore, well-developed sensing technologies are frequently used in the lab, which cannot meet the critical need of on-site determination, which necessitates portable sensors to obtain the result instantly [117]. As a result, simple, rapid, economical, and portable analytical approaches for determining different food contaminations have received considerable attention [113,118].

5.1.2. New Technological Approaches

The evaluation of quality in the food sector is a significant problem due to the need for high-cost equipment and lengthy analysis to ensure that products reaching customers are safe and of the highest quality. Existing technologies often require significant resources, skilled individuals, and sophisticated analytical techniques, which has generated a desire for quick and cost-effective solutions [111]. The use of current analytical systems ensures that the utilized technique is sensitive, linear, and repeatable. This is particularly important for method development and validation, as it provides the foundation for future studies (Table 2).

Table 2. New technological approaches and their demonstrated potential.

Typology	Techniques	Supported Analysis
Spectroscopy	UV/VIS Spectroscopy	Grape-must caramel in vinegar; edible oils degradation; antioxidant compounds
	FTIR-ATR	Detect mycotoxins; detect adulteration in dairy products; authenticity of meat and meat products; identify pathogenic bacteria; quantify sugars; characterize polysaccharides
	NIRS	Detect meat adulteration; fish quality; analyze constituents in fruits, oils and milk products; quantify adulterated oils; analyze rheological parameters; determine protein concentration and nutritional composition
Electrochemical	E-Nose	Detect food spoilage due to bacterial and fungal infections; monitor fermentation processes; detect fraud in food products; monitor the oxidation process; determine the shelf life of macroalgae
	E-Eye	Classification of olive oil; identify food fraud
	E-tongue	Detection of tetracycline residues in milk; microbiological quality of fish samples; sensory analysis of vegetable milk; quality parameters; characteristics of aqueous extracts from seaweed
Imaging	Hyperspectral imaging	Detect fecal contamination; detect chemical residues and contaminants; detection of foodborne pathogens
Fluorescence	X-ray Fluorescence	Seaweed biomass products; mineral composition of milk; analyze plant material; determination of cadmium; evaluate algae-based supplements; heavy metal contamination

UV/VIS Spectroscopy

Absorption spectroscopy in the ultraviolet and visible (UV/VIS) range is a rapid method used for the qualitative and quantitative assessment of sample substances. This approach has been applied in numerous fields of study in food science and food processing companies due to its simplicity and dependability [119]. UV/VIS spectroscopy is a technique used to monitor and measure the interactions of UV and visible light with various chemical substances in the wavelength range of 200 to 780 nm. The approach takes advantage of the physical reactions of light and analytes in the sample, such as absorption, scattering, diffraction, refraction, and reflection. UV and visible light absorption are limited to certain chromophores and chemical species with specified molecular functional groups. When the electrons within chromophores are stimulated, they produce distinct absorption spectra for individual molecules. UV/VIS spectroscopy applies to solid, liquid, and gaseous materials. However, UV absorption solute analysis is only feasible in homogeneous solutions. Nonhomogeneous samples often show significant interference within the spectra, especially when solid particles are present. This is due to the absorption and light-scattering effects of individual particles. The benefits and limitations of this technique include the sensitivity, equipment cost, remote-sampling capabilities, and the ability to analyze liquid, solid, and slurry samples with chemical resolution. The path length can also have an influence [120,121].

The utilization of UV/VIS spectroscopy stands out prominently in ensuring food safety, as evidenced by several recent studies across diverse food industries [122].

In the assessment of grape-must caramel in balsamic vinegar from Modena and Spanish PDO wine vinegar, UV/VIS spectroscopy emerges as a rapid analytical method, offering the potential for the quantification of the caramel content below legal limits. This underscores its crucial role in verifying compliance and authenticity within regulated food products [123].

Similarly, UV/VIS spectroscopy, coupled with chemometrics, demonstrates its efficacy in evaluating the impact of heating on various edible oils. By discerning the characteristic spectral changes indicative of oil degradation, UV/VIS spectroscopy enables the precise determination of the acid value, thus ensuring the quality and safety of heated oils [124].

Moreover, in the quantitative analysis of edible blend oils, UV/VIS spectroscopy is integrated with weighted multiscale support vector regression, providing enhanced accuracy in predicting the oil composition. This innovative approach showcases the instrumental role of UV/VIS spectroscopy in addressing complex compositional challenges, thereby bolstering food safety standards [125].

In one study, the Irish seaweed *Himanthalia elongata* (Phaeophyceae) was investigated as a natural source of antioxidant compounds [126]. The ethyl acetate subfraction showed high scavenging capacities for DPPH and lipid peroxidation, indicating strong antioxidant activity. LC-DAD-ESI-MS/MS analysis identified eight phenolic compounds, including some not previously reported in *H. elongata*. These results suggest the potential use of purified subfractions in food, pharmaceutical and cosmetic applications for health promotion due to their antioxidant properties [126]. Overall, this study highlights the importance of UV/VIS spectroscopy in the characterization of compounds from seaweeds and their potential contribution to food safety and health.

UV/VIS spectroscopy plays a vital role in food safety, enabling rapid and accurate analysis of the compounds present in both conventional foods and marine algae. By providing insight into the chemical composition of these products, UV/VIS spectroscopy helps to identify and quantify substances that may affect their safety and quality. This analytical technique makes a significant contribution to ensuring the integrity of food products and marine algae-based dietary supplements, thereby protecting public health and enhancing consumer confidence.

FTIR-ATR

Fourier transform infrared (FTIR) spectroscopy is a very effective method of analysis that does not damage the material and provides a "fingerprint" of the compounds present. The spectral peaks correspond to the vibration frequencies between the bonds of each atom in each chemical present in the sample, and they indicate the absorption of the IR beam. FTIR procedures are efficient, dependable, and easy to use, requiring no sample pre-treatment. Such approaches offer a simple and consistent way to handle a wide range of foods with non-destructive analysis, with the entire sample and analysis process often taking less than five minutes from start to finish. Previous research has demonstrated that FTIR-based methods, when combined with other chemometric techniques, can be effectively utilized in food industry processes to identify chemicals that may compromise the quality of the food or have been added to falsely claim that a food item is something other than what it is [127].

FTIR-ATR (Fourier-transform infrared spectroscopy–attenuated total reflectance) is a low-cost technique that analyses the chemical bonds of dried materials. This technique is an improvement over previous techniques that required a liquid extract solution [128]. It employs infrared light to vibrate chemical bonds and can be used to analyze the polysaccharides, pigments, phenolic fractions, compound oxidation, and microplastics in seaweeds before commercialization [128–131]. Compared to chromatography, the FTIR approach is less expensive and easier to use. However, it is less sensitive to biochemical quantification and quality analysis [107,132].

Food safety is a growing concern worldwide as consumers become increasingly aware of the risks associated with food contamination, adulteration, and deterioration. In this context, the use of FTIR-ATR is proving to be a powerful and versatile tool for ensuring the quality and authenticity of food.

The application of FTIR-ATR in food safety has been demonstrated in several recent studies. For example, researchers have used the technique to detect the presence of mycotoxins in foods such as sultanas, helping to prevent risks to human health [133]. FTIR-ATR has also been used to detect adulteration in dairy products such as cheese, ensuring the authenticity and quality of these foods [134].

Another important aspect is monitoring the quality and authenticity of meat and meat products, where FTIR-ATR has proven to be a valuable tool. This technique allows the detection of adulteration and the assessment of the deterioration status, and it ensures compliance with food safety standards quickly and efficiently [135].

In addition, FTIR-ATR has been used to identify pathogenic bacteria rapidly and accurately in a wide range of foods, providing an effective alternative to traditional methods of microbiological analysis. This rapid identification capability is critical in preventing outbreaks of foodborne illness [136].

The technique has also been used to quantify the sugars in products such as honey, monitor pesticide residues in crops and to detect adulteration in a wide range of foods [137,138].

In this study, FTIR-ATR was used to extract and characterize polysaccharides from macroalgae including *Eucheuma denticulatum*, *Solieria chordalis* (Rhodophyta) and *Sargassum muticum* (Phaeophyceae) [139]. The research focused on the identification of cell-wall polysaccharides such as carrageenans, fucoidans and alginates, highlighting the potential of these compounds for various applications in food and biotechnology.

Comparison of purified extracts with commercial solutions of polysaccharides showed strong similarities in the spectra, validating the extraction methods and confirming FTIR-ATR as a reliable technique for polysaccharide analysis. In addition, the study investigated seasonal variations in the polysaccharide composition, revealing differences in fucoidans, alginates and carrageenans depending on the time of harvest [139]. This information can guide the optimization of extraction processes and improve the utilization of macroalgal polysaccharides in different economic sectors.

The research also provided insights into the structural characteristics of carrageenans in specific algal species, such as the presence of iota-carrageenan at mature stages of development [139]. This knowledge can facilitate better control of the extraction methods and enable targeted applications of these compounds.

In a related study, ulvan, an edible sulphated polysaccharide extracted from *Ulva lactuca* (Chlorophyta), was used for the biosynthesis of silver nanoparticles (Ag-NPs) to produce bionanocomposite films for active food packaging. FTIR-ATR spectroscopy was used to confirm the formation of these films, which exhibited antimicrobial and antioxidant properties, making them potential replacements for conventional food-packaging materials [140].

FTIR-ATR is proving to be a valuable tool for ensuring food safety, particularly in the analysis of both traditional and novel food sources such as marine algae.

NIRS

Near-infrared spectroscopy (NIRS) is a promising approach for nondestructive and easy food safety inspection and control. It offers great benefits, such as speed, noninvasive measurement, ease of use, and low sample preparation requirements [141]. NIRS has been widely demonstrated to be effective in this field.

When radiation enters the sample, it is either reflected, absorbed by molecular bonds, or transmitted, resulting in changes in the light energy. These changes may reflect certain chemical bonds and, therefore, the properties of the tested items. To conduct scientific studies, it is important to select appropriate equipment and apparatus with better configurations, such as high spectral resolution, extensive scanning ranges, and adjustable scanning speed. Although NIRS systems are used to monitor manufacturing lines, they should be explicitly developed for commercial use as the application is typically specified [142]. However, it is important to consider the impact of the characteristics of NIRS instrumentation components on the overall performance when creating such a system. In recent years, advancements in hardware and software have made NIRS sensors more portable and practical [143]. The method of sample collection that is most suitable for the elements to be examined can have a significant impact on the overall outcome of the endeavor. The IRS for evaluating food quality involves collecting the spectra of tested items and constructing calibration models to correlate the spectral fingerprints with the sample attributes [143,144]. Nonetheless, NIR spectra contain enormous amounts of data that need to be processed, so chemometrics (a mixture of statistical and quantitative sciences) is often used to extract usable information that can significantly improve the potential applications of NIRS. Furthermore, to prevent fraudulent situations, NIRS can be used to perform qualitative and quantitative analyses of harmful compounds in food. The food industry has effectively demonstrated the potential of NIRS. Although more research has been conducted on food quality analysis, the use of NIRS in food safety evaluation and control is also increasing. This will be explored in detail in the following sections. The following sections show the number of recently published publications cited in this study, covering several application disciplines (such as freshness assessment, authenticity and adulteration, toxin detection and unlawful treatments) [141].

The NIRS technique has been widely studied in the food industry, particularly in detecting meat adulteration and assessing fish quality and fishery products [145,146]. However, its potential is not limited to these sectors alone, as it offers numerous possibilities in other areas of the food industry. For example, Kurz et al. [147] identified and analyzed the constituents in fruits, Kuligowski et al. [148] in oils, and Balabin and Smirnov [149] in milk products. Studies have demonstrated that NIRS can accurately discriminate and quantify adulterated oils [150].

NIR technology has been successfully applied in various areas, such as evaluating beef freshness, determining tomato maturity and textural properties, and analyzing rheological parameters in wheat grains [151–153]. These examples highlight the versatility and effectiveness of NIRS in the food industry, providing a valuable tool to ensure food quality, authenticity, and safety [141].

NIRS is a promising tool for ensuring food safety and quality in traditional food products and macroalgae. Recent studies have shown that NIRS can be applied to various aspects of seaweed analysis, including monitoring microbiological growth, determining the protein concentration, and assessing the nutritional composition [154–156]. For example, NIRS has proven effective in predicting the microbial counts in stored seaweed samples, providing rapid and real-time assessment of the product quality [154]. Furthermore, NIRS has been successfully used for on-site detection of the protein concentration in red seaweed, offering a non-destructive and precise alternative to traditional laboratory-based methods [155]. Furthermore, NIRS has demonstrated potential in evaluating the nutritional value and digestibility of brown seaweeds [156]. However, further validation on larger datasets may be required to assess its robustness.

NIRS shows potential as a versatile and efficient tool for enhancing food safety and quality assurance in conventional and emerging food sources like macroalgae. Its non-destructive nature, rapid analysis capabilities, and real-time assessments make it valuable for monitoring and ensuring food safety throughout the production and supply chain. Although NIRS shows considerable potential, further research and validation are required to fully establish its reliability and accuracy in different food matrices, including macroalgae. Nevertheless, with ongoing advancements and refinement, NIRS is poised to play a crucial role in advancing food safety and quality standards in the future.

E-Nose

Electrochemical (EC) methods, including impedance spectroscopy, voltammetry, potentiometry, and coulometry, have made significant contributions to food analysis. It is important to note that these methods provide objective and precise measurements of food components. EC techniques directly convert chemical processes occurring at the electrode/electrolyte interface into quantifiable electronic signals, such as altered conductive properties (conductometric), current (amperometry), and potential or charge accumulation (potentiometric) [111,157].

The electronic nose, as a non-invasive technology for identifying volatile substances, has been applied to food safety and quality analysis. The use of the E-Nose for pathogen identification has been proven to be successful and superior to traditional approaches. The E-Nose is a non-invasive and rapid approach that requires little or no sample preparation, making it perfect for use as an online monitoring tool. An E-Nose is a device that combines a chemical sensor array system with partial specificity and an appropriate pattern recognition system to distinguish simple or complex odors. It can analyze volatile organic compounds (VOCs) generated by microorganisms and is used as an alternative approach to identify and classify various chemicals and bacteria [110].

In the food industry, the E-Nose is used to detect food spoilage bacteria, total volatile basic nitrogen, trimethylamine, and fungal infections [158–164]. The E-Nose has several advantages over traditional and non-invasive approaches, such as vibrational spectroscopy and hyperspectral imaging. Despite these benefits, E-Noses are currently used in a limited number of applications in the food industry. The growth of E-Nose systems in the food industry has been rapid despite challenges in terms of sensor selection and the difficulty of implementing pattern recognition algorithms in low-cost hardware components. However, the necessity for such algorithms has hindered progress [111]. It is important to maintain objectivity when discussing these limitations.

E-Nose technology has a significant application in classifying the degree of contamination in leftover cooked foods, as demonstrated by research conducted in Malaysia. The odor characteristics of local leftover cooked meals are analyzed by E-Nose devices equipped with sensor arrays, which effectively classify the contamination levels with high accuracy rates ranging from 90% to 100%. This article highlights the potential of E-Nose technology in ensuring the safety and quality of prepared foods worldwide [165].

E-Nose technology has proven to be valuable in monitoring fermentation processes. Studies on the fermentation of *Tremella aurantialba* (fungi) have exemplified this. By detecting volatile compounds associated with fermentation, E-Nose devices allow for real-time analysis and prediction of the fermentation phases. This technology offers precise monitoring and control, ensuring the quality and safety of fermented food products [166].

In postharvest scenarios, E-Nose technology coupled with machine vision provides a rapid and non-destructive method for detecting the freshness and spoilage of perishable food items such as spinach [167]. By analyzing odor and image data, E-Nose devices accurately classify the freshness levels, facilitating timely quality assessments during storage [168]. E-Nose technology has been used to detect microbial spoilage in canned foods by analyzing volatile organic compounds (VOCs), providing early indicators of spoilage [169].

E-Nose technology is being used to detect fraud in food products, specifically extra virgin olive oil, by analyzing the VOCs emitted by olive oil samples [170]. This technology helps distinguish between authentic and adulterated or fraudulent products. Additionally, E-Nose devices are used to monitor the oxidation process of frying oils, ensuring quality control and preventing the consumption of degraded or rancid oils [171].

The various applications of E-Nose technology in food safety highlight its importance in modern food production and distribution systems. E-Nose devices offer versatile solutions for enhancing food safety and quality control practices, from detecting contamination and spoilage to monitoring fermentation processes and ensuring product authenticity. As technology advances, research and implementation of E-Nose technology are poised to redefine food safety standards. This will benefit both consumers and the food industry.

Electronic nose (E-Nose) technology provides a quick and efficient method for evaluating the quality and shelf life of food products, including macroalgae. In a recent study, the shelf life of different types of macroalgae was evaluated using an E-Nose over a storage period of 150 days. The E-Nose system detected the release of volatile organic compounds (VOCs) from the macroalgae, which are indicative of food deterioration [172].

The E-Nose recorded significant changes in the values for certain types of macroalgae during the storage period. These changes in the sensor values were positively correlated with the physical, microbiological, and Fourier-transform infrared (FTIR) spectroscopy parameters, providing comprehensive insights into the quality and shelf life of the macroalgae samples [172].

E-Nose technology has the potential to enhance food safety and quality assurance in both traditional food products and emerging food sources such as macroalgae. Its ability to detect changes in food quality quickly and accurately makes it a valuable tool for the food industry, facilitating timely interventions to maintain product freshness and safety.

E-Eye

Color is a crucial aspect of food quality as it is closely linked to perceptions of freshness, maturity, attractiveness, and safety. Customers often examine the color of food when making purchases [173]. Color is the perceptual response to the visible spectrum of the light that is reflected or emitted by an object. This response is generated by the interaction of light with receptors in the retina, which then sends a signal to the brain via the optic nerve. Color perception is influenced not only by the object itself but also by the surrounding lighting conditions [174]. Therefore, color analysis is crucial for categorizing items such as meat, peas, maize, canola, rice, and wheat for human and animal consumption [173].

The E-Eye is a detection system that utilizes visual information identification and analysis to assess food quality [175]. Color differentiation is achieved by comparing wavelengths. There are various color spaces available, such as HSI (hue, saturation, intensity), HSV (hue, saturation, value), HSL (hue, saturation, lightness), and HSB (hue, saturation, brightness), which differ in sensitivity [174]. The E-Eye has advantages for use in food quality evaluation due to its low cost, portability, and ease of implementation on a large scale [176].

This technology can be powered by colorimetry, spectrophotometry, or computer vision. Using appropriate equipment can result in a more accurate color description [177]. The E-Eye can detect the appearance-related characteristics of samples but not the flavor- or aroma-related components [178]. Therefore, combining multiple technologies such as the E-Nose and E-Tongue has become an alternative to detect the various characteristics of samples. Spectrophotometers are devices used to measure color that analyze the spectrum distribution of a sample's transmittance or reflectance. They offer a spectral analysis of the reflectance wavelength and/or transmission qualities of objects [179]. Near-infrared reflectance (NIR) spectrophotometers are commonly used in the food industry to evaluate the chemical composition of items, particularly in cases where color is a crucial factor. This includes assessing the levels of proteins, oil, starch, fiber, and moisture. It is important to maintain objectivity in the evaluation of these elements [173,180].

Several research studies have demonstrated the applicability of E-Eye technology to various food products and safety scenarios. For example, in the characterization of edible olive oils, E-Eye devices were used alongside an electronic nose (E-Nose) and electronic tongue (E-Tongue) to assess the oil quality deterioration during storage [181]. Through innovative data-processing techniques and mid-level data fusion approaches, E-Eye technology facilitated the classification of olive oil samples based on freshness, demonstrating high classification accuracy rates.

E-Eye technology has also been instrumental in the fight against food fraud, particularly in the case of Italian lentils. Using red–green–blue (RGB) imaging and discriminant classifiers, E-Eye devices were able to classify lentils according to the harvest year and origin, as well as to discriminate between expired and edible samples. These results highlight the effectiveness of E-Eye technology in detecting subtle visual changes associated with food quality and authenticity [182].

The integration of E-Eye technology into food safety practices offers significant benefits, enabling rapid and non-destructive assessment of external food characteristics. From assessing food coloring in Chinese medicine to distinguishing between expired and edible lentils, E-Eye devices show immense potential for improving food safety measures. As research continues to explore the capabilities of E-Eye technology, its adoption is expected to grow, providing valuable tools to ensure the integrity and safety of food products worldwide [175,182].

Although macroalgae are vulnerable to spoilage, they offer numerous health-promoting compounds, making them a valuable food source. Due to the increasing demand for food quality and safety, sensitive and rapid analytical technologies are needed in the seafood industry. When applying the E-Eye technology to macroalgae safety, it is expected to yield similar benefits. The E-Eye system can effectively assess the quality and authenticity of macroalgae products by utilizing rapid and non-destructive techniques, such as spectroscopic methods. This approach offers advantages such as speed, minimal sample preparation, and the ability to monitor products in real time, which enhances food safety practices [183].

E-Eye technology shows potential for enhancing food safety and quality assessment for both conventional food products and sources such as macroalgae. Its quick and non-invasive nature makes it a valuable tool for monitoring freshness and authenticity, thereby contributing to consumer confidence and industry standards.

E-Tongue

Taste is the perception of chemicals that stimulate the taste buds on the tongue [184]. The E-Tongue has advanced in terms of the sensitivity, selectivity, and multiplexing capabilities of current biosensors, enabling accurate and rapid quality prediction of samples. This has led to its use in various industries, including pharmaceuticals, cosmetics, and food [185].

The term "E-Tongue", which is an analogy to the human tongue, was first introduced in the 1990s and has since been extensively researched. It is a technology that uses a collection of sensors to detect chemical solutions. The system consists of an electrochemical cell (sensor array), a measuring module, and a pattern recognition algorithm capable of distinguishing between the simple and complex chemical systems that make up the flavor [180,184].

By examining various components of a sample, the electronic tongue extracts a signal signature [186]. The E-Tongue can describe the flavor of complex liquids or samples that have been transformed into liquid form [187]. The main purpose of this technology is to analyze meals using a collection of sensors, such as ion-selective electrodes with specific features, followed by statistical analysis. This allows for the collection of information on the freshness and maturity levels [187,188]. An E-Tongue sensor array provides multi-dimensional information in a short amount of time [189]. This technology distinguishes distinct patterns of classes of molecules responsible for flavor. The complicated information obtained by E-Tongue measurement is processed using multivariate statistical analysis. In contrast, human taste and flavor perception involves matching signals from taste receptors with memories to determine taste and flavor [187]. Additionally, an E-Tongue's sensor suite can include a wide range of chemical sensors, such as electrochemical, optical, mass, and enzymatic sensors [180,189]. The combination of these techniques can improve seaweed quality characterization and safety.

One application of the E-Tongue is the detection of tetracycline residues in milk samples [190]. The E-Tongue successfully identified the presence of tetracycline residues in milk samples, providing a rapid and reliable method for contaminant detection without the need for sample pre-treatment.

Similarly, studies have demonstrated the superiority of E-Tongue technology coupled with advanced machine-learning algorithms such as back-propagation neural networks (BP-NN) in assessing the microbiological quality of fish samples [191]. These approaches outperformed traditional microbiological plating methods in predicting the total viable counts (TVCs), demonstrating the high accuracy and reliability of E-Tongue-based models in assessing microbiological quality parameters and improving food safety assurance measures.

In addition, the E-Tongue technology was instrumental in the sensory analysis of vegetable milk [192]. The results underlined the potential feasibility of using electronic tongues for simple, rapid, and effective sensory evaluation to ensure quality and consumer satisfaction.

E-Tongue technology was also explored for the express evaluation of quality parameters in vegetable oils [193]. Investigations into the use of multi-sensor systems have shown promising results. The application of these tools has demonstrated the potential for the development of rapid and simple methods for express quality assessment of vegetable oils, further highlighting the versatility and effectiveness of E-Tongue technology in improving food safety and quality assurance measures.

The diverse applications of E-Tongue technology underscore its importance in ensuring food safety, quality, and consumer satisfaction in various sectors of the food industry. From contaminant detection to sensory analysis and quality assessment, E-Tongue technology continues to play a key role in advancing food safety measures and maintaining high food quality standards.

The use of E-Tongue technology in food safety, specifically in relation to macroalgae, presents potential for quality assessment and product development. By utilizing the E-Tongue, researchers can examine the taste profile and chemical composition of macroalgae-based products, aiding in the creation of innovative food formulations and substitutes.

In the development of a low-sodium salt substitute, the E-Tongue can evaluate the taste characteristics of aqueous extracts from seaweed and other marine sources. By assessing taste parameters such as saltiness, bitterness, and umami, the E-Tongue can identify extracts with desirable flavor profiles. This helps researchers identify potential candidates for salt substitute formulations [194]. In addition, the E-Tongue can aid in comprehending the intricate interplay of salty constituents in macroalgae extracts. This can lead to the creation of improved salt substitutes that emulate the flavor of conventional sodium-based salts. By scrutinizing the sensory characteristics of macroalgae extracts, researchers can refine formulations to attain the desired taste profiles while diminishing the sodium content. This can help address health issues linked to excessive salt consumption [194].

The use of E-Tongue technology shows potential to improve food safety and quality assessment for both traditional food products and innovative macroalgae-based formulations. The E-Tongue provides rapid and accurate sensory analysis, allowing researchers to evaluate taste characteristics, identify flavor compounds, and optimize product formulations. This contributes to the development of healthier and more sustainable food alternatives.

Hyperspectral Imaging

Hyperspectral imaging (HSI) is a non-destructive technology used in food safety to identify adulterations, as well as microbiological, chemical, and physical contamination. HSI combines conventional imaging with spectroscopy to obtain a spectrum for each point (i.e., pixel) in an area of interest in the material being examined [195]. HSI's spectroscopic component can cover spectral ranges from ultraviolet (UV) to terahertz. However, the most commonly used ranges are visible (VIS) and near-infrared (NIR). Studies have shown that when integrated with chemometrics and machine learning, HSI can identify fungal contamination in food. HSI detected fungal infections on grains at an early stage, indicating the possibility of early identification and removal of affected sections to minimize or prevent the development of fungal diseases. The use of HSI has been documented in identifying pollutants in fruits, vegetables, and meat products in the VIS-NIR or NIR range [195]. This method has also been found feasible in detecting faults and features related to the composition of meat products [196]. Physical contamination of food poses two primary difficulties, namely fecal contamination and the presence of foreign elements in food matrices. Fecal contamination is a significant food safety concern as it can introduce harmful germs into fruits, vegetables, and meat, leading to potential cases of food poisoning. It is associated with various fungal and bacterial pollutants [197,198].

Hyperspectral imaging has been investigated by researchers for its potential to ensure food safety. Studies have shown that hyperspectral-imaging systems are effective in detecting fecal contamination on fresh produce surfaces and in poultry-processing facilities [199]. These systems can differentiate between fecal contaminants and food surfaces, providing valuable insights for preventing foodborne illnesses.

Investigations have shown that hyperspectral imaging is highly versatile in various food safety applications. It can detect chemical residues and contaminants, as well as assess food quality attributes such as freshness and ripeness [198]. Hyperspectral imaging provides a non-destructive and efficient means of monitoring food safety throughout the supply chain.

Additionally, studies have highlighted its potential for rapid and accurate detection of foodborne pathogens. Hyperspectral-imaging systems can identify specific microbial strains and assess their presence in food products, enabling proactive measures to prevent foodborne outbreaks and ensure consumer safety [198].

The research conducted in this field demonstrates the significant potential of hyperspectral imaging in enhancing food safety practices and standards. As technology advances, hyperspectral-imaging systems are expected to play a crucial role in shaping the future of food safety in the food industry.

Hyperspectral imaging shows potential for improving food safety, especially in the context of macroalgae. By using advanced imaging techniques and innovative algorithms, researchers can enhance the accuracy and efficiency of foreign object detection, ensuring the quality and safety of seaweed products [200]. Furthermore, hyperspectral imaging can be applied to a range of food products, making it a versatile and powerful tool for ensuring food quality across multiple industries.

X-ray Fluorescence

Mineral analysis is a critical aspect of assessing seaweed's impact on human health. However, it is a complicated and time-consuming technique. To reduce the time and expense associated with Inductively Coupled Plasma (ICP) techniques, a faster methodology suitable for frequent and numerous sample testing is required. A beam of primary X-rays, produced by an X-ray tube in X-ray fluorescence (XRF), bombards the material. The collimator restricts the cross-section of the main beam, allowing for specific excitation of the sample. This stimulation results in the release of fluorescence radiation. An energy-dispersive detector is used to measure the energy distribution of the fluorescence light. It allows for the identification and calculation of the relative concentrations of elements present in a sample. X-ray fluorescence spectroscopy (XRF) is a non-destructive and continuous technique that has become viable for generating high-resolution elemental data. The adjusted operating conditions improve the minimal detection limits and detection efficiency for several evaluated items. X-ray fluorescence spectrometry offers advantages over other multi-element methods, such as ICP-MS/ICP-OES. It requires limited sample preparation, provides non-destructive analysis, has a higher total speed, generates less hazardous waste, and has lower operating costs [201,202].

The gamma X-ray (XRF) technology was tested and calibrated against wet samples of seaweed to determine the relative and quantitative levels of elements. XRF is commonly used in geology and material sectors to evaluate the amounts of valuable or dangerous metals in solid or manufactured materials. However, it has not been widely accepted in the biological sciences, despite its potential [203].

The use of XRF has enabled frequent testing of seaweed biomass products, resulting in a developing record of range measurements for each element. This has allowed for the development of a profile of elemental ranges for nutrition and safety purposes for a specific type of seaweed. The nutritional value and safety of seaweed are determined by its elemental profile, which varies across different species. Therefore, it is crucial to thoroughly test all seaweeds and provide the results to the user. It is important to note that some seaweeds, such as Kombu, contain quantities of trace elements that exceed the recommended daily intake and even the upper limit. Hence, Kombu should only be consumed in very small doses [203].

A recent research review examined the application of XRF in the elemental analysis of milk [204]. Different configurations of XRF spectrometers were evaluated for their ability to quantify minerals and trace elements in milk samples with the accuracy and reproducibility required for food products. The study highlighted practical examples of the use of XRF techniques to determine the mineral composition of milk, emphasizing the importance of instrumentation, sample preparation, calibration and quantification procedures to ensure reliable results [204].

Another study investigated the use of low-power XRF systems, including energy-dispersive X-ray fluorescence (EDXRF) and micro-XRF (μ-XRF), for multi-element analysis and chemical imaging of various edible plant species [205]. The research demonstrated the feasibility of these XRF methods in analyzing plant material without the need for complex sample preparation, offering advantages in terms of simplicity and cost-effectiveness compared to traditional methods. The results underlined the potential of XRF techniques to increase knowledge of the impact of environmental factors, such as irrigation with treated wastewater, on the elemental composition of food products [205].

Another study focused on the rapid determination of cadmium in rice using energy-dispersive X-ray fluorescence spectrometry. The research evaluated the detection efficiency, periodic stability, and detection limits of different XRF instruments and demonstrated their applicability for the detection of cadmium in rice samples. The study highlighted the sensitivity of XRF spectrometers in meeting the rapid detection requirements with minimal impact from variations in the rice morphology. However, the research also identified challenges related to false negative and false positive results, highlighting the importance of careful calibration and validation procedures in XRF analysis for food safety applications [206].

The elemental composition of algae-based dietary supplements is crucial for assessing their safety and nutritional value. Energy-dispersive X-ray fluorescence (XRF) was used in a study evaluating 15 algae-based supplements commonly sold on the Portuguese market [207]. Despite the beneficial iodine content found in most of the kelp samples, there were concerns about potential excess iodine intake, which can lead to thyroid dysfunction. In addition, the presence of lead in various seaweed supplements posed significant health risks due to the potential for high daily doses. There was also considerable variability in the arsenic levels, raising further concerns about the safety of these supplements. These findings highlight the importance of informed consumption and regulatory oversight to mitigate the health risks associated with algae-based products.

Another study focused on XRF analysis of kelp supplement powders and capsules for heavy metal contamination [208]. While no significant levels of lead or mercury were found in the 17 samples analyzed, all the samples had detectable levels of arsenic exceeding the recommended reference doses. This highlights the importance of stringent quality control measures to ensure the safety of dietary supplements containing algae-based ingredients.

In Kenya, where seaweed is not widely consumed, there is growing interest in its potential as a food and medicine [209]. However, concerns remain about marine pollution and heavy metal contamination. In a study, XRF was used to analyze the trace metal concentrations in seaweed samples collected from different sites along the Kenyan coast. The results showed relatively high concentrations of essential trace elements such as calcium, manganese, iron, copper, and zinc, while toxic elements such as arsenic and lead were within acceptable limits according to the EPA/WHO regulations. This suggests that direct consumption of edible seaweed can be encouraged as it poses minimal health risks. In addition, due to their ability to accumulate trace metals, seaweeds serve as valuable indicators of marine pollution, making them essential for environmental health monitoring.

Overall, X-ray fluorescence spectroscopy plays a crucial role in assessing the safety and quality of algae-based dietary supplements and edible seaweeds, thereby ensuring consumer protection and environmental sustainability in the food industry.

5.2. Bioavailability

After conducting mandatory biochemical profiling of commercially available food, a major question remains: "The nutritional profile of seaweed and its true nutritional benefit to humans?". Scientific evidence supports the fact that the nutritional value of vegetables differs from their real nutritional value due to the synergistic behavior of the vegetable matrix compounds [4,13,52,210]. Therefore, it is extremely necessary to conduct bioavailability assays to fully understand the true nutritional power of seaweed.

Bioavailability is a two-stage process consisting of bioaccessibility and bioactivity. Bioaccessibility refers to the amount of a consumed component that is released from its food matrix and available for absorption in the colon [211]. The biological activity of a medicine or food component involves transporting the component to the target tissue, interacting with other biomolecules, undergoing biotransformation and/or metabolism, and inducing a physiological response [108,212].

Thus, bioavailability is closely associated with the human digestive system and the food matrix. In vitro digestion methods reveal that food matrix effects have a significant impact on bioaccessibility, which is the fraction of the ingested substance available for absorption and is frequently the rate-limiting factor determining the overall bioavailability. Ideally, test methodologies should accurately predict the behavior of loaded food matrices in real-world settings, such as human consumption, and produce reproducible findings across laboratories [213]. In vivo human-feeding experiments provide the most precise approximation of real-world settings, but they are time-consuming and expensive to conduct. In vivo animal experiments can be used as an alternative, but they are time-consuming, expensive, and raise ethical concerns about animal confinement and sacrifice. Additionally, due to differences between the gastrointestinal tract of test animals and humans, the data obtained cannot be accurately extrapolated. As a result, researchers are developing in vitro digestion procedures to simulate the human gastrointestinal system. These approaches are especially useful as screening tools for determining the influence of certain food matrix effects on the bioavailability [214–216].

Over the last decade, several in vitro digestive models have been developed to investigate the bioavailability of bioactive compounds in foods and medicines [217–219]. However, in recent years, various approaches have been created to understand the nutritional and nutraceutical potential of seaweed. The INFOGEST static digestive simulator has become the most extensively used in vitro digestion model for foods due to its ability to precisely replicate the human gut while being simple to operate [219]. This method was developed through a COST action to promote a more accurate model and has gained popularity for its effectiveness [213]. This approach exposes food samples to sequential oral, gastric, and intestinal digestion. Factors such as the electrolytes, enzymes, bile, dilution, pH, and digestion time are dependent on the available physiological data. The method has been modified and enhanced from the INFOGEST 2.0 digesting technique by including the oral phase and using gastric lipase. The method described here assesses the food digestion endpoints by analyzing digestion products such as peptides, amino acids, fatty acids, and simple sugars, as well as quantifying micronutrient release from the food matrix. It is important to note that this method is not static and can be improved to provide more information based on real human digestion and its complexity. The procedure is divided into three steps that simulate the oral, gastric, and small intestine phases of digestion in vivo. At each step, the text describes the length, physical and biochemical environment, and rationale for the selection of the steps [220].

5.2.1. Seaweed Bioavailability Analysis

Understanding the kinetics of seaweed vs. stability may be critical for gaining important insights into the impacts of seaweed in the digestive process, ensuring improved food safety levels, and retaining a comparable product with stable chemicals that can be bioavailable to humans, thereby promoting human welfare. However, there is a general lack of studies on the bioavailability of seaweeds, and most of them use isolated compounds, mainly seaweed polysaccharides [52,53].

Raw seaweed biomass contains proteins, polysaccharides, lipids, and minerals. However, due to the variable intrinsic composition of seaweeds, some of these components may not be accessible for biological functions in the human body. In vitro research includes laboratory-based mimics/models for one or more circumstances faced by these substances in the digestive system, such as the mouth cavity, esophagus, stomach, and small and large intestines. In vivo investigations aim to evaluate the precise digestion and fermentation pathways of seaweed-isolated proteins, polysaccharides, and lipids by the gut microbiota [31,221].

Seaweed Proteins

Protein digestibility is a crucial factor in determining protein availability. The protein content of seaweed varies significantly. Digestive enzymes can easily break down seaweed proteins, making them readily absorbable by humans. However, the high soluble fiber content of seaweed can lead to inadequate protein digestibility, limiting protein digestion and reducing the accessibility of proteolytic enzymes. There is an inverse correlation between protein digestibility and the polysaccharide content in seaweed [222].

A recent study determined the in vitro protein digestibility values of six different Irish seaweeds using a rapid method. This methodology assessed the protein quality and bioavailability in seaweeds based on their amino acid profiles and protein content [223]. Furthermore, the study found that seaweeds are rich in essential amino acids and taurine, which are integral to human health. This suggests that seaweeds have the potential to serve as valuable protein sources in human nutrition, providing a diverse array of essential and non-essential amino acids to consumers [223].

The study emphasizes the significance of evaluating the protein bioavailability in seaweeds. This enables the identification of appropriate alternative protein sources for human diets, contributing to sustainable food systems and addressing the increasing demand for protein in a resource-efficient manner.

Seaweed Polysaccharides

Seaweed polysaccharides and fibers typically range from 33% to 62% (dry weight basis) and are generally indigestible by the human stomach and intestinal enzymes. They act as prebiotic food in the human body [224]. Seaweed is abundant in soluble fibers, which are known for their ability to retain water and their hydrocolloid properties. Hydrocolloids are used as functional additives in food formulation to improve food consistency. They act as thickening, stabilizing and emulsifying agents, and they enhance the sensory quality of food products [225].

Most seaweed polysaccharides are resistant to digestion in the stomach and small intestine. Oral digestion of marine algae has shown no breakdown of marine polysaccharides or release of oligosaccharides due to saliva enzymatic reactions. In vitro evidence suggests that no monosaccharides are identified in the gastric juice, indicating that they are not digested in the stomach. Similarly, no monosaccharides or oligosaccharides were detected in the small intestine, indicating that seaweed polysaccharides were not degraded there. Seaweed polysaccharides are carbohydrate polymers that are neither digested nor absorbed in the small intestine. Instead, they are fermented by bacteria in the colon, which influences and modifies bacterial populations in the gut. The human large intestine contains a diverse range of bacterial species that play a crucial role in various metabolic and regulatory processes related to human health [31,221].

Recent studies have investigated the in vitro digestibility and fermentability of polysaccharides extracted from two types of seaweed: *Ascophyllum nodosum* (AnPs) (Phaeophyceae) [226] and *Pyropia haitanensis* (formerly *Porphyra haitanensis*) (PHPs) (Rhodophyta) [227]. The findings suggest that these polysaccharides have potential as functional food components with implications for human health.

AnPs showed significant resistance to enzymatic digestion by salivary amylase, gastric juice, and intestinal juice. However, fermentation by the gut microbiota resulted in a notable reduction in the molecular weight and reduced the sugar content of AnPs. This process also caused significant changes in the composition of the gut microbiota, particularly by increasing the abundance of Bacteroidetes and Firmicutes, which are associated with metabolic health benefits. Furthermore, the fermentation of AnPs led to a significant rise in the overall content of short-chain fatty acids (SCFAs), suggesting potential advantages for gut health and overall well-being [226].

Polysaccharides extracted from PHPs using water extraction and alcohol precipitation methods showed a promising yield with a distinctive molar ratio of galactose, glucose, and fucose. These polysaccharides were resistant to digestion by alpha-amylase, indicating their potential as prebiotics that can selectively stimulate the growth and activity of beneficial gut microbiota. Furthermore, PHPs have demonstrated significant antioxidant activity, making them attractive candidates for applications in the food and cosmetic industries [227].

The investigation into the digestibility and bioavailability of polysaccharides from macroalgae highlights their potential as functional food ingredients with diverse health-promoting properties. Further research is needed to elucidate their physiological effects and explore their applications in human nutrition and well-being.

Seaweed Lipids

Macroalgae, also known as seaweeds, are being considered as potential candidates for improving the nutritional value of various food products. This is due to their rich composition of lipids and diverse bioactive compounds. Studies examining the lipid content and availability of lipids in different species of seaweeds provide valuable insights into their potential as functional food ingredients [228].

Seaweeds have a diverse nutritional composition, with lipids being a key component. The lipid content of seaweeds usually ranges from 0.2% to 2.5%, with variations observed among different species and environmental factors. Seaweeds are valued for their unique lipid profiles and bioactive lipid compounds, despite their relatively low lipid content compared to other food sources [228,229].

Analysis of the fatty acid profiles in seaweeds reveals that they contain a predominance of polyunsaturated fatty acids (PUFAs), including omega-3 and omega-6 fatty acids, which are essential for human health. Seaweeds have higher levels of PUFAs compared to saturated fatty acids (SFAs) and monounsaturated fatty acids (MUFAs), making them a valuable source of healthy fats [228].

The study conducted by Demarco et al. [222] investigated the lipid and fatty acid contents of brown seaweed *Fucus spiralis* (Phaeophyceae) before and after in vitro modeling of the human digestive process. The results showed that *F. spiralis* has a low lipid content of 3.5% dry weight, with a lipid bioaccessibility of about 12.1%. The most abundant −3 fatty acid found was eicosatetraenoic acid, with a bioaccessibility percentage of 13.0%. The low lipid bioaccessibility may be because most of the internal lipids in seaweed are phospholipids and glycolipids, which are associated with cell membranes [222].

An investigation found that different seaweed species have a wide range of total lipid content, ranging from $0.7 \pm 0.1\%$ in *Chondrus crispus* (Rhodophyta) to $3.8 \pm 0.6\%$ in *Ulva* spp. (Chlorophyta). This variability highlights the diverse lipid compositions present in seaweeds.

Analysis of the fatty acid profiles revealed varying proportions of polyunsaturated fatty acids (PUFAs), monounsaturated fatty acids (MUFAs), and saturated fatty acids (SFAs) across different species. *Ulva* species exhibited the highest content of polyunsaturated fatty acids (PUFAs), while *C. crispus* and *Gracilaria* species were rich in saturated fatty acids (SFAs) [230].

Seaweed Minerals

Minerals are chemical elements that organisms require as essential nutrients for survival. They are primarily obtained through the consumption of plants, animals, or water. The human body contains five primary minerals: calcium (Ca), phosphorus (P), potassium (K), sodium (Na), and magnesium (Mg). Sulfur (S), iron (Fe), chloride (Cl), cobalt (Co), copper (Cu), zinc (Zn), manganese (Mn), molybdenum (Mo), iodine (I), and selenium (Se) are trace elements with unique metabolic functions in the human body. The bioavailability of these metals is strongly related to the carbohydrate and dietary fiber concentrations [22]. Samples with high carbohydrate or dietary fiber concentrations have high metal bioavailability ratios. A negative association was found between the bioavailability percentages of certain metals and proteins. This is because during in vitro digestion, proteins are hydrolyzed into amino acids. Most soluble amino acids carry positive or negative charges at a physiological pH, which increases the ionic strength in the aqueous phase. This results in lower metal solubility via the salting-out effect [231]. Metals and lipids have a negative correlation, which is due to metals not being emulsified by bile extracts during in vitro digestion and being introduced into the liquid phase before becoming available for absorption [22,222].

One study has shown that certain seaweeds, such as nori and sea lettuce, contain high levels of bioavailable iron. However, the absorption efficiency of iron from seaweeds varies, with some species exhibiting lower absorption rates compared to spinach. Factors such as the vitamin C content can enhance the iron bioavailability in specific seaweeds, but high levels of arsenic or other minerals may limit their regular consumption as a safe iron source [232].

The absorption of magnesium from seaweeds differs based on their fiber content and solubility. Seaweeds like Aosa (Chlorophyta) and Kombu (Phaeophyceae) have been identified as good sources of magnesium due to their high absorbable magnesium concentration. Despite variations in the magnesium solubility after in vitro digestion, factors beyond solubility alone influence magnesium absorption from seaweeds [233].

The bioavailability of minerals in seaweeds can be affected by cooking methods. Some metals, such as chromium, iron, cobalt, nickel, and zinc, are released into the cooking water during heat treatment, while others, such as copper, iron, and zinc, are retained in the seaweed [210].

Understanding the bioavailability and digestibility of minerals in edible seaweeds is crucial for maximizing their nutritional benefits. Seaweeds offer a diverse range of minerals, but variations in the absorption rates and interactions with food constituents highlight the need to consider these factors when incorporating seaweeds into the diet. Further research is needed to optimize the bioavailability of minerals from seaweeds and ensure their safe and effective use as valuable sources of essential nutrients [52,53,95,107].

The seaweed food matrix can be optimized to improve the bioaccessibility by modifying its content and structure. Seaweed compounds can interact positively or negatively, affecting the bioavailability of nutrients. In vitro digestion models are powerful tools for modifying the food matrix and improving the bioavailability of bioactive compounds, resulting in potentially healthier food products [52,53,95,107].

Furthermore, there is a need to further develop and study this topic due to its vital importance for human health. The nutritional value of seaweed is currently undervalued, despite being a key factor in human welfare and a potential tool for future food nutraceuticals. With the evolution of technology, it is possible to develop a food safety system using new technologies such as the E-Nose, FTIR-ATR, and NIRS, coupled with the INFOGEST system. This can enhance food safety and improve understanding of the real nutrient availability through pro analysis component analysis and logarithms. It is important to develop a new technological safety system to improve human welfare quality while reducing the costs associated with classical assays.

6. Future Road into Seaweed Food Safety and Nutraceutical Potential

Even though each biochemical and biological study has both positive and negative implications for seaweed quality certification, these analytical techniques are currently in development due to the complexity of seaweeds' composition. The additional expenses that these studies may incur for businesses might be reflected in the final product pricing. In contrast, seaweed characterization is required to provide quality control and customer safety. Thus, the framework and regulations for regulating seaweed food quality management are continuously evolving, with specialized analysis for seaweeds considering the chemical diversity and compound complexity [4]. It is still unknown how the gut microbiota influences dietary component metabolism and how much this differs across people (because humans differ as to their digestive system, with different microbiomes). Many kinds of food substances, such as complex polysaccharides, lipids, proteins, and phytochemicals, are transformed by gut microorganisms, and these metabolic processes are associated with a variety of health advantages as well as disease susceptibility. Furthermore, the gut microbiota changes the toxicities and lives of industrial chemicals and pollutants in the body [234]. There is currently a scarcity of information on the interactions and metabolic regulation of the gut microbiota and seaweeds' components. Due to the inherent heterogeneity of the gut microbiome from individual to individual, it is critical to understand the functions of gut bacteria in the processing of components acquired from macroalgae sources, most creating a robust gastrointestinal model (INFOGEST). The gut microbiota's direct and indirect metabolic effect on the chemical alterations of a wide variety of chemicals might possibly have consequences for host health [107,235].

As a novel food product, the seaweed sector for human consumption confronts a number of obstacles, including a lack of standard quality and safety processes, as well as a lack of regulatory rules. There are now various analytical techniques available for evaluating the nutritional value of seaweeds, which can guarantee a better quality and safety check of seaweed-based products; however, advancements to standardize the methodology and findings are urgently needed. Actually, a desire to collaborate with the scientific community to establish rules and laws applicable to seaweed farmers and industry for the food sector is rising throughout the Asian, European, and American continents [107].

Seaweeds are a promising novel food source that can help address global food security challenges. However, ensuring the safety of seaweed consumption is of the utmost importance. A systematic review was conducted to assess the microbiological, chemical, physical, and allergenic risks associated with seaweed consumption. The review found no significant hazards for microbiological, allergenic, and physical risks. However, there are concerns regarding chemical risks, particularly the accumulation of heavy metals, when harvesting seaweed in polluted areas. To mitigate this risk, seaweed cultivation in controlled environments is recommended. It is also necessary to periodically monitor the heavy metal levels in final products. Establishing safety standards for seaweed consumption is crucial as its popularity continues to grow worldwide [236].

The seaweed industry has great potential for sustainable ocean economies and aligns with the UN Decade of Ocean Science for Sustainable Development. Seaweed production accounts for over 50% of the total global marine production and supports the livelihoods of millions. However, the industry faces significant challenges, including the need to enhance biosecurity, traceability, and pest management, as well as to promote technological innovation and policy coordination, to ensure its long-term sustainability. Paying attention to small-scale farmers and processors is crucial to ensure inclusive growth and resilience while addressing key Sustainable Development Goals [237].

The expansion of the seaweed market in Australia provides a valuable example for Europe as it incorporates seaweed into its food systems. Australia's initiatives, including significant investments in research, development, and commercialization, serve as a blueprint for Europe to follow [238].

In Australia, concerted efforts have been made to sustainably cultivate native seaweed species, tapping into the country's rich biodiversity. This approach is in line with Europe's increasing interest in using local resources and promoting biodiversity conservation. European countries can explore the cultivation of indigenous seaweed species suited to their coastal environments, thus reducing their dependence on imported seaweeds by adopting similar strategies [238].

Additionally, Australia's emphasis on innovation and technology in seaweed cultivation and processing provides valuable insights for Europe. By embracing advancements in cultivation techniques, biorefinery processes, and product development, Europe's emerging seaweed industry can enhance its efficiency and sustainability [238].

Collaboration between scientists, government agencies, industry stakeholders and local communities has been instrumental in the growth of the seaweed sector in Australia. Similarly, to overcome regulatory hurdles, address food safety concerns, and ensure consumer acceptance of seaweed-based products, fostering interdisciplinary collaborations and stakeholder engagement will be crucial for Europe [238].

To ensure the safety of seaweed for food and feed purposes, a comprehensive assessment of the potential hazards is necessary. Major hazards, such as heavy metal accumulation and microbial contamination, pose significant risks to human health. It is essential to address data gaps and prioritize monitoring programs to identify and mitigate hazards. The hazard presence is influenced by factors such as the seaweed type, cultivation environment, and processing methods. Prioritizing monitoring efforts based on the hazard severity and knowledge gaps can inform risk management strategies and ensure the safety of seaweed-based products [97,239]. There needs to be developed a new system based on the HACCP, GMP and ISO 2200 and linked to the new characterization technologies and bioavailability analysis to evolve human food into a new step toward human safe nutraceutical food [102,104,105].

Despite the need for responses to growing health concerns and a secure and stable food supply, it is critical to adopt new rules that improve the quality and safety of seaweeds and their sub-products for human use. Furthermore, promoting and improving the required human daily nutrition dosage is critical for improving overall health conditions. Thus, there is a need to observe if the different types of seaweed will need a different type of safety analysis protocol or can an intermediate protocol be used for all. Furthermore, there is need to develop a control of seaweed quality and a handbook of best practices for the culture that enhances sustainably and healthy seaweed aquaculture. A handbook focusing on the methods to analyze the quality and safety of seaweeds and their sub-products for human use, including microbial, heavy metals and nutritional methods, is required, as is bioavailability assays to be applied around the globe, and clarifying if different types of algae need different approaches due to brown seaweeds having different compounds from red and green seaweed [4,23,52,240].

7. Conclusions

There is a new wave of research to find a new natural food source and ingredient, which is being setting on seaweeds. However, there still a lot to do to verify seaweeds' real potential at a food solution for humankind's future. Still, the principal features and discoveries demonstrated that seaweeds are a future key for human nutrition and healthcare solution.

Food safety is becoming an international problem for human health worldwide. Common food contaminants, including but not limited to antibiotics, veterinary medications, pesticides, mycotoxins, heavy metals, foodborne diseases, unlawful additives, and allergies, can constitute a substantial hazard to human health. Therefore, to have a positive impact on human health, there is need to correctly certify seaweeds' quality as a food ingredient by the regulations and principles with which food raw sources are normally accessed. Thus, a variety of analytical approaches were used to check food safety, but each had practical limits or analytical flaws. There are still margins to develop even more technological tools to have more data to check and enhance seaweed in relations to food safety and security.

More assays concerning seaweed compounds' bioavailability are needed to promote seaweed foods and their safety for human welfare, using standard methods which can be validated by diverse food agencies. This is one of the most important topics, which continues without a theoretical and practical recommended and applied method for the studies of nutraceutical compounds and foods. Thus, bioavailability analysis of seaweeds and their compounds/products requires an integrated and comprehensive approach that includes in vitro, in silico and in vivo studies, new modern analytical methods (connect to the classical techniques), and evaluation of the many affecting variables. Understanding the bioavailability of seaweed-derived nutrients and bioactive compounds will help with their best application in functional and healthy meals, nutritional supplements, and therapeutic uses, being a keystone to the future of seaweeds as a safe nutraceutical source.

Author Contributions: Writing—original draft preparation, J.C., J.O.T. and R.S.; writing—review and editing, J.C., J.O.T., R.S. and L.P.; visualization, J.C. and J.O.T.; supervision, L.P. All authors have read and agreed to the published version of the manuscript.

Funding: This work was supported by FCT—Fundação para a Ciência e Tecnologia, I.P. by project reference UIDB/04004/2020 and DOI identifier 10.54499/UIDB/04004/2020 (https://doi.org/10.54499/UIDB/04004/2020). This work was supported by FCT—Fundação para a Ciência e Tecnologia, I.P. by project reference UIDP/04004/2020 and DOI identifier 10.54499/UIDP/04004/2020 (https://doi.org/10.54499/UIDP/04004/2020).

Institutional Review Board Statement: Not applicable.

Informed Consent Statement: Not applicable.

Data Availability Statement: Not applicable.

Conflicts of Interest: The authors declare no conflicts of interest.

Abbreviations

Due to the extensive work, we created a list of abbreviations:

%	Percentage
Ag-NPs	Silver nanoparticles
AnPs	*Ascophyllum nodosum*
AOAC	Association of Official Analytical Chemists
BP-NN	Back-propagation neural networks
Ca	Calcium
Cl	Chloride
Co	Cobalt
CO_2	Carbon dioxide
Cu	Copper
DPPH	2,2-diphenyl-1-picrylhydrazyl
E401	Sodium alginate
E407a	Transformed Eucheuma algae
EAAs	Essential amino acids
EC	Commission Regulation
EC	Electrochemical
EDXRF	Energy-dispersive X-ray fluorescence
ELISA	Enzyme-linked immunosorbent assay
EPA	Environmental Protection Agency
EU	European Union
FA	Fatty acids
FAO	Food and Agriculture Organization of the United Nations
Fe	Iron
FSMS	Food Safety Management System

FTIR-ATR	Fourier-transform infrared spectroscopy
GC-MS	Gas chromatography-mass spectrometer
GMP	Good Manufacturing Practices
HACCP	Hazard Analysis Critical Control Point
HIS	Hue, saturation, intensity
HLS	High-level structure
HPLC	High-performance liquid chromatography
HSB	Hue, saturation, brightness
HSI	Hyperspectral imaging
HSL	Hue, saturation, lightness
HSV	Hue, saturation, value
I	Iodine
ICP	Inductively coupled plasma
ICP-MS/ICP-OES	Inductively coupled plasma mass/spectroscopy inductively coupled plasma atomic emission spectroscopy
INFOGEST	COST INFOGEST network standardized protocol for human digestion assay
K	Potassium
LC-DAD-ESI-MS/MS	Liquid chromatography coupled to diode array detection and electrospray ionization tandem mass spectrometry
Mg	Magnesium
Mn	Manganese
Mo	Molybdenum
MS	Mass spectrometry
MUFAs	Monounsaturated fatty acids
Na	Sodium
NEAAs	Non-essential amino acids
NIRS	Near-infrared spectroscopy
P	Phosphorus
PCR	Polymerase chain reaction
PDO	Protected designation of origin
pH	Potential of hydrogen
PHPs	*Pyropia haitanensis*
PUFAs	Polyunsaturated fatty acids
RGB	Red–green–blue
S	Sulfur
SCFAs	Short-chain fatty acids
Se	Selenium
SFAs	Saturated fatty acids
TVC	Total viable count
UV-Vis	Absorption spectroscopy in ultraviolet and visible
VOCs	Volatile organic compounds
WHO	World Health Organization
WWII	World War II
XRF	X-ray fluorescence spectroscopy
Zn	Zinc
μ-XRF	Micro X-ray fluorescence spectroscopy
ω	Ômega

References

1. Geada, P.; Moreira, C.; Silva, M.; Nunes, R.; Madureira, L.; Rocha, C.M.R.; Pereira, R.N.; Vicente, A.A.; Teixeira, J.A. Algal Proteins: Production Strategies and Nutritional and Functional Properties. *Bioresour. Technol.* **2021**, *332*, 125125. [CrossRef] [PubMed]
2. Dopelt, K.; Radon, P.; Davidovitch, N. Environmental Effects of the Livestock Industry: The Relationship between Knowledge, Attitudes, and Behavior among Students in Israel. *Int. J. Environ. Res. Public Health* **2019**, *16*, 1359. [CrossRef] [PubMed]
3. Parodi, A.; Leip, A.; De Boer, I.J.M.; Slegers, P.M.; Ziegler, F.; Temme, E.H.M.; Herrero, M.; Tuomisto, H.; Valin, H.; Van Middelaar, C.E.; et al. The Potential of Future Foods for Sustainable and Healthy Diets. *Nat. Sustain.* **2018**, *1*, 782–789. [CrossRef]
4. Leandro, A.; Pacheco, D.; Cotas, J.; Marques, J.C.; Pereira, L.; Gonçalves, A.M.M. Seaweed's Bioactive Candidate Compounds to Food Industry and Global Food Security. *Life* **2020**, *10*, 140. [CrossRef]

5. Sultana, F.; Wahab, M.A.; Nahiduzzaman, M.; Mohiuddin, M.; Iqbal, M.Z.; Shakil, A.; Mamun, A.-A.; Khan, M.S.R.; Wong, L.; Asaduzzaman, M. Seaweed Farming for Food and Nutritional Security, Climate Change Mitigation and Adaptation, and Women Empowerment: A Review. *Aquac. Fish.* **2023**, *8*, 463–480. [CrossRef]
6. Ross, F.W.R.; Boyd, P.W.; Filbee-Dexter, K.; Watanabe, K.; Ortega, A.; Krause-Jensen, D.; Lovelock, C.; Sondak, C.F.A.; Bach, L.T.; Duarte, C.M.; et al. Potential Role of Seaweeds in Climate Change Mitigation. *Sci. Total Environ.* **2023**, *885*, 163699. [CrossRef] [PubMed]
7. Klnc, B.; Cirik, S.; Turan, G.; Tekogul, H.; Koru, E. Seaweeds for Food and Industrial Applications. In *Food Industry*; InTech: London, UK, 2013.
8. Shannon, E.; Abu-Ghannam, N. Seaweeds as Nutraceuticals for Health and Nutrition. *Phycologia* **2019**, *58*, 563–577. [CrossRef]
9. Xu, J.; Liao, W.; Liu, Y.; Guo, Y.; Jiang, S.; Zhao, C. An Overview on the Nutritional and Bioactive Components of Green Seaweeds. *Food Prod. Process. Nutr.* **2023**, *5*, 18. [CrossRef]
10. Cherry, P.; O'Hara, C.; Magee, P.J.; McSorley, E.M.; Allsopp, P.J. Risks and Benefits of Consuming Edible Seaweeds. *Nutr. Rev.* **2019**, *77*, 307–329. [CrossRef]
11. da Costa, E.; Silva, J.; Mendonça, S.; Abreu, M.; Domingues, M. Lipidomic Approaches towards Deciphering Glycolipids from Microalgae as a Reservoir of Bioactive Lipids. *Mar. Drugs* **2016**, *14*, 101. [CrossRef]
12. Ścieszka, S.; Klewicka, E. Algae in Food: A General Review. *Crit. Rev. Food Sci. Nutr.* **2019**, *59*, 3538–3547. [CrossRef] [PubMed]
13. Mendes, M.; Navalho, S.; Ferreira, A.; Paulino, C.; Figueiredo, D.; Silva, D.; Gao, F.; Gama, F.; Bombo, G.; Jacinto, R.; et al. Algae as Food in Europe: An Overview of Species Diversity and Their Application. *Foods* **2022**, *11*, 1871. [CrossRef] [PubMed]
14. Alba, K.; Kontogiorgos, V. Seaweed Polysaccharides (Agar, Alginate Carrageenan). In *Encyclopedia of Food Chemistry*; Elsevier: Amsterdam, The Netherlands, 2019; pp. 240–250.
15. Ferrara, L. Seaweeds: A Food for Our Future. *J. Food Chem. Nanotechnol.* **2020**, *6*, 56–64. [CrossRef]
16. FAO; WHO. *Report of the Expert Meeting on Food Safety for Seaweed—Current Status and Future Perspectives*; FAO: Rome, Italy, 2022; ISBN 9789251365908.
17. FAO. New Report Urges Food Safety Guidance on Seaweed. In *Inocuidad y Calidad de los Alimentos*; FAO: Rome, Italy, 2022; Available online: https://www.fao.org/food-safety/news/news-details/es/c/1607078/ (accessed on 20 November 2023).
18. Peñalver, R.; Lorenzo, J.M.; Ros, G.; Amarowicz, R.; Pateiro, M.; Nieto, G. Seaweeds as a Functional Ingredient for a Healthy Diet. *Mar. Drugs* **2020**, *18*, 301. [CrossRef] [PubMed]
19. Giercksky, E.; Doumeizel, V. *Seaweed Revolution: A Manifesto for a Sustainable Future*; Lloyd's Register Foundation: London, UK, 2020.
20. Sloth, J.J.; Holdt, S.L. Setting the Standards for Seaweed Analysis. In *New Food*; Russell Publishing Ltd.: Kent, UK, 2021.
21. Cian, R.; Drago, S.; De Medina, F.; Martínez-Augustin, O. Proteins and Carbohydrates from Red Seaweeds: Evidence for Beneficial Effects on Gut Function and Microbiota. *Mar. Drugs* **2015**, *13*, 5358–5383. [CrossRef]
22. Moreda-Piñeiro, J.; Moreda-Piñeiro, A.; Romarís-Hortas, V.; Domínguez-González, R.; Alonso-Rodríguez, E.; López-Mahía, P.; Muniategui-Lorenzo, S.; Prada-Rodríguez, D.; Bermejo-Barrera, P. Trace Metals in Marine Foodstuff: Bioavailability Estimation and Effect of Major Food Constituents. *Food Chem.* **2012**, *134*, 339–345. [CrossRef]
23. Morais, T.; Inácio, A.; Coutinho, T.; Ministro, M.; Cotas, J.; Pereira, L.; Bahcevandziev, K. Seaweed Potential in the Animal Feed: A Review. *J. Mar. Sci. Eng.* **2020**, *8*, 559. [CrossRef]
24. Guiry, M.D.M.; Guiry, M.D.M. *AlgaeBase*; World-Wide Electronic Publication; National Univerity of Ireland: Galway, Ireland, 2021.
25. Thiviya, P.; Gamage, A.; Gama-Arachchige, N.S.; Merah, O.; Madhujith, T. Seaweeds as a Source of Functional Proteins. *Phycology* **2022**, *2*, 216–243. [CrossRef]
26. Webb, P.; Somers, N.K.; Thilsted, S.H. Seaweed's Contribution to Food Security in Low- and Middle-Income Countries: Benefits from Production, Processing and Trade. *Glob. Food Sec* **2023**, *37*, 100686. [CrossRef]
27. Boukid, F.; Rosell, C.M.; Rosene, S.; Bover-Cid, S.; Castellari, M. Non-Animal Proteins as Cutting-Edge Ingredients to Reformulate Animal-Free Foodstuffs: Present Status and Future Perspectives. *Crit. Rev. Food Sci. Nutr.* **2022**, *62*, 6390–6420. [CrossRef]
28. Garcia-Vaquero, M.; Hayes, M. Red and Green Macroalgae for Fish and Animal Feed and Human Functional Food Development. *Food Rev. Int.* **2016**, *32*, 15–45. [CrossRef]
29. van der Heide, M.E.; Stødkilde, L.; Værum Nørgaard, J.; Studnitz, M. The Potential of Locally-Sourced European Protein Sources for Organic Monogastric Production: A Review of Forage Crop Extracts, Seaweed, Starfish, Mussel, and Insects. *Sustainability* **2021**, *13*, 2303. [CrossRef]
30. Torres, M.D.; Flórez-Fernández, N.; Domínguez, H. Integral Utilization of Red Seaweed for Bioactive Production. *Mar. Drugs* **2019**, *17*, 314. [CrossRef] [PubMed]
31. Monteiro, P.; Lomartire, S.; Cotas, J.; Pacheco, D.; Marques, J.C.; Pereira, L.; Gonçalves, A.M.M. Seaweeds as a Fermentation Substrate: A Challenge for the Food Processing Industry. *Processes* **2021**, *9*, 1953. [CrossRef]
32. Pereira, L. *Edible Seaweeds of the World*; CRC Press: Boca Raton, FL, USA, 2016; ISBN 9780429154041.
33. Pereira, L. *Therapeutic and Nutritional Uses of Algae*; A Science Publishers Book; CRC Press: Boca Raton, FL, USA, 2018; ISBN 9781315152844.
34. Menaa, F.; Wijesinghe, U.; Thiripuranathar, G.; Althobaiti, N.A.; Albalawi, A.E.; Khan, B.A.; Menaa, B. Marine Algae-Derived Bioactive Compounds: A New Wave of Nanodrugs? *Mar. Drugs* **2021**, *19*, 484. [CrossRef] [PubMed]

35. Rebours, C.; Marinho-Soriano, E.; Zertuche-González, J.A.; Hayashi, L.; Vásquez, J.A.; Kradolfer, P.; Soriano, G.; Ugarte, R.; Abreu, M.H.; Bay-Larsen, I.; et al. Seaweeds: An Opportunity for Wealth and Sustainable Livelihood for Coastal Communities. *J. Appl. Phycol.* **2014**, *26*, 1939–1951. [CrossRef] [PubMed]
36. Tanna, B.; Mishra, A. Metabolites Unravel Nutraceutical Potential of Edible Seaweeds: An Emerging Source of Functional Food. *Compr. Rev. Food Sci. Food Saf.* **2018**, *17*, 1613–1624. [CrossRef]
37. Brown, E.M.; Allsopp, P.J.; Magee, P.J.; Gill, C.I.; Nitecki, S.; Strain, C.R.; McSorley, E.M. Seaweed and Human Health. *Nutr. Rev.* **2014**, *72*, 205–216. [CrossRef]
38. Tibbetts, S.M.; Milley, J.E.; Lall, S.P. Nutritional Quality of Some Wild and Cultivated Seaweeds: Nutrient Composition, Total Phenolic Content and In Vitro Digestibility. *J. Appl. Phycol.* **2016**, *28*, 3575–3585. [CrossRef]
39. Salehi, B.; Sharifi-Rad, J.; Seca, A.M.L.; Pinto, D.C.G.A.; Michalak, I.; Trincone, A.; Mishra, A.P.; Nigam, M.; Zam, W.; Martins, N. Current Trends on Seaweeds: Looking at Chemical Composition, Phytopharmacology, and Cosmetic Applications. *Molecules* **2019**, *24*, 4182. [CrossRef]
40. Collins, K.; Fitzgerald, G.; Stanton, C.; Ross, R. Looking Beyond the Terrestrial: The Potential of Seaweed Derived Bioactives to Treat Non-Communicable Diseases. *Mar. Drugs* **2016**, *14*, 60. [CrossRef] [PubMed]
41. Galland-Irmouli, A.-V.; Fleurence, J.; Lamghari, R.; Luçon, M.; Rouxel, C.; Barbaroux, O.; Bronowicki, J.-P.; Villaume, C.; Guéant, J.-L. Nutritional Value of Proteins from Edible Seaweed Palmaria Palmata (Dulse). *J. Nutr. Biochem.* **1999**, *10*, 353–359. [CrossRef] [PubMed]
42. Dawczynski, C.; Schubert, R.; Jahreis, G. Amino Acids, Fatty Acids, and Dietary Fibre in Edible Seaweed Products. *Food Chem.* **2007**, *103*, 891–899. [CrossRef]
43. Wong, K.H.; Cheung, P.C.K. Nutritional Evaluation of Some Subtropical Red and Green Seaweeds. *Food Chem.* **2000**, *71*, 475–482. [CrossRef]
44. Holdt, S.L.; Kraan, S. Bioactive Compounds in Seaweed: Functional Food Applications and Legislation. *J. Appl. Phycol.* **2011**, *23*, 543–597. [CrossRef]
45. Kendel, M.; Wielgosz-Collin, G.; Bertrand, S.; Roussakis, C.; Bourgougnon, N.; Bedoux, G. Lipid Composition, Fatty Acids and Sterols in the Seaweeds Ulva Armoricana, and Solieria Chordalis from Brittany (France): An Analysis from Nutritional, Chemotaxonomic, and Antiproliferative Activity Perspectives. *Mar. Drugs* **2015**, *13*, 5606–5628. [CrossRef] [PubMed]
46. Rohani-Ghadikolaei, K.; Abdulalian, E.; Ng, W.-K. Evaluation of the Proximate, Fatty Acid and Mineral Composition of Representative Green, Brown and Red Seaweeds from the Persian Gulf of Iran as Potential Food and Feed Resources. *J. Food Sci. Technol.* **2012**, *49*, 774–780. [CrossRef]
47. Polat, S.; Ozogul, Y. Biochemical Composition of Some Red and Brown Macro Algae from the Northeastern Mediterranean Sea. *Int. J. Food Sci. Nutr.* **2008**, *59*, 566–572. [CrossRef]
48. McHugh, D.J. *A Guide to the Seaweed Industry*; FAO: Rome, Italy, 2003; ISBN 9251049580.
49. MacArtain, P.; Gill, C.I.R.; Brooks, M.; Campbell, R.; Rowland, I.R. Nutritional Value of Edible Seaweeds. *Nutr. Rev.* **2008**, *65*, 535–543. [CrossRef]
50. Narayan, B.; Kumar, C.S.; Sashima, T.; Maeda, H.; Hosokawa, M.; Miyashita, K. Composition, Functionality and Potential Applications of Seaweed Lipids. In *Biocatalysis and Bioenergy*; Wiley: Hoboken, NJ, USA, 2008; pp. 463–490.
51. Valverde, M.E.; Hernández-Pérez, T.; Paredes-López, O. Edible Mushrooms: Improving Human Health and Promoting Quality Life. *Int. J. Microbiol.* **2015**, *2015*, 376387. [CrossRef]
52. Tavares, J.O.; Cotas, J.; Valado, A.; Pereira, L. Algae Food Products as a Healthcare Solution. *Mar. Drugs* **2023**, *21*, 578. [CrossRef]
53. Cotas, J.; Lomartire, S.; Gonçalves, A.M.M.; Pereira, L. From Ocean to Medicine: Harnessing Seaweed's Potential for Drug Development. *Int. J. Mol. Sci.* **2024**, *25*, 797. [CrossRef]
54. Circuncis, A.R.; Catarino, M.D.; Cardoso, S.M.; Silva, A.M.S. Minerals from Macroalgae Origin: Health Benefits and Risks for Consumers. *Mar. Drugs* **2018**, *16*, 400. [CrossRef] [PubMed]
55. Jadeja, R.N.; Batty, L. Metal Content of Seaweeds in the Vicinity of Acid Mine Drainage in Amlwch, North Wales, U.K. *Indian. J. Mar. Sci.* **2013**, *42*, 16–20.
56. Vellinga, R.E.; Sam, M.; Verhagen, H.; Jakobsen, L.S.; Ravn-Haren, G.; Sugimoto, M.; Torres, D.; Katagiri, R.; Thu, B.J.; Granby, K.; et al. Increasing Seaweed Consumption in the Netherlands and Portugal and the Consequences for the Intake of Iodine, Sodium, and Exposure to Chemical Contaminants: A Risk-Benefit Study. *Front. Nutr.* **2022**, *8*, 792923. [CrossRef] [PubMed]
57. Zhou, Q.; Xue, S.; Zhang, L.; Chen, G. Trace Elements and the Thyroid. *Front. Endocrinol.* **2022**, *13*, 904889. [CrossRef]
58. American Cancer Society. Arsenic and Cancer Risk. Available online: https://www.cancer.org/cancer/risk-prevention/chemicals/arsenic.html#:~:t (accessed on 15 December 2023).
59. Davis, T.A.; Volesky, B.; Mucci, A. A Review of the Biochemistry of Heavy Metal Biosorption by Brown Algae. *Water Res.* **2003**, *37*, 4311–4330. [CrossRef] [PubMed]
60. Ibrahim, W.M. Biosorption of Heavy Metal Ions from Aqueous Solution by Red Macroalgae. *J. Hazard. Mater.* **2011**, *192*, 1827–1835. [CrossRef]
61. Banach, J.L.; van den Burg, S.W.K.; van der Fels-Klerx, H.J. Food Safety during Seaweed Cultivation at Offshore Wind Farms: An Exploratory Study in the North Sea. *Mar. Policy* **2020**, *120*, 104082. [CrossRef]
62. Grebe, G.S.; Byron, C.J.; St. Gelais, A.; Kotowicz, D.M.; Olson, T.K. An Ecosystem Approach to Kelp Aquaculture in the Americas and Europe. *Aquac. Rep.* **2019**, *15*, 100215. [CrossRef]

63. Lähteenmäki-Uutela, A.; Rahikainen, M.; Camarena-Gómez, M.T.; Piiparinen, J.; Spilling, K.; Yang, B. European Union Legislation on Macroalgae Products. *Aquac. Int.* **2021**, *29*, 487–509. [CrossRef]
64. Gianello, D.; Ávila-Hernández, E.; Aguer, I.; Crettaz-Minaglia, M.C. Water Quality Assessment of a Temperate Urban Lagoon Using Physico-Chemical and Biological Indicators. *SN Appl. Sci.* **2019**, *1*, 470. [CrossRef]
65. Rahikainen, M.; Yang, B. *Macroalgae as Food and Feed Ingredients in the Baltic Sea Region—Regulation by the European Union*; Growing Algae Sustainably in the Baltic Sea (GRASS) Project, University of Turku: Turku, Finland, 2020.
66. Augyte, S.; Yarish, C.; Redmond, S.; Kim, J.K. Cultivation of a Morphologically Distinct Strain of the Sugar Kelp, Saccharina Latissima Forma Angustissima, from Coastal Maine, USA, with Implications for Ecosystem Services. *J. Appl. Phycol.* **2017**, *29*, 1967–1976. [CrossRef]
67. Concepcion, A.; DeRosia-Banick, K.; Balcom, N. *Seaweed Production and Processing in Connecticut: A Guide to Understanding and Controlling Potential Food Safety Hazards*; NOAA: Washington, DC, USA, 2020.
68. Campbell, I.; Kambey, C.S.B.; Mateo, J.P.; Rusekwa, S.B.; Hurtado, A.Q.; Msuya, F.E.; Stentiford, G.D.; Cottier-Cook, E.J. Biosecurity Policy and Legislation for the Global Seaweed Aquaculture Industry. *J. Appl. Phycol.* **2020**, *32*, 2133–2146. [CrossRef]
69. WHO. *Food Safety*; WHO: Geneva, Switzerland, 1999.
70. Suutari, M.; Leskinen, E.; Spilling, K.; Kostamo, K.; Seppälä, J. Nutrient Removal by Biomass Accumulation on Artificial Substrata in the Northern Baltic Sea. *J. Appl. Phycol.* **2017**, *29*, 1707–1720. [CrossRef]
71. Biancarosa, I.; Espe, M.; Bruckner, C.G.; Heesch, S.; Liland, N.; Waagbø, R.; Torstensen, B.; Lock, E.J. Amino Acid Composition, Protein Content, and Nitrogen-to-Protein Conversion Factors of 21 Seaweed Species from Norwegian Waters. *J. Appl. Phycol.* **2017**, *29*, 1001–1009. [CrossRef]
72. Guo, J.; Qi, M.; Chen, H.; Zhou, C.; Ruan, R.; Yan, X.; Cheng, P. Macroalgae-Derived Multifunctional Bioactive Substances: The Potential Applications for Food and Pharmaceuticals. *Foods* **2022**, *11*, 3455. [CrossRef] [PubMed]
73. Desideri, D.; Cantaluppi, C.; Ceccotto, F.; Meli, M.A.; Roselli, C.; Feduzi, L. Essential and Toxic Elements in Seaweeds for Human Consumption. *J. Toxicol. Environ. Health A* **2016**, *79*, 112–122. [CrossRef] [PubMed]
74. Hamdy, A.A. Biosorption of Heavy Metals by Marine Algae. *Curr. Microbiol.* **2000**, *41*, 232–238. [CrossRef]
75. Negara, B.F.S.P.; Sohn, J.H.; Kim, J.-S.; Choi, J.-S. Effects of Phlorotannins on Organisms: Focus on the Safety, Toxicity, and Availability of Phlorotannins. *Foods* **2021**, *10*, 452. [CrossRef]
76. Nutrition and Food Safety (NFS). *Basic Food Safety for Health Workers*; Nutrition and Food Safety (NFS): Geneva, Switzerland, 1999.
77. Lozano Muñoz, I.; Díaz, N.F. Minerals in Edible Seaweed: Health Benefits and Food Safety Issues. *Crit. Rev. Food Sci. Nutr.* **2022**, *62*, 1592–1607. [CrossRef] [PubMed]
78. Wells, M.L.; Potin, P.; Craigie, J.S.; Raven, J.A.; Merchant, S.S.; Helliwell, K.E.; Smith, A.G.; Camire, M.E.; Brawley, S.H. Algae as Nutritional and Functional Food Sources: Revisiting Our Understanding. *J. Appl. Phycol.* **2017**, *29*, 949–982. [CrossRef] [PubMed]
79. Goecke, F.; Klemetsdal, G.; Ergon, Å. Cultivar Development of Kelps for Commercial Cultivation—Past Lessons and Future Prospects. *Front. Mar. Sci.* **2020**, *8*, 110. [CrossRef]
80. Lomartire, S.; Cotas, J.; Pacheco, D.; Marques, J.C.; Pereira, L.; Gonçalves, A.M.M. Environmental Impact on Seaweed Phenolic Production and Activity: An Important Step for Compound Exploitation. *Mar. Drugs* **2021**, *19*, 245. [CrossRef] [PubMed]
81. Roleda, M.Y.; Hurd, C.L. Seaweed Nutrient Physiology: Application of Concepts to Aquaculture and Bioremediation. *Phycologia* **2019**, *58*, 552–562. [CrossRef]
82. Hafting, J.T.; Critchley, A.T.; Cornish, M.L.; Hubley, S.A.; Archibald, A.F. On-Land Cultivation of Functional Seaweed Products for Human Usage. *J. Appl. Phycol.* **2012**, *24*, 385–392. [CrossRef]
83. García-Poza, S.; Leandro, A.; Cotas, C.; Cotas, J.; Marques, J.C.; Pereira, L.; Gonçalves, A.M.M. The Evolution Road of Seaweed Aquaculture: Cultivation Technologies and the Industry 4.0. *Int. J. Environ. Res. Public Health* **2020**, *17*, 6528. [CrossRef] [PubMed]
84. Hafting, J.T.; Craigie, J.S.; Stengel, D.B.; Loureiro, R.R.; Buschmann, A.H.; Yarish, C.; Edwards, M.D.; Critchley, A.T. Prospects and Challenges for Industrial Production of Seaweed Bioactives. *J. Phycol.* **2015**, *51*, 821–837. [CrossRef]
85. Snethlage, J.S.; de Koning, S.; Giesbers, E.; Veraart, J.A.; Debrot, A.O.; Harkes, I.; van den Burg, S.W.K.; Hamon, K.G. Knowledge Needs in Realising the Full Potential of Seaweed for World Food Provisioning. *Glob. Food Secur.* **2023**, *37*, 100692. [CrossRef]
86. Farkas, J. Physical Methods of Food Preservation. In *Food Microbiology: Fundamentals and Frontiers*, 3rd ed.; American Society of Microbiology: Washington, DC, USA, 2007; pp. 685–712.
87. Tapia, M.S.; Alzamora, S.M.; Chirife, J. Effects of Water Activity (a_w) on Microbial Stability as a Hurdle in Food Preservation. In *Water Activity in Foods*; Wiley: Hoboken, NJ, USA, 2020; pp. 323–355.
88. Skonberg, D.I.; Fader, S.; Perkins, L.B.; Perry, J.J. Lactic Acid Fermentation in the Development of a Seaweed Sauerkraut-style Product: Microbiological, Physicochemical, and Sensory Evaluation. *J. Food Sci.* **2021**, *86*, 334–342. [CrossRef]
89. Santhoshkumar, P.; Yoha, K.S.; Moses, J.A. Drying of Seaweed: Approaches, Challenges and Research Needs. *Trends Food Sci. Technol.* **2023**, *138*, 153–163. [CrossRef]
90. Regal, A.L.; Alves, V.; Gomes, R.; Matos, J.; Bandarra, N.M.; Afonso, C.; Cardoso, C. Drying Process, Storage Conditions, and Time Alter the Biochemical Composition and Bioactivity of the Anti-Greenhouse Seaweed Asparagopsis Taxiformis. *Eur. Food Res. Technol.* **2020**, *246*, 781–793. [CrossRef]
91. Gupta, S.; Cox, S.; Abu-Ghannam, N. Effect of Different Drying Temperatures on the Moisture and Phytochemical Constituents of Edible Irish Brown Seaweed. *LWT—Food Sci. Technol.* **2011**, *44*, 1266–1272. [CrossRef]

92. Cherry, P.; Yadav, S.; Strain, C.R.; Allsopp, P.J.; McSorley, E.M.; Ross, R.P.; Stanton, C. Prebiotics from Seaweeds: An Ocean of Opportunity? *Mar. Drugs* **2019**, *17*, 327. [CrossRef] [PubMed]
93. Moreira-Leite, B.; Antunes, R.; Cotas, J.; Martins, N.; Costa, N.; Noronha, J.P.; Mata, P.; Diniz, M. Modified Atmosphere Packaging (MAP) for Seaweed Conservation: Impact on Physicochemical Characteristics and Microbiological Activity. *Foods* **2023**, *12*, 2736. [CrossRef]
94. Moreira Leite, B.S. Novas Alternativas Para o Uso de Macroalgas Da Costa Portuguesa Em Alimentação. Master's Thesis, Universidade Nova de Lisboa, Lisboa, Portugal, 2017.
95. Cascais, M.; Monteiro, P.; Pacheco, D.; Cotas, J.; Pereira, L.; Marques, J.C.; Gonçalves, A.M.M. Effects of Heat Treatment Processes: Health Benefits and Risks to the Consumer. *Appl. Sci.* **2021**, *11*, 8740. [CrossRef]
96. Vucko, M.J.; Magnusson, M.; Kinley, R.D.; Villart, C.; de Nys, R. The Effects of Processing on the in Vitro Antimethanogenic Capacity and Concentration of Secondary Metabolites of Asparagopsis Taxiformis. *J. Appl. Phycol.* **2017**, *29*, 1577–1586. [CrossRef]
97. Banach, J.L.; Hoek-van den Hil, E.F.; van der Fels-Klerx, H.J. Food Safety Hazards in the European Seaweed Chain. *Compr. Rev. Food Sci. Food Saf.* **2020**, *19*, 332–364. [CrossRef]
98. Nitschke, U.; Stengel, D.B. Quantification of Iodine Loss in Edible Irish Seaweeds during Processing. *J. Appl. Phycol.* **2016**, *28*, 3527–3533. [CrossRef]
99. WHO. *Cyanobacterial Toxins: Anatoxin-a and Analogues*; WHO: Geneva, Switzerland, 2020.
100. Wright, A. What Are Good Manufacturing Practices in the Food Industry? Available online: https://www.imsm.com/gb/blogs/what-are-good-manufacturing-practices-in-the-food-industry/ (accessed on 20 December 2023).
101. NQA. GMP: Food Safety Management. Available online: https://www.nqa.com/en-gb/certification/standards/gmp (accessed on 20 December 2023).
102. Chon, J.; Koo, R.; Song, K.; Kang, I.; Kim, D.; Bae, D.; Kim, H.; Kim, S.; Seo, K. Strategies for Expanding HACCP Certification Rate Using an Awareness Survey of Dairy Farmers. *Int. J. Dairy. Technol.* **2021**, *74*, 453–461. [CrossRef]
103. Suherman, S.; Janitra, A.A.; Budhiary, K.N.S.; Pratiwi, W.Z.; Idris, F.A. Review on Hazard Analysis and Critical Control Point (HACCP) in the Dairy Product: Cheese. *IOP Conf. Ser. Mater. Sci. Eng.* **2021**, *1053*, 012081. [CrossRef]
104. Food Safety Management—ISO 22000:2018. Available online: https://www.iso.org/publication/PUB100430.html (accessed on 20 December 2023).
105. NQA. Guide to ISO 22000. Available online: https://www.nqa.com/en-ca/resources/blog/february-2019/guide-to-iso-22000 (accessed on 20 December 2023).
106. ISO 22000:2018—Food Safety Management Systems, a Practical Guide. Available online: https://www.iso.org/publication/PUB100454.html (accessed on 20 December 2023).
107. Monteiro, P.; Cotas, J.; Pacheco, D.; Figueirinha, A.; da Silva, G.J.; Pereira, L.; Gonçalves, A.M.M. Seaweed as Food: How to Guarantee Their Quality? In *Sustainable Global Resources of Seaweeds Volume 2*; Springer International Publishing: Cham, Switzerland, 2022; pp. 309–321.
108. Shannon, E.; Conlon, M.; Hayes, M. Seaweed Components as Potential Modulators of the Gut Microbiota. *Mar. Drugs* **2021**, *19*, 358. [CrossRef] [PubMed]
109. AOAC Scientific Standards & Methods—AOAC International. Available online: https://www.aoac.org/scientific-solutions/ (accessed on 20 November 2023).
110. Bonah, E.; Huang, X.; Aheto, J.H.; Osae, R. Application of Electronic Nose as a Non-Invasive Technique for Odor Fingerprinting and Detection of Bacterial Foodborne Pathogens: A Review. *J. Food Sci. Technol.* **2020**, *57*, 1977–1990. [CrossRef]
111. Anwar, H.; Anwar, T.; Murtaza, S. Review on Food Quality Assessment Using Machine Learning and Electronic Nose System. *Biosens. Bioelectron. X* **2023**, *14*, 100365. [CrossRef]
112. Zhang, J.; Huang, H.; Song, G.; Huang, K.; Luo, Y.; Liu, Q.; He, X.; Cheng, N. Intelligent Biosensing Strategies for Rapid Detection in Food Safety: A Review. *Biosens. Bioelectron.* **2022**, *202*, 114003. [CrossRef]
113. Mahmudiono, T.; Olegovich Bokov, D.; Abdalkareem Jasim, S.; Kamal Abdelbasset, W.; Dinora, M. Khashirbaeva State-of-the-Art of Convenient and Low-Cost Electrochemical Sensor for Food Contamination Detection: Technical and Analytical Overview. *Microchem. J.* **2022**, *179*, 107460. [CrossRef]
114. Kaufmann, A. The Current Role of High-Resolution Mass Spectrometry in Food Analysis. *Anal. Bioanal. Chem.* **2012**, *403*, 1233–1249. [CrossRef]
115. Nollet, L.M.L.; Toldra, F. (Eds.) *Food Analysis by HPLC*; CRC Press: Boca Raton, FL, USA, 2012; ISBN 9780429151620.
116. Kholafazad Kordasht, H.; Hasanzadeh, M. Biomedical Analysis of Exosomes Using Biosensing Methods: Recent Progress. *Anal. Methods* **2020**, *12*, 2795–2811. [CrossRef]
117. Huang, X.; Chalmers, A.N. Review of Wearable and Portable Sensors for Monitoring Personal Solar UV Exposure. *Ann. Biomed. Eng.* **2021**, *49*, 964–978. [CrossRef]
118. Kholafazad-Kordasht, H.; Hasanzadeh, M.; Seidi, F. Smartphone Based Immunosensors as next Generation of Healthcare Tools: Technical and Analytical Overview towards Improvement of Personalized Medicine. *TrAC Trends Anal. Chem.* **2021**, *145*, 116455. [CrossRef]
119. Haque, F.; Bubli, S.Y.; Khan, M.S. UV–Vis Spectroscopy for Food Analysis. In *Techniques to Measure Food Safety and Quality*; Springer International Publishing: Cham, Switzerland, 2021; pp. 169–193.

120. Karoui, R. Spectroscopic Technique: Fluorescence and Ultraviolet-Visible (UV-Vis) Spectroscopies. In *Modern Techniques for Food Authentication*; Elsevier: Amsterdam, The Netherlands, 2018; pp. 219–252.
121. Claßen, J.; Aupert, F.; Reardon, K.F.; Solle, D.; Scheper, T. Spectroscopic Sensors for In-Line Bioprocess Monitoring in Research and Pharmaceutical Industrial Application. *Anal. Bioanal. Chem.* **2017**, *409*, 651–666. [CrossRef] [PubMed]
122. Kharbach, M.; Alaoui Mansouri, M.; Taabouz, M.; Yu, H. Current Application of Advancing Spectroscopy Techniques in Food Analysis: Data Handling with Chemometric Approaches. *Foods* **2023**, *12*, 2753. [CrossRef] [PubMed]
123. Ríos-Reina, R.; Azcarate, S.M.; Camiña, J.; Callejón, R.M. Assessment of UV–Visible Spectroscopy as a Useful Tool for Determining Grape-Must Caramel in High-Quality Wine and Balsamic Vinegars. *Food Chem.* **2020**, *323*, 126792. [CrossRef] [PubMed]
124. Zhang, W.; Li, N.; Feng, Y.; Su, S.; Li, T.; Liang, B. A Unique Quantitative Method of Acid Value of Edible Oils and Studying the Impact of Heating on Edible Oils by UV–Vis Spectrometry. *Food Chem.* **2015**, *185*, 326–332. [CrossRef] [PubMed]
125. Wu, X.; Bian, X.; Lin, E.; Wang, H.; Guo, Y.; Tan, X. Weighted Multiscale Support Vector Regression for Fast Quantification of Vegetable Oils in Edible Blend Oil by Ultraviolet-Visible Spectroscopy. *Food Chem.* **2021**, *342*, 128245. [CrossRef] [PubMed]
126. Rajauria, G.; Foley, B.; Abu-Ghannam, N. Identification and Characterization of Phenolic Antioxidant Compounds from Brown Irish Seaweed Himanthalia Elongata Using LC-DAD–ESI-MS/MS. *Innov. Food Sci. Emerg. Technol.* **2016**, *37*, 261–268. [CrossRef]
127. Barnes, M.; Sulé-Suso, J.; Millett, J.; Roach, P. Fourier Transform Infrared Spectroscopy as a Non-Destructive Method for Analysing Herbarium Specimens. *Biol. Lett.* **2023**, *19*, 20220546. [CrossRef] [PubMed]
128. Pereira, L.; Sousa, A.; Coelho, H.; Amado, A.M.; Ribeiro-Claro, P.J.A. Use of FTIR, FT-Raman and 13C-NMR Spectroscopy for Identification of Some Seaweed Phycocolloids. *Biomol. Eng.* **2003**, *20*, 223–228. [CrossRef]
129. Bogolitsyn, K.; Parshina, A.; Druzhinina, A.; Ovchinnikov, D.; Krasikov, V.; Khviyuzov, S. Physicochemical Characteristics of the Active Fractions of Polyphenols from Arctic Macrophytes. *J. Appl. Phycol.* **2020**, *32*, 4277–4287. [CrossRef]
130. Cotas, J.; Figueirinha, A.; Pereira, L.; Batista, T. The Effect of Salinity on Fucus Ceranoides (Ochrophyta, Phaeophyceae) in the Mondego River (Portugal). *J. Oceanol. Limnol.* **2019**, *37*, 881–891. [CrossRef]
131. Li, Q.; Feng, Z.; Zhang, T.; Ma, C.; Shi, H. Microplastics in the Commercial Seaweed Nori. *J. Hazard. Mater.* **2020**, *388*, 122060. [CrossRef] [PubMed]
132. Sim, S.F.; Lee, T.Z.E.; Mohd Irwan Lu, N.A.L.; Samling, B. Synchronized Analysis of FTIR Spectra and GCMS Chromatograms for Evaluation of the Thermally Degraded Vegetable Oils. *J. Anal. Methods Chem.* **2014**, *2014*, 271970. [CrossRef] [PubMed]
133. Galvis-Sánchez, A.C.; Barros, A.; Delgadillo, I. FTIR-ATR Infrared Spectroscopy for the Detection of Ochratoxin A in Dried Vine Fruit. *Food Addit. Contam.* **2007**, *24*, 1299–1305. [CrossRef] [PubMed]
134. Leite, A.I.N.; Pereira, C.G.; Andrade, J.; Vicentini, N.M.; Bell, M.J.V.; Anjos, V. FTIR-ATR Spectroscopy as a Tool for the Rapid Detection of Adulterations in Butter Cheeses. *LWT* **2019**, *109*, 63–69. [CrossRef]
135. Candoğan, K.; Altuntas, E.G.; İğci, N. Authentication and Quality Assessment of Meat Products by Fourier-Transform Infrared (FTIR) Spectroscopy. *Food Eng. Rev.* **2021**, *13*, 66–91. [CrossRef]
136. Al-Deen, R.B.; Azizieh, A.; Al-Ameer, L. Identification of Enterobacteriaceae Foodborne Bacteria in Syrian Foods by PCR and FTIR-ATR Techniques. *Adv. Environ. Biol.* **2014**, *2014*, 1233.
137. Keshavarzi, Z.; Barzegari Banadkoki, S.; Faizi, M.; Zolghadri, Y.; Shirazi, F.H. Comparison of Transmission FTIR and ATR Spectra for Discrimination between Beef and Chicken Meat and Quantification of Chicken in Beef Meat Mixture Using ATR-FTIR Combined with Chemometrics. *J. Food Sci. Technol.* **2020**, *57*, 1430–1438. [CrossRef] [PubMed]
138. Anjos, O.; Campos, M.G.; Ruiz, P.C.; Antunes, P. Application of FTIR-ATR Spectroscopy to the Quantification of Sugar in Honey. *Food Chem.* **2015**, *169*, 218–223. [CrossRef] [PubMed]
139. Vandanjon, L.; Burlot, A.-S.; Zamanileha, E.F.; Douzenel, P.; Ravelonandro, P.H.; Bourgougnon, N.; Bedoux, G. The Use of FTIR Spectroscopy as a Tool for the Seasonal Variation Analysis and for the Quality Control of Polysaccharides from Seaweeds. *Mar. Drugs* **2023**, *21*, 482. [CrossRef]
140. Amin, H.H. Safe Ulvan Silver Nanoparticles Composite Films for Active Food Packaging. *Am. J. Biochem. Biotechnol.* **2021**, *17*, 28–39. [CrossRef]
141. Qu, J.-H.; Liu, D.; Cheng, J.-H.; Sun, D.-W.; Ma, J.; Pu, H.; Zeng, X.-A. Applications of Near-Infrared Spectroscopy in Food Safety Evaluation and Control: A Review of Recent Research Advances. *Crit. Rev. Food Sci. Nutr.* **2015**, *55*, 1939–1954. [CrossRef] [PubMed]
142. Lin, H.; Ying, Y. Theory and Application of near Infrared Spectroscopy in Assessment of Fruit Quality: A Review. *Sens. Instrum. Food Qual. Saf.* **2009**, *3*, 130–141. [CrossRef]
143. Nicolaï, B.M.; Defraeye, T.; De Ketelaere, B.; Herremans, E.; Hertog, M.L.A.T.M.; Saeys, W.; Torricelli, A.; Vandendriessche, T.; Verboven, P. Nondestructive Measurement of Fruit and Vegetable Quality. *Annu. Rev. Food Sci. Technol.* **2014**, *5*, 285–312. [CrossRef]
144. Pojić, M.M.; Mastilović, J.S. Near Infrared Spectroscopy—Advanced Analytical Tool in Wheat Breeding, Trade, and Processing. *Food Bioproc Tech.* **2013**, *6*, 330–352. [CrossRef]
145. Morsy, N.; Sun, D.-W. Robust Linear and Non-Linear Models of NIR Spectroscopy for Detection and Quantification of Adulterants in Fresh and Frozen-Thawed Minced Beef. *Meat Sci.* **2013**, *93*, 292–302. [CrossRef]
146. Tito, N.B.; Rodemann, T.; Powell, S.M. Use of near Infrared Spectroscopy to Predict Microbial Numbers on Atlantic Salmon. *Food Microbiol.* **2012**, *32*, 431–436. [CrossRef]

147. Kurz, C.; Leitenberger, M.; Carle, R.; Schieber, A. Evaluation of Fruit Authenticity and Determination of the Fruit Content of Fruit Products Using FT-NIR Spectroscopy of Cell Wall Components. *Food Chem.* **2010**, *119*, 806–812. [CrossRef]
148. Kuligowski, J.; Carrión, D.; Quintás, G.; Garrigues, S.; de la Guardia, M. Direct Determination of Polymerised Triacylglycerides in Deep-Frying Vegetable Oil by near Infrared Spectroscopy Using Partial Least Squares Regression. *Food Chem.* **2012**, *131*, 353–359. [CrossRef]
149. Balabin, R.M.; Smirnov, S.V. Melamine Detection by Mid- and near-Infrared (MIR/NIR) Spectroscopy: A Quick and Sensitive Method for Dairy Products Analysis Including Liquid Milk, Infant Formula, and Milk Powder. *Talanta* **2011**, *85*, 562–568. [CrossRef]
150. Mignani, A.G.; Ciaccheri, L.; Ottevaere, H.; Thienpont, H.; Conte, L.; Marega, M.; Cichelli, A.; Attilio, C.; Cimato, A. Visible and Near-Infrared Absorption Spectroscopy by an Integrating Sphere and Optical Fibers for Quantifying and Discriminating the Adulteration of Extra Virgin Olive Oil from Tuscany. *Anal. Bioanal. Chem.* **2011**, *399*, 1315–1324. [CrossRef]
151. Prieto, N.; Andrés, S.; Giráldez, F.J.; Mantecón, A.R.; Lavín, P. Discrimination of Adult Steers (Oxen) and Young Cattle Ground Meat Samples by near Infrared Reflectance Spectroscopy (NIRS). *Meat Sci.* **2008**, *79*, 198–201. [CrossRef] [PubMed]
152. Flores, K.; Sánchez, M.-T.; Pérez-Marín, D.; Guerrero, J.-E.; Garrido-Varo, A. Feasibility in NIRS Instruments for Predicting Internal Quality in Intact Tomato. *J. Food Eng.* **2009**, *91*, 311–318. [CrossRef]
153. Yuan, W.; Xiang, B.; Yu, L.; Xu, J. A Non-Invasive Method for Screening Sodium Hydroxymethanesulfonate in Wheat Flour by Near-Infrared Spectroscopy. *Food Anal. Methods* **2011**, *4*, 550–558. [CrossRef]
154. Lytou, A.E.; Tsakanikas, P.; Lymperi, D.; Nychas, G.-J.E. Rapid Assessment of Microbial Quality in Edible Seaweeds Using Sensor Techniques Based on Spectroscopy, Imaging Analysis and Sensors Mimicking Human Senses. *Sensors* **2022**, *22*, 7018. [CrossRef] [PubMed]
155. Tadmor Shalev, N.; Ghermandi, A.; Tchernov, D.; Shemesh, E.; Israel, A.; Brook, A. NIR Spectroscopy and Artificial Neural Network for Seaweed Protein Content Assessment In-Situ. *Comput. Electron. Agric.* **2022**, *201*, 107304. [CrossRef]
156. Campbell, M.; Ortuño, J.; Koidis, A.; Theodoridou, K. The Use of Near-Infrared and Mid-Infrared Spectroscopy to Rapidly Measure the Nutrient Composition and the in Vitro Rumen Dry Matter Digestibility of Brown Seaweeds. *Anim. Feed. Sci. Technol.* **2022**, *285*, 115239. [CrossRef]
157. Niu, Y.; Sun, F.; Xu, Y.; Cong, Z.; Wang, E. Applications of Electrochemical Techniques in Mineral Analysis. *Talanta* **2014**, *127*, 211–218. [CrossRef] [PubMed]
158. Ampuero, S.; Zesiger, T.; Gustafsson, V.; Lundén, A.; Bosset, J. Determination of Trimethylamine in Milk Using an MS Based Electronic Nose. *Eur. Food Res. Technol.* **2002**, *214*, 163–167. [CrossRef]
159. Lippolis, V.; Cervellieri, S.; Damascelli, A.; Pascale, M.; Di Gioia, A.; Longobardi, F.; De Girolamo, A. Rapid Prediction of Deoxynivalenol Contamination in Wheat Bran by MOS-based Electronic Nose and Characterization of the Relevant Pattern of Volatile Compounds. *J. Sci. Food Agric.* **2018**, *98*, 4955–4962. [CrossRef]
160. Liu, Q.; Zhao, N.; Zhou, D.; Sun, Y.; Sun, K.; Pan, L.; Tu, K. Discrimination and Growth Tracking of Fungi Contamination in Peaches Using Electronic Nose. *Food Chem.* **2018**, *262*, 226–234. [CrossRef]
161. Li, M.; Wang, H.; Sun, L.; Zhao, G.; Huang, X. Application of Electronic Nose for Measuring Total Volatile Basic Nitrogen and Total Viable Counts in Packaged Pork During Refrigerated Storage. *J. Food Sci.* **2016**, *81*, M906–M912. [CrossRef] [PubMed]
162. Pallottino, F.; Costa, C.; Antonucci, F.; Strano, M.C.; Calandra, M.; Solaini, S.; Menesatti, P. Electronic Nose Application for Determination of Penicillium Digitatum in Valencia Oranges. *J. Sci. Food Agric.* **2012**, *92*, 2008–2012. [CrossRef]
163. Pattarapon, P.; Zhang, M.; Bhandari, B.; Gao, Z. Effect of Vacuum Storage on the Freshness of Grass Carp (Ctenopharyngodon Idella) Fillet Based on Normal and Electronic Sensory Measurement. *J. Food Process Preserv.* **2018**, *42*, e13418. [CrossRef]
164. Wang, D.; Wang, X.; Liu, T.; Liu, Y. Prediction of Total Viable Counts on Chilled Pork Using an Electronic Nose Combined with Support Vector Machine. *Meat Sci.* **2012**, *90*, 373–377. [CrossRef] [PubMed]
165. Azman, W.N.F.S.W.; Azir, K.N.F.b.K.; Khairuddin, A.b.M. E-Nose: Spoiled Food Detection Embedded Device Using Machine Learning for Food Safety Application. In *Computing and Informatics*; Springer: Singapore, 2024; pp. 221–234.
166. Dai, C.; Huang, X.; Lv, R.; Zhang, Z.; Sun, J.; Aheto, J.H. Analysis of Volatile Compounds of Tremella Aurantialba Fermentation via Electronic Nose and HS-SPME-GC-MS. *J. Food Saf.* **2018**, *38*, e12555. [CrossRef]
167. Huang, X.; Yu, S.; Xu, H.; Aheto, J.H.; Bonah, E.; Ma, M.; Wu, M.; Zhang, X. Rapid and Nondestructive Detection of Freshness Quality of Postharvest Spinaches Based on Machine Vision and Electronic Nose. *J. Food Saf.* **2019**, *39*, e12708. [CrossRef]
168. Shen, H.; Tao, J. Applying Electronic Nose Based on Odour Classification and Identification Technology in Detecting the Shelf Life of Fresh Fruits. *Chem. Eng. Trans.* **2018**, *68*, 217–222. [CrossRef]
169. Chongthanaphisut, P.; Seesaard, T.; Kerdcharoen, T. Monitoring of Microbial Canned Food Spoilage and Contamination Based on E-Nose for Smart Home. In Proceedings of the 2015 12th International Conference on Electrical Engineering/Electronics, Computer, Telecommunications and Information Technology (ECTI-CON), Hua Hin, Thailand, 24–27 June 2015; IEEE: New York, NY, USA, 2015; pp. 1–5.
170. Zarezadeh, M.R.; Aboonajmi, M.; Varnamkhasti, M.G.; Azarikia, F. Olive Oil Classification and Fraud Detection Using E-Nose and Ultrasonic System. *Food Anal. Methods* **2021**, *14*, 2199–2210. [CrossRef]
171. Xu, L.; Li, X.; Huang, J.; Gao, P.; Jin, Q.; Wang, X. Rapid Measuring Flavor Quality Changes of Frying Rapeseed Oils Using a Flash Gas Chromatography Electronic Nose. *Eur. J. Lipid Sci. Technol.* **2019**, *121*, 1800260. [CrossRef]

172. Singh, S.; Gaur, S. Development of Rapid and Non-Destructive Electric Nose (E-Nose) System for Shelf Life Evaluation of Different Edible Seeds. *Food Chem.* **2023**, *426*, 136562. [CrossRef]
173. McCaig, T.N. Extending the Use of Visible/near-Infrared Reflectance Spectrophotometers to Measure Colour of Food and Agricultural Products. *Food Res. Int.* **2002**, *35*, 731–736. [CrossRef]
174. Wu, D.; Sun, D.-W. Colour Measurements by Computer Vision for Food Quality Control—A Review. *Trends Food Sci. Technol.* **2013**, *29*, 5–20. [CrossRef]
175. YongXia, C.; RuiXin, L.; ZhaoZhou, L.; PengJu, C.; LiLi, W.; YanLi, W.; SuiQing, C. Quality Evaluation Based on Color Grading: Quality Discrimination of the Chinese Medicine Corni Fructus by an E-Eye. *Sci. Rep.* **2019**, *9*, 17006. [CrossRef] [PubMed]
176. Li, Z.; Yang, Z.; Song, C.; Li, C.; Peng, Z.; Xu, W. E-Eye. In Proceedings of the 16th ACM Conference on Embedded Networked Sensor Systems, New York, NY, USA, 4–7 November 2018; ACM: New York, NY, USA, 2018; pp. 68–81.
177. Calvini, R.; Pigani, L. Toward the Development of Combined Artificial Sensing Systems for Food Quality Evaluation: A Review on the Application of Data Fusion of Electronic Noses, Electronic Tongues and Electronic Eyes. *Sensors* **2022**, *22*, 577. [CrossRef] [PubMed]
178. Wang, Y.; Liu, Y.; Cui, Q.; Li, L.; Ning, J.; Zhang, Z. Monitoring the Withering Condition of Leaves during Black Tea Processing via the Fusion of Electronic Eye (E-Eye), Colorimetric Sensing Array (CSA), and Micro-near-Infrared Spectroscopy (NIRS). *J. Food Eng.* **2021**, *300*, 110534. [CrossRef]
179. Pathare, P.B.; Opara, U.L.; Al-Said, F.A.-J. Colour Measurement and Analysis in Fresh and Processed Foods: A Review. *Food Bioproc Tech.* **2013**, *6*, 36–60. [CrossRef]
180. Munekata, P.E.S.; Finardi, S.; de Souza, C.K.; Meinert, C.; Pateiro, M.; Hoffmann, T.G.; Domínguez, R.; Bertoli, S.L.; Kumar, M.; Lorenzo, J.M. Applications of Electronic Nose, Electronic Eye and Electronic Tongue in Quality, Safety and Shelf Life of Meat and Meat Products: A Review. *Sensors* **2023**, *23*, 672. [CrossRef] [PubMed]
181. Buratti, S.; Malegori, C.; Benedetti, S.; Oliveri, P.; Giovanelli, G. E-Nose, e-Tongue and e-Eye for Edible Olive Oil Characterization and Shelf Life Assessment: A Powerful Data Fusion Approach. *Talanta* **2018**, *182*, 131–141. [CrossRef] [PubMed]
182. Foschi, M.; Di Maria, V.; D'Archivio, A.A.; Marini, F.; Biancolillo, A. E-Eye-Based Approach for Traceability and Annuality Compliance of Lentils. *Appl. Sci.* **2023**, *13*, 1433. [CrossRef]
183. Hassoun, A.; Karoui, R. Quality Evaluation of Fish and Other Seafood by Traditional and Nondestructive Instrumental Methods: Advantages and Limitations. *Crit. Rev. Food Sci. Nutr.* **2017**, *57*, 1976–1998. [CrossRef]
184. Titova, T.; Nachev, V. "Electronic Tongue" in the Food Industry. *Food Sci. Appl. Biotechnol.* **2020**, *3*, 71. [CrossRef]
185. Surányi, J.; Zaukuu, J.-L.Z.; Friedrich, L.; Kovacs, Z.; Horváth, F.; Németh, C.; Kókai, Z. Electronic Tongue as a Correlative Technique for Modeling Cattle Meat Quality and Classification of Breeds. *Foods* **2021**, *10*, 2283. [CrossRef]
186. Banerjee, R.; Tudu, B.; Bandyopadhyay, R.; Bhattacharyya, N. A Review on Combined Odor and Taste Sensor Systems. *J. Food Eng.* **2016**, *190*, 10–21. [CrossRef]
187. Kaya, Z.; Koca, İ. Gıda Mühendisliğinde Elektronik Dil Uygulamaları. *Turk. J. Agric.—Food Sci. Technol.* **2020**, *8*, 1463–1471. [CrossRef]
188. Tahara, Y.; Toko, K. Electronic Tongues—A Review. *IEEE Sens. J.* **2013**, *13*, 3001–3011. [CrossRef]
189. Di Rosa, A.R.; Leone, F.; Chiofalo, V. Electronic Noses and Tongues. In *Chemical Analysis of Food*; Elsevier: Amsterdam, The Netherlands, 2020; pp. 353–389.
190. Scagion, V.P.; Mercante, L.A.; Sakamoto, K.Y.; Oliveira, J.E.; Fonseca, F.J.; Mattoso, L.H.C.; Ferreira, M.D.; Correa, D.S. An Electronic Tongue Based on Conducting Electrospun Nanofibers for Detecting Tetracycline in Milk Samples. *RSC Adv.* **2016**, *6*, 103740–103746. [CrossRef]
191. Han, F.; Huang, X.; Teye, E.; Gu, H. Quantitative Analysis of Fish Microbiological Quality Using Electronic Tongue Coupled with Nonlinear Pattern Recognition Algorithms. *J. Food Saf.* **2015**, *35*, 336–344. [CrossRef]
192. Pascual, L.; Gras, M.; Vidal-Brotóns, D.; Alcañiz, M.; Martínez-Máñez, R.; Ros-Lis, J.V. A Voltammetric E-Tongue Tool for the Emulation of the Sensorial Analysis and the Discrimination of Vegetal Milks. *Sens. Actuators B Chem.* **2018**, *270*, 231–238. [CrossRef]
193. Semenov, V.; Volkov, S.; Khaydukova, M.; Fedorov, A.; Lisitsyna, I.; Kirsanov, D.; Legin, A. Determination of Three Quality Parameters in Vegetable Oils Using Potentiometric E-Tongue. *J. Food Compos. Anal.* **2019**, *75*, 75–80. [CrossRef]
194. Zhang, Y.; Cheng, Q.; Yao, Y.; Guo, X.; Wang, R.; Peng, Z. A Preliminary Study: Saltiness and Sodium Content of Aqueous Extracts from Plants and Marine Animal Shells. *Eur. Food Res. Technol.* **2014**, *238*, 565–571. [CrossRef]
195. Gowen, A.; Odonnell, C.; Cullen, P.; Downey, G.; Frias, J. Hyperspectral Imaging—An Emerging Process Analytical Tool for Food Quality and Safety Control. *Trends Food Sci. Technol.* **2007**, *18*, 590–598. [CrossRef]
196. Al-Sarayreh, M.; Reis, M.M.; Yan, W.Q.; Klette, R. A Sequential CNN Approach for Foreign Object Detection in Hyperspectral Images. In *Computer Analysis of Images and Patterns*; Springer: Cham, Switzerland, 2019; pp. 271–283.
197. Feng, Y.-Z.; Sun, D.-W. Application of Hyperspectral Imaging in Food Safety Inspection and Control: A Review. *Crit. Rev. Food Sci. Nutr.* **2012**, *52*, 1039–1058. [CrossRef]
198. Soni, A.; Dixit, Y.; Reis, M.M.; Brightwell, G. Hyperspectral Imaging and Machine Learning in Food Microbiology: Developments and Challenges in Detection of Bacterial, Fungal, and Viral Contaminants. *Compr. Rev. Food Sci. Food Saf.* **2022**, *21*, 3717–3745. [CrossRef]

199. Kim, M.S.; Chao, K.; Chen, Y.-R.; Chan, D.; Mehl, P.M. Hyperspectral Imaging System for Food Safety: Detection of Fecal Contamination on Apples. In *Proc. SPIE 4206, Photonic Detection and Intervention Technologies for Safe Food*; Chen, Y.-R., Tu, S.-I., Eds.; SPIE: Bellingham, WA, USA, 2001; pp. 174–184.
200. Kwak, D.-H.; Son, G.-J.; Park, M.-K.; Kim, Y.-D. Rapid Foreign Object Detection System on Seaweed Using VNIR Hyperspectral Imaging. *Sensors* **2021**, *21*, 5279. [CrossRef]
201. El Hosry, L.; Sok, N.; Richa, R.; Al Mashtoub, L.; Cayot, P.; Bou-Maroun, E. Sample Preparation and Analytical Techniques in the Determination of Trace Elements in Food: A Review. *Foods* **2023**, *12*, 895. [CrossRef]
202. McComb, J.Q.; Rogers, C.; Han, F.X.; Tchounwou, P.B. Rapid Screening of Heavy Metals and Trace Elements in Environmental Samples Using Portable X-Ray Fluorescence Spectrometer, A Comparative Study. *Water Air Soil. Pollut.* **2014**, *225*, 2169. [CrossRef]
203. Winberg, P.C. *Best Practices for the Emerging Australian Seaweed Industry: Seaweed Quality Control Systems*; AgriFutures: Wagga Wagga, NSW, Australia, 2017.
204. Pashkova, G.V.; Smagunova, A.N.; Finkelshtein, A.L. X-Ray Fluorescence Analysis of Milk and Dairy Products: A Review. *TrAC Trends Anal. Chem.* **2018**, *106*, 183–189. [CrossRef]
205. Gallardo, H.; Queralt, I.; Tapias, J.; Guerra, M.; Carvalho, M.L.; Marguí, E. Possibilities of Low-Power X-Ray Fluorescence Spectrometry Methods for Rapid Multielemental Analysis and Imaging of Vegetal Foodstuffs. *J. Food Compos. Anal.* **2016**, *50*, 1–9. [CrossRef]
206. Li, F.; Wang, J.; Xu, L.; Wang, S.; Zhou, M.; Yin, J.; Lu, A. Rapid Screening of Cadmium in Rice and Identification of Geographical Origins by Spectral Method. *Int. J. Environ. Res. Public. Health* **2018**, *15*, 312. [CrossRef]
207. Reboredo, F.H.; Junior, W.; Pessoa, M.F.; Lidon, F.C.; Ramalho, J.C.; Leitão, R.G.; Silva, M.M.; Alvarenga, N.; Guerra, M. Elemental Composition of Algae-Based Supplements by Energy Dispersive X-Ray Fluorescence. *Plants* **2021**, *10*, 2041. [CrossRef] [PubMed]
208. Garshott, D.M.; MacDonald, E.A.; Murray, M.N.; Benvenuto, M.A.; Roberts-Kirchhoff, E.S. Elemental Analysis of a Variety of Dried, Powdered, Kelp Food Supplements for the Presence of Heavy Metals via Energy-Dispersive X-ray Fluorescence Spectrometry. In *ACS Symposium Series*; ACS Publications: Washington, DC, USA, 2011; pp. 123–133.
209. Chesori, C.R. Determination of Elemental Concentrations in Edible Seaweeds, Sea Sediments and Seawater Samples from the Kenyan Coast Using X-Ray Fluorescence Techniques. Master's Thesis, University of Nairobi, Nairobi, Kenya, 2015.
210. García-Sartal, C.; Barciela-Alonso, M.d.C.; Moreda-Piñeiro, A.; Bermejo-Barrera, P. Study of Cooking on the Bioavailability of As, Co, Cr, Cu, Fe, Ni, Se and Zn from Edible Seaweed. *Microchem. J.* **2013**, *108*, 92–99. [CrossRef]
211. Thakur, N.; Raigond, P.; Singh, Y.; Mishra, T.; Singh, B.; Lal, M.K.; Dutt, S. Recent Updates on Bioaccessibility of Phytonutrients. *Trends Food Sci. Technol.* **2020**, *97*, 366–380. [CrossRef]
212. Alegría, A.; Garcia-Llatas, G.; Cilla, A. Static Digestion Models: General Introduction. In *The Impact of Food Bioactives on Health*; Springer International Publishing: Cham, Switzerland, 2015; pp. 3–12.
213. Brodkorb, A.; Egger, L.; Alminger, M.; Alvito, P.; Assunção, R.; Ballance, S.; Bohn, T.; Bourlieu-Lacanal, C.; Boutrou, R.; Carrière, F.; et al. INFOGEST Static in Vitro Simulation of Gastrointestinal Food Digestion. *Nat. Protoc.* **2019**, *14*, 991–1014. [CrossRef] [PubMed]
214. Tan, Y.; Li, R.; Zhou, H.; Liu, J.; Muriel Mundo, J.; Zhang, R.; McClements, D.J. Impact of Calcium Levels on Lipid Digestion and Nutraceutical Bioaccessibility in Nanoemulsion Delivery Systems Studied Using Standardized INFOGEST Digestion Protocol. *Food Funct.* **2020**, *11*, 174–186. [CrossRef] [PubMed]
215. Zhang, C.; Xu, W.; Jin, W.; Shah, B.R.; Li, Y.; Li, B. Influence of Anionic Alginate and Cationic Chitosan on Physicochemical Stability and Carotenoids Bioaccessibility of Soy Protein Isolate-Stabilized Emulsions. *Food Res. Int.* **2015**, *77*, 419–425. [CrossRef]
216. Zheng, B.; Peng, S.; Zhang, X.; McClements, D.J. Impact of Delivery System Type on Curcumin Bioaccessibility: Comparison of Curcumin-Loaded Nanoemulsions with Commercial Curcumin Supplements. *J. Agric. Food Chem.* **2018**, *66*, 10816–10826. [CrossRef] [PubMed]
217. Bohn, T.; Carriere, F.; Day, L.; Deglaire, A.; Egger, L.; Freitas, D.; Golding, M.; Le Feunteun, S.; Macierzanka, A.; Menard, O.; et al. Correlation between in Vitro and in Vivo Data on Food Digestion. What Can We Predict with Static in Vitro Digestion Models? *Crit. Rev. Food Sci. Nutr.* **2018**, *58*, 2239–2261. [CrossRef] [PubMed]
218. McClements, D.J.; Li, Y. Review of in Vitro Digestion Models for Rapid Screening of Emulsion-Based Systems. *Food Funct.* **2010**, *1*, 32. [CrossRef]
219. Tan, Y.; Zhou, H.; McClements, D.J. Application of Static in Vitro Digestion Models for Assessing the Bioaccessibility of Hydrophobic Bioactives: A Review. *Trends Food Sci. Technol.* **2022**, *122*, 314–327. [CrossRef]
220. Mackie, A.; Rigby, N. InfoGest Consensus Method. In *The Impact of Food Bioactives on Health*; Springer International Publishing: Cham, Switzerland, 2015; pp. 13–22.
221. Satoor, S.N.; Patil, D.P.; Kristensen, H.D.; Joglekar, M.V.; Shouche, Y.; Hardikar, A.A. Manipulation and Assessment of Gut Microbiome for Metabolic Studies. *Methods Mol. Biol.* **2014**, *1194*, 449–469.
222. Demarco, M.; Oliveira de Moraes, J.; Matos, Â.P.; Derner, R.B.; de Farias Neves, F.; Tribuzi, G. Digestibility, Bioaccessibility and Bioactivity of Compounds from Algae. *Trends Food Sci. Technol.* **2022**, *121*, 114–128. [CrossRef]
223. De Bhowmick, G.; Hayes, M. In Vitro Protein Digestibility of Selected Seaweeds. *Foods* **2022**, *11*, 289. [CrossRef]
224. Zheng, L.-X.; Chen, X.-Q.; Cheong, K.-L. Current Trends in Marine Algae Polysaccharides: The Digestive Tract, Microbial Catabolism, and Prebiotic Potential. *Int. J. Biol. Macromol.* **2020**, *151*, 344–354. [CrossRef]

225. Qin, Y. Seaweed Hydrocolloids as Thickening, Gelling, and Emulsifying Agents in Functional Food Products. In *Bioactive Seaweeds for Food Applications*; Elsevier: Amsterdam, The Netherlands, 2018; pp. 135–152.
226. Chen, L.; Xu, W.; Chen, D.; Chen, G.; Liu, J.; Zeng, X.; Shao, R.; Zhu, H. Digestibility of Sulfated Polysaccharide from the Brown Seaweed Ascophyllum Nodosum and Its Effect on the Human Gut Microbiota in Vitro. *Int. J. Biol. Macromol.* **2018**, *112*, 1055–1061. [CrossRef] [PubMed]
227. Dong, M.; Jiang, Y.; Wang, C.; Yang, Q.; Jiang, X.; Zhu, C. Determination of the Extraction, Physicochemical Characterization, and Digestibility of Sulfated Polysaccharides in Seaweed—Porphyra Haitanensis. *Mar. Drugs* **2020**, *18*, 539. [CrossRef] [PubMed]
228. Peñalver, R.; Lorenzo, J.M.; Nieto, G. Bioaccessibility, Digestibility and Nutritional Properties of Algae and Cyanophyceae as Basis of Their Potential as Functional Food Ingredients. *Appl. Food Res.* **2024**, *4*, 100404. [CrossRef]
229. Rupérez, P.; Saura-Calixto, F. Dietary Fibre and Physicochemical Properties of Edible Spanish Seaweeds. *Eur. Food Res. Technol.* **2001**, *212*, 349–354. [CrossRef]
230. Soares, C.; Sousa, S.; Machado, S.; Vieira, E.; Carvalho, A.P.; Ramalhosa, M.J.; Morais, S.; Correia, M.; Oliva-Teles, T.; Domingues, V.F.; et al. Bioactive Lipids of Seaweeds from the Portuguese North Coast: Health Benefits versus Potential Contamination. *Foods* **2021**, *10*, 1366. [CrossRef]
231. Kramer, R.M.; Shende, V.R.; Motl, N.; Pace, C.N.; Scholtz, J.M. Toward a Molecular Understanding of Protein Solubility: Increased Negative Surface Charge Correlates with Increased Solubility. *Biophys. J.* **2012**, *102*, 1907–1915. [CrossRef] [PubMed]
232. Flores, S.R.L.; Dobbs, J.; Dunn, M.A. Mineral Nutrient Content and Iron Bioavailability in Common and Hawaiian Seaweeds Assessed by an in Vitro Digestion/Caco-2 Cell Model. *J. Food Compos. Anal.* **2015**, *43*, 185–193. [CrossRef]
233. Nakamura, E.; Yokota, H.; Matsui, T. The in Vitro Digestibility and Absorption of Magnesium in Some Edible Seaweeds. *J. Sci. Food Agric.* **2012**, *92*, 2305–2309. [CrossRef] [PubMed]
234. Tu, P.; Chi, L.; Bodnar, W.; Zhang, Z.; Gao, B.; Bian, X.; Stewart, J.; Fry, R.; Lu, K. Gut Microbiome Toxicity: Connecting the Environment and Gut Microbiome-Associated Diseases. *Toxics* **2020**, *8*, 19. [CrossRef] [PubMed]
235. Clarke, G.; Sandhu, K.V.; Griffin, B.T.; Dinan, T.G.; Cryan, J.F.; Hyland, N.P. Gut Reactions: Breaking Down Xenobiotic–Microbiome Interactions. *Pharmacol. Rev.* **2019**, *71*, 198–224. [CrossRef] [PubMed]
236. Cavallo, G.; Lorini, C.; Garamella, G.; Bonaccorsi, G. Seaweeds as a "Palatable" Challenge between Innovation and Sustainability: A Systematic Review of Food Safety. *Sustainability* **2021**, *13*, 7652. [CrossRef]
237. Cottier-Cook, E.J.; Nagabhatla, N.; Asri, A.; Beveridge, M.; Bianchi, P.; Bolton, J.; Bondad-Reantaso, M.G.; Brodie, J.; Buschmann, A.; Cabarubias, J.; et al. *Ensuring the Sustainable Future of the Rapidly Expanding Global Seaweed Aquaculture Industry—A Vision*; UNU Institute on Comparative Regional Integration Studies: Bruges, Belgium, 2021.
238. Deepika, C.; Wolf, J.; Moheimani, N.; Hankamer, B.; von Herzen, B.; Rao, A.R. Utilisation of Seaweeds in the Australian Market—Commercialisation Strategies: Current Trends and Future Prospects. In *Sustainable Global Resources Of Seaweeds Volume 1*; Springer International Publishing: Cham, Switzerland, 2022; pp. 265–294.
239. Blikra, M.J.; Altintzoglou, T.; Løvdal, T.; Rognså, G.; Skipnes, D.; Skåra, T.; Sivertsvik, M.; Noriega Fernández, E. Seaweed Products for the Future: Using Current Tools to Develop a Sustainable Food Industry. *Trends Food Sci. Technol.* **2021**, *118*, 765–776. [CrossRef]
240. Salido, M.; Soto, M.; Seoane, S. Seaweed: Nutritional and Gastronomic Perspective. A Review. *Algal Res.* **2024**, *77*, 103357. [CrossRef]

nutraceuticals

Review

The Marine Alga *Sargassum horneri* Is a Functional Food with High Bioactivity

Masayoshi Yamaguchi

Cancer Biology Program, University of Hawaii Cancer Center, University of Hawaii at Manoa, 701 Ilalo Street, Hawaii, HI 96813, USA; yamamasa11555@yahoo.co.jp

Abstract: Functional food factors can play a preventive and therapeutic role in several human diseases. The marine alga *Sargassum horneri* (*S. horneri*) has restorative effects in several types of metabolic disorders, including osteoporosis, diabetes, inflammatory conditions, and cancer cell growth. Osteoporosis is widely recognized as a major public health problem. Bone loss associated with ageing and diabetic states was prevented through the intake of bioactive compounds from *S. horneri* water extract in vivo by stimulating osteoblastic bone formation and inhibiting osteoclastic bone resorption in vitro. The intake of *S. horneri* water extract was found to have preventive effects on diabetic findings with an increase in serum glucose and lipid components. Furthermore, the *S. horneri* component has been shown to suppress adipogenesis from rat bone marrow cells and inflammatory conditions in vitro. Notably, the growth of bone metastatic human breast cancer MDA-MB-231 cells, which induce bone loss with osteolytic effects, was suppressed through culturing with the *S. horneri* water extract component in vitro. The *S. horneri* component, which has a molecular weight of less than 1000, was found to suppress the activation of NF-κB signaling by tumor necrosis factor-α, a cytokine associated with inflammation, in osteoblastic cells and macrophage RAW264.7 cells in vitro, suggesting a molecular mechanism. The bioactive component of *S. horneri* may play a multifunctional role in the prevention and treatment of metabolic disorders. This review outlines the advanced knowledge of the biological activity of the aqueous extract components of *S. horneri* and discusses the development of health supplements using this material.

Keywords: *Sargassum horneri*; osteoporosis; diabetes; obesity; cancer; NF-κB signaling; dietary supplement

Citation: Yamaguchi, M. The Marine Alga *Sargassum horneri* Is a Functional Food with High Bioactivity. *Nutraceuticals* **2024**, *4*, 181–189. https://doi.org/10.3390/nutraceuticals4020012

Academic Editors: Ivan Cruz-Chamorro and Guillermo Santos Sánchez

Received: 24 February 2024
Revised: 24 March 2024
Accepted: 2 April 2024
Published: 8 April 2024

1. Introduction

Functional food factors can play a preventive and therapeutic role in several human diseases. The knowledge that dietary factors regulate biological functions provides a wealth of information about food and health. In recent years, the marine alga *Sargassum horneri* (*S. horneri*) has been shown to have restorative effects in several types of metabolic disorders, including osteoporosis, diabetes, inflammatory conditions, and cancer cell growth. Osteoporosis, the bone loss associated with ageing, is widely recognized as a major public health problem. Therefore, it may be of great interest to elucidate the role of food-derived factors in the regulation of bone metabolism and to investigate the prevention and repair of osteoporosis [1–3].

Bone tissue is formed by the destruction of old bone by osteoclasts, which break down bone minerals, and the formation of new bone tissue by osteoblasts, which make new bone. The bone then continues to become flexible and elastic [4–6], which is called bone remodeling [4]. It has a mechanism to hold it in place. When this balance is disturbed by ageing or many pathological conditions, bone mass decreases, leading to osteoporosis. This bone disease makes people susceptible to fractures, causes them to be bedridden with fractures, disrupts their daily lives, and accelerates their death. It can play an important role in maintaining health and helping to prevent and repair disease. The incidence of bone disease, including osteoporosis, has been increasing annually with the recent increase in the ageing population, and there has been much interest in its prevention.

The author has a strong interest in elucidating the role of food-derived factors in the regulation of bone metabolism and research related to the prevention and repair of osteoporosis [1–3]. Alongside fisheries researchers with morphological characteristics, we collected seaweeds from the sea in Shizuoka Prefecture (Shimoda City, Japan), such as Undaria pinnatifida, *Sargassum horneri*, Eisenia bicyclis, and Cryptonemia schmitziana. Gelidium amansii and Ulva ulvaceae from Lake Hamana (Shizuoka, Japan) were used to study the effects of their extracts on bone calcium content [4]. As a result, we initially discovered that, among edible algae, the water-extracted components of *Sargassum horneri* (Japanese name, Akamoku) have a strong effect on increasing bone calcium levels in rat femoral tissue in vitro [7]. After that, the components of several seaweeds have been reported to express osteogenic effects [8–14]. *S. horneri* is found along the coasts of Japan and China, growing from winter to spring on rocky areas of the ocean floor where the waves are relatively calm. This seaweed grows rapidly and becomes entangled in fishing nets and ship propellers, so it is rarely used. Although it has been used as food in some areas, much of it has been discarded into the environment. This seaweed is a brown alga belonging to the kelp, mozuku (Japanese), and wakame (Japanese; *Undaria pinnatifida*) families. It has just the right amount of sliminess and a unique chewy texture, and has been used as a delicious and healthy seaweed since ancient times in areas such as Akita, Iwate, Toyama, and the Noto Peninsula in Japan. It was eaten raw. In addition, *S. horneri* (Japanese name, Akamoku) is known to contain significantly higher amounts of nutritional components than other seaweeds, such as carotene, vitamin C, vitamin B2, dietary fiber, minerals, and trace elements.

Subsequently, we found that in addition to promoting bone mass, the water extract components of *S. horneri* have the effects of ameliorating hyperglycemia and hyperlipidemia in diabetes in vivo, and suppressing the formation of fat cells associated with obesity both ex vivo and in vivo. [15]. It has been shown that these effects are based on a mechanism by which the *S. horneri* components suppress the expression and activity of nuclear factor kappa B (NF-κB), an intracellular information transduction molecule that is important for the expression of the above pathological conditions [16]. Thus, the water extract component of *S. horneri* was found to be effective in preventing and ameliorating a variety of pathological conditions, and it is expected to be effective as a traditional Chinese herbal medicine with a variety of physiological activities that help improve health.

This mini review outlines the advanced knowledge of the biological activity of the aqueous extract components of *S. horneri* and discusses the development of health supplements using this material.

2. *S. horneri* Component May Prevent Osteoporosis

2.1. Unique Anabolic Effect of S. horneri Component in Bone Tissue

Seaweed extract components were prepared to clarify the biological effects of various seaweeds on bone tissue in vitro. *S. horneri* extract was proven to increase the amount of calcium in the femoral tissue in a dose-dependent manner [7].

The experiment used a lyophilized sample of the extract dissolved in the purified distilled water of various seaweeds. Rats were given a 5% aqueous solution of aqueous extracts of wakame, red moss, arame, occidental ibis, maculata, and sea lettuce orally for 7 consecutive days. Among the different seaweeds, the intake of *S. horneri* aqueous extract uniquely increased the amount of calcium in femoral metaphyseal tissue (spongy bone tissue) in a dose-dependent manner [7]. In addition, the activity of alkaline phosphatase, an enzyme that promotes bone mineralization, and the amount of DNA (evaluated as an index of the number of cells in bone tissue) were shown to be significantly increased [17]. Such an effect of *S. horneri* extract was also induced in the diaphyseal (cortical) tissues of rat femurs. The anabolic effects of the *S. horneri* extract component were based on the enhancement of bone protein synthesis in bone cells [7,17]. The aqueous extract component of *S. horneri* may contain a bioactive factor that enhances osteoblast function, which promotes bone formation [7,17]. Brown algae contain several active compounds such as fucoidan, fucosterol and fucoxanthin. We have confirmed that these substances do not have an anabolic effect on bone formation.

It is noteworthy that the aqueous extract component of *S. horneri* has also been shown to suppress bone resorption through the action of osteoclasts causing osteolysis [18]. When the diaphyseal and metaphyseal tissues of rat femurs were cultured in the presence of bone resorption-promoting factors such as parathyroid hormone and prostaglandin E2, bone calcium content decreased [18]. These effects were almost completely suppressed in the presence of the aqueous extract component of *S. horneri*. Thus, the aqueous extract of *S. horneri* was found to have an inhibitory effect on bone resorption in vitro.

2.2. Characterisation of the Active Ingredient in S. horneri Aqueous Extract

An attempt was made to identify the active compounds in the aqueous extract of *S. horneri*. Firstly, the molecular weight of the active compounds in *S. horneri* water extract was determined using membrane fractionation [19]. The compounds that suppress bone resorption have been shown to have a molecular weight of 50,000 or more. To further separate the active components, the active fraction of the Sephadex G-25 (Sigma-Aldrich, St. Louis, MO, USA) column elution of the red mosquito extract was separated by high performance liquid chromatography (Asahipack GF-310HQ (Resonac, New York, NY, USA)) and GS-220HQ (Resonac, New York, NY, USA). The final fraction was found to contain several new and known substances with molecular weights (MW) of approximately 350–400 [19]. Finally, we found the presence of 4 chemicals in the *S. horneri* components (less than MW 3000) by analysis using a liquid chromatography–mass spectrophotometry system (LCMS-IT-TOF; Shimadzu, Kyoto, Japan). These chemicals were identified as 1,3,5-tris(oxolan-2-ylmethyl)-1,3,5-triazinane (MW 339), 5-phenyl-2-[2-(5-phenyltetrazol-2-ethyl)] tetrazole (MW 318), 3-(hexadecylamine)propane-1,2-diol (MW 316), and 2-(2-hydroxyethyl-tridecylamino)ethanol (MW 288). These chemicals may affect osteoblastogenesis and/or osteoclastogenesis. The combination of these compounds has a more potent anabolic effect on bone than either component alone. The manifestation of this potent combined effect may be important in the expression of multi-functionality.

2.3. S. horneri Component Has an Inhibitory Effect on the Activation of the NF-κB Signalling Pathway

Endogenous tumor necrosis factor-alpha (TNF-α) decreases peak bone mass and inhibits osteoblastic Smad activation through NF-κB [20–22]. The *S. horneri* component has been shown to have anabolic effects on bone tissue [7,19]. To elucidate the mechanistic property, we used the *S. horneri* components with molecular weights below 3000. The inflammatory cytokine TNF-α plays a central role in the molecular and cellular mechanisms of skeletal pathology [20,21]. We focused on the involvement of TNF-α in the expression of the anabolic effects of *S. horneri* on bone tissue. Inhibition of NF-κB, which is associated with TNF-α signaling, promotes osteoblastic bone formation and suppresses osteoclastic bone resorption in vitro. Notably, the aqueous extract component of *S. horneri* was found to block the impairment of osteoblast function by TNF-α and the control of osteoclast activation by the receptor activator of NF-κB (RANK) ligand (RANKL) [16]. This result suggests that the suppression of TNF-α signaling plays a critical role in exerting the anabolic effect of the *S. horneri* extract component on bone. We found that there are several new and known chemical substances with molecular weights between 200 and 400 in the aqueous extract components of *S. horneri* (MW3000 or less) [19]. These molecules may contribute to the inhibition of NF-κB signaling activation. Mechanistically, this inhibition may be important in exerting the bone anabolic effects of the *S. horneri* aqueous extract component.

The cellular mechanism by which the *S. horneri* component exerts an anabolic effect on bone metabolism, leading to increased bone mass, is described in Figure 1.

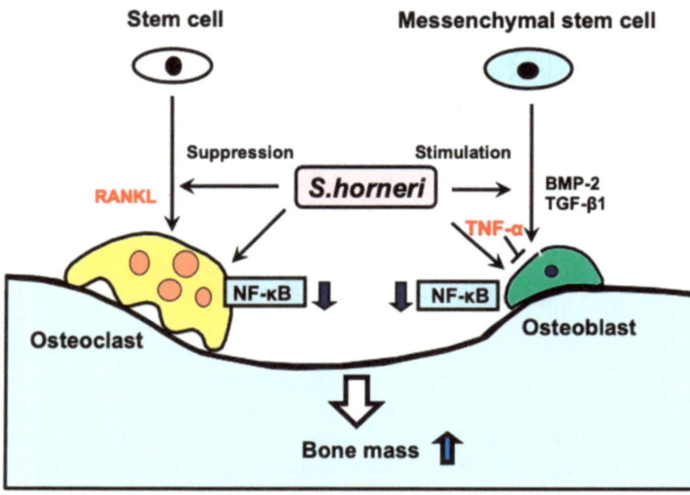

Figure 1. The cellular mechanism by which the *S. horneri* component exerts its osteogenic effects. Osteoclasts are differentiated from stem cells. RANKL stimulates osteoclastogenesis. Osteoblastic cells are differentiated from bone marrow mesenchymal stem cells and are stimulated by bone growth factors (TGF-β1 and BMP-2). The *S. horneri* component stimulates osteoblastic bone formation and suppresses osteoclastic bone resorption, thereby increasing bone mass. The *S. horneri* component suppresses TNF-α- and RANKL-enhanced NF-κB activation in osteoblasts and osteoclasts, suggesting a possible molecular mechanism for the osteogenic effects of the *S. horneri* component. In addition, bone marrow mesenchymal stem cells are differentiated into adipocytes. The *S. horneri* component suppresses adipogenesis from bone marrow mesenchymal stem cells. This may lead to the prevention of obesity and associated bone loss.

2.4. Involvement of *S. horneri* Water Extract Component in Osteoporosis Prevention

As mentioned above, *S. horneri* aqueous extract has been shown to promote bone formation and inhibit bone resorption [7,15–19]. In addition, we investigated its effect on promoting bone constituents and preventing osteoporosis in growing and ageing rats [23]. First, growing (4-week-old male) and ageing (50-week-old female) rats were given 2.5, 5 and 10 mg of *S. horneri* water extract (lyophilized product) per 100 g of rat body weight once a day [16]. After continuous oral administration for 7 and 14 days, changes in bone constituents (calcium content, alkaline phosphatase activity and DNA content) in the diaphyseal (bone) and metaphyseal (cancellous bone) tissues of the femur were examined [23]. Significant increases in all bone components, including calcium content, alkaline phosphatase activity, and DNA content, were found in the diaphyseal and metaphyseal tissues of growing and ageing rats [23]. Thus, it has been demonstrated that the intake of *S. horneri* components has the effect of promoting bone mass during both growth and ageing.

Osteoporosis with bone loss as a complication of diabetes has attracted clinical attention as an intractable disease [24,25]. Streptozotocin (STZ) is used as a model animal for diabetes because it destroys the pancreas and induces type 1 diabetes due to impaired insulin secretion [26]. In STZ-treated rats, bone constituents (calcium, alkaline phosphatase activity and DNA content) in the diaphysis and metaphysis of the femur were significantly reduced [15]. Notably, STZ-induced bone loss was significantly suppressed by continuous oral administration of *S. horneri* extract for 14 and 21 days in vivo [15]. Thus, the water extract component of *S. horneri* may have an improved effect on diabetic osteoporosis and may be useful in the supplemental prevention of osteoporosis.

2.5. Complementary Effects of S. horneri Water Extract Component in Humans

Based on the above knowledge, a new functional food material was developed using *S. horneri* water extract as the active ingredient to prevent osteoporosis. The material of the final product was named Hormax OT, after the name *S. horneri*. Note that OT is an abbreviation for osteon (meaning bone). Generally, healthy volunteers (24 men and women, 11 to 13 in each group) took 1 tablet (containing 300 mg of Formax OT) or 3 tablets (containing 900 mg of Formax OT) a day for 4 or 8 weeks [27].

This study was analyzed by measuring the serum concentration of bone turnover markers developed for the clinical evaluation of osteoporosis. Note that a starch tablet was used as a placebo. Bone turnover markers included bone-type alkaline phosphatase [28] and osteocalcin, markers of bone formation (proteins specifically produced by osteoblasts) [29], tartrate-resistant acid phosphatase, a marker of bone resorption (a protein produced by osteoclasts) [30], and type 1 collagen degradation product N-telopeptide (a specific degradation product of type 1 collagen protein present in bone tissue during bone resorption) [31].

Specifically, Formax OT was found to significantly reduce tartrate-resistant acid phosphatase, a serum marker of bone resorption, and suppress bone resorption (bone mineral dissolution) [24]. It was concluded that *S. horneri* extract constituents exert their bone resorption inhibiting effect early in the intake and then promote bone formation, thus helping to maintain bone mass.

In addition, general blood biochemistry tests and blood cell counts did not change significantly after taking 1 or 3 tablets of Formax daily for 8 weeks, and no toxicity was observed [27]. The functional material "Formax OT", which was first developed at Maruhachi Muramatsu Co., Ltd. (Shizuoka, Japan), is a powdered version of the water extract component of *S. horneri*, which has been shown to be effective in humans and may be effective in preventing osteoporosis.

3. *S. horneri* Water Extract Component Has a Preventive Effect on Diabetes

Whether the *S. horneri* component ameliorates diabetic states has been investigated using model rats treated with STZ, which destroys the pancreas and induces type 1 diabetes due to impaired insulin secretion [8]. In this rat model, suppressed body weight gain, elevated serum glucose and triglyceride concentrations, and elevated serum calcium and inorganic phosphorus concentrations were observed [15]. These diabetes-related fluctuations were significantly improved through continuous oral administration of the *S. horneri* water extract component for 14 and 21 days [15]. This result suggests that there are factors in the *S. horneri* water extract component that exert anti-diabetic effects and that its intake is effective in improving diabetic conditions. In addition, it is believed that there are two factors in the *S. horneri* water extract component, one that improves diabetic conditions and one that repairs diabetic osteoporosis, as described previously [15]. However, the same factor may be effective in improving both conditions, suggesting that they may be related.

4. *S. horneri* Extract Component Suppresses Adipogenesis in Bone Marrow Cells

Bone marrow stem cells are pluripotent and differentiate into osteoblasts, chondrocytes, cardiomyocytes, and adipocytes [32–34]. We have also found that factors in the aqueous extract component of *S. horneri* suppress the formation of adipocytes from bone marrow cells [35]. The question of whether bone marrow stem cells differentiate into adipocytes or osteoblasts has attracted much interest in the progression of osteoporosis. Therefore, we investigated the formation of adipocytes by culturing mouse bone marrow cells in a culture medium containing a component of *S. horneri* water extract (molecular weight less than 3000) in vitro [35]. We found that the *S. horneri* water extract component enhanced the insulin-stimulated differentiation of bone marrow stem cells into osteoblasts and suppressed the formation of adipocytes [35]. *S. horneri* extract also has inhibitory effects on lipid metabolism in 3T3-L1 adipocytes in vitro [36]. Thus, the *S. horneri* component may have the effect of suppressing the differentiation of bone marrow cells into adipocytes.

TNF-α secreted by mature adipocytes may be involved in the pathogenesis of obesity-induced osteoporosis [37,38]. In addition, the factors in the aqueous extract component of *S. horneri* were found to block the inhibition of osteoblast function and osteoclast hyperplasia caused by the activation of intracellular NF-κB signaling stimulated by TNF-α [16]. Note that TNF-α is upregulated in bone marrow adipocytes and mature adipocytes and is an important factor in causing inflammation and insulin resistance [35]. The *S. horneri* component has been shown to improve the status of type 1 diabetes [15]. The *S. horneri* component may also be useful in preventing type 2 diabetes, which is associated with obesity and insulin resistance.

5. Suppressive Effects of the *S. horneri* Component on the Inflammatory State Associated with TNF-α in Various Cell Types

The *S. horneri* component has been shown to exert suppressive effects on the inflammatory state of various cell types, including macrophages, retinal cells, and dermal fibroblasts [39–51]. In this mini review, the author has focused on inflammatory macrophages. The *S. horneri* component with 70% EtOH extracts, such as phenolics and flavonoids, has been shown to have excellent anti-inflammatory and antioxidant activities [39]. Levels of several cytokines, including prostaglandin E2 (PGE2), TNF-α, and interleukin (IL)-6, which mediate pro-inflammatory effects, were found to be reduced by treatment with *S. horneri* extract [39]. *S. horneri* extract has been shown to downregulate pro-inflammatory cytokines, PGE2 and nitric oxide secretion by blocking the downstream activation of Toll-like receptor (TLR)-mediated NF-κB and mitogen-activated kinase (MAPK) phosphorylation [40,50]. Furthermore, *S. horneri* extract inhibited chronic lung inflammation induced by particulate matter by blocking TLR/NF-κB/MAPK signaling in lung macrophages [42]. TNF-α, which stimulates TLR signaling, is important in the development of inflammatory diseases and cancer, and is a contributing factor in a variety of diseases. It may be critical as a common mechanism by which the *S. horneri* component has suppressive effects on the activation of NF-κB signaling.

6. Suppressive Effects of the *S. horneri* Component on Cancer Cells

Among various types of cancer, bone metastasis is a very serious and deadly disease. Breast cancer cells metastasize to bone, and approximately 70% of patients with advanced breast cancer will develop bone metastases [52–57]. Breast cancer promotes the formation of osteoclasts by secreting osteoporotic cytokines, including TNF-α [58,59]. Treatment includes drugs and antibodies that inhibit the action of RANKL, a cytokine that activates osteoclast function leading to bone resorption [20,21]. We found that the aqueous extract component of *S. horneri* stimulates osteoblastic bone formation and osteoclastic bone resorption, leading to bone loss [7,16,17], and investigated whether the *S. horneri* component affects breast cancer cells.

MDA-MB-231 human breast cancer cells have bone metastatic potential and lack the expression of estrogen receptor alpha, progesterone, and epidermal growth factor receptor 2. MDA-MB-231 triple negative breast cancer cells are well established as a model of human breast cancer that is difficult to treat with drugs [52,53,57,59]. We investigated whether the *S. horneri* component would be useful in controlling breast cancer. In the culture of human breast cancer MDA-MB-231 cells, the *S. horneri* component with a molecular weight of less than 3000 was found to suppress proliferation with cell cycle arrest and stimulate apoptotic cell death of MDA-MB-231 cells, resulting in a reduced numbers of cancer cells [60]. This study may suggest that the *S. horneri* compound is a useful tool for cancer prevention and therapy without side effects.

In addition, a polysaccharide fraction obtained from *S. horneri* by hot water extraction was shown to inhibit the growth of human colon cancer DLD cells in a dose-dependent manner by inducing apoptosis of DLD cells [61]. The polysaccharide component obtained from *S. horneri* caused the accumulation of cells in G0/G1 and S phase and affected the expression of apoptosis-associated genes such as Bcl-2 and Bax [62]. This may explain the growth inhibition of DLD cells. These studies suggest that sulphate content and molecular weight may influence antioxidant and antitumor activities. Thus, the *S. horneri* component has been shown to have anticancer activity. Further studies are anticipated.

7. Conclusions

Research into the biological activities of *S. horneri* has been fruitful, with the discovery in 2001 that it has osteogenic effects [7]. This is popular because it helps prevent osteoporosis, which leads to fractures and is common in an ageing population. This disease is common in bed-ridden people and is considered a socially important disease wherein prevention is important. The aqueous extract component of *S. horneri* has been shown to have an osteogenic effect by stimulating osteoblastic bone formation and suppressing osteoclastic bone resorption, leading to the prevention of age-related bone loss. Among the many edible seaweeds, *S. horneri* was found to have a unique osteogenic effect. Subsequently, the *S. horneri* component was found to have suppressive effects against diabetes, adipogenesis, obesity, inflammation, and cancer cell growth. Mechanistically, the *S. horneri* component was shown to suppress the TNF-α-activated transcription factor NF-κB signaling, which is associated with several diseases. *S. horneri* may be a functional food with multifunctional bioactivities. In addition, these bioactive components may be needed to identify the detailed chemical structure and associated biological activity. This study may lead to the development of new pharmaceutical drugs. Thus, the *S. horneri* marine algae component has a variety of physiological activities and is expected to be used as a material for supplements with Chinese herbal medicine effects.

Funding: This research received no external funding.

Institutional Review Board Statement: This article does not include any human or animal studies conducted by any of the authors.

Informed Consent Statement: Not applicable.

Data Availability Statement: The datasets used in this study are available from the respective authors upon reasonable request.

Conflicts of Interest: The author declares no conflicts of interest.

References

1. Yamaguchi, M. Regulatory mechanism of food factors in bone metabolism and prevention of osteoporosis. *Yakugaku Zasshi* **2006**, *126*, 1117–1137. [CrossRef]
2. Yamaguchi, M. *Nutritional Factors and Osteoporosis Prevention*; Nova Science Publishers, Inc.: New York, NY, USA, 2010.
3. Yamaguchi, M. *Biomedical Osteoporosis Treatment*; Nova Science Publishers, Inc.: New York, NY, USA, 2013.
4. Raggatt, L.J.; Partridge, N.C. Cellular and Molecular Mechanisms of Bone Remodeling. *J. Biol. Chem.* **2010**, *285*, 25103–25108. [CrossRef]
5. Xu, J.; Yu, L.; Liu, F.; Wan, L.; Deng, Z. The effect of cytokines on osteoblasts and osteoclasts in bone remodeling in osteoporosis: A review. *Front. Immunol.* **2023**, *14*, 1222129. [CrossRef]
6. Chin, K.-Y.; Ng, B.N.; Rostam, M.K.I.; Fadzil, N.F.D.M.; Raman, V.; Yunus, F.M.; Hashim, S.A.S.; Ekeuku, S.O. A Mini Review on Osteoporosis: From biology to pharmacological management of bone loss. *J. Clin. Med.* **2022**, *11*, 6434. [CrossRef]
7. Yamaguchi, M.; Hachiya, S.; Hiratuka, S.; Suzuki, T. Effect of marine algae extract on bone calcification in the femoral-metaphyseal tissues of rats: Anabolic effect of *Sargassum horneri*. *J. Health Sci.* **2001**, *47*, 533–538. [CrossRef]
8. Park, M.H.; Kim, S.; Cheon, J.; Lee, J.; Kim, B.K.; Lee, S.H.; Kong, C.; Kim, Y.Y.; Kim, M. Effects of Scytosiphon lomentaria on osteoblastic proliferation and differentiation of MC3T3- E1 cells. *Nutr. Res. Pract.* **2016**, *10*, 148–153. [CrossRef]
9. Koshi, R.; Nakai, K.; Tanaka, H.; Kato, K.; Charlestoncoad, T.; Matsuike, R.; Nakasugi, T. Original an extract of Eisenia bicyclis stimulates mineralized nodule formation by osteoblasts. *J. Hard Tissue Biol.* **2019**, *28*, 359–364. [CrossRef]
10. Kim, M.; Kang, J.H.; Oh, G.H.; Park, M.H. *Ishige sinicola* extract stimulates osteoblast proliferation and differentiation via the bone morphogenetic protein 2/runt-related gene 2 signalling pathway. *Z. Für Naturforschung C* **2019**, *74*, 167–174. [CrossRef]
11. Minetti, M.; Bernardini, G.; Biazzo, M.; Gutierrez, G.; Geminiani, M.; Petrucci, T.; Santucci, A. *Padina pavonica* extract promotes in vitro differentiation and functionality of human primary osteoblasts. *Mar. Drugs* **2019**, *17*, 473. [CrossRef]
12. Walsh, P.J.; McGrath, S.; McKelvey, S.; Ford, L.; Sheldrake, G.; Clarke, S.A. The osteogenic potential of brown seaweed extracts. *Mar. Drugs* **2019**, *17*, 141. [CrossRef]

13. Gyu, D. Floridoside from Laurencia undulata promotes osteogenic differentiation in murine bone marrow mesenchymal cells. *J. Funct. Foods* **2015**, *19*, 505–511. [CrossRef]
14. Saleng, S.; Hendra, F.N.; Ruslin, M.; Forouzanfar, T.; Helder, M.N. The osteogenic potential of seaweed: A systematic review and meta-analysis. *Algal Res.* **2024**, *79*, 103445. [CrossRef]
15. Uchiyama, S.; Yamaguchi, M. Preventive effect of marine alga *Sargassum horneri* extract on bone loss in streptozotocin-diabetic rats in vivo. *J. Health Sci.* **2003**, *49*, 149–155. [CrossRef]
16. Yamaguchi, M.; Matsumoto, T. Marine algae *Sargassum horneri* bioactive factor stimulates osteoblastogenesis and suppresses osteoclastogenesis in vitro. *OA Biotechnol.* **2012**, *1*, 3. [CrossRef]
17. Uchiyama, S.; Yamaguchi, M. Stimulatory effect of *Sargassum horneri* extract on bone formation in rat femoral-diaphyseal and -metaphyseal tissues in vitro. *J. Health Sci.* **2002**, *48*, 148–153. [CrossRef]
18. Uchiyama, S.; Yamaguchi, M. Inhibitory effect of marine alga *Sargassum horneri* extract on bone resorption in tissue culture in vitro. *J. Health Sci.* **2002**, *48*, 154–160. [CrossRef]
19. Uchiyama, S.; Hashizume, M.; Hokari, Y.; Nakagawa, T.; Igarashi, A.; Yamaguchi, M. Characterization of active component in marine alga *Sargassum horneri* extract in stimulating bone calcification in vitro. *J. Health Sci.* **2004**, *50*, 634–639. [CrossRef]
20. Nanes, M.S. Tumor necrosis factor-α: Molecular and cellular mechanisms in skeletal pathology. *Gene* **2003**, *321*, 1–15. [CrossRef] [PubMed]
21. Li, Y.; Li, A.; Strait, K.; Zhang, H.; Nanes, M.S.; Weitzmann, M.N. Endogenous TNFα lowers maximum peak bone mass and inhibits osteoblastic smad activation through NF-κB. *J. Bone Miner. Res.* **2007**, *22*, 646–655. [CrossRef]
22. Yamaguchi, M.; Weitzmann, M.N. Vitamin K2 stimulates osteoblastogenesis and suppresses osteoclastogenesis by suppressing NF-κB activation. *Int. J. Mol. Med.* **2011**, *27*, 3–14. [CrossRef]
23. Uchiyama, S.; Yamaguchi, M. Anabolic Effect of Marine Alga *Sargassum horneri* Extract on Bone Components in the Femoral-diaphyseal and -metaphyseal Tissues of Young and Aged Rats in Vivo. *J. Health Sci.* **2002**, *48*, 325–330. [CrossRef]
24. Li, Z.; Zhao, W.; Lin, X.; Li, F. AI algorithms for accurate prediction of osteoporotic fractures in patients with diabetes: An up-to-date review. *J. Orthop. Surg. Res.* **2023**, *18*, 956. [CrossRef] [PubMed]
25. Guimarães, G.C.; Coelho, J.B.C.; Silva, J.G.O.; de Sant'ana, A.C.C.; de Sá, C.A.C.; Moreno, J.M.; Reis, L.M.; Guimarães, C.S.d.O. Obesity, diabetes and risk of bone fragility: How BMAT behavior is affected by metabolic disturbances and its influence on bone health. *Osteoporos. Int.* **2023**, *35*, 1–14. [CrossRef] [PubMed]
26. Ghasemi, A.; Jeddi, S. Streptozotocin as a tool for induction of rat models of diabetes: A practical guide. *EXCLI J.* **2023**, *22*, 274–294. [CrossRef]
27. Matsumoto, T.; Hokari, Y.; Hashizume, M.; Yamaguchi, M. Effect of *Sargassum horneri* extract on circulating bone metabolic markers: Supplemental intake has an effect in healthy humans. *J. Health Sci.* **2008**, *54*, 50–55. [CrossRef]
28. Gomez, B.; Ardakani, S.; Ju, J.; Jenkins, D.; Cerelli, M.J.; Daniloff, G.Y.; Kung, V.T. Monoclonal antibody assay for measuring bone-specific alkaline phosphatase activity in serum. *Clin. Chem.* **1995**, *41*, 1560–1566. [CrossRef] [PubMed]
29. Koyama, N.; Ohara, K.; Yokota, H.; Kurome, T.; Katayama, M.; Hino, F.; Kato, I.; Akai, T. A one step sandwich enzyme immunoassay for γ-carboxylated osteocalcin using monoclonal antibodies. *J. Immunol. Methods* **1991**, *139*, 17–23. [CrossRef] [PubMed]
30. Halleen, J.M.; Alatalo, S.L.; Suominen, H.; Cheng, S.; Janckila, A.J.; Väänänen, H.K. Tartrate-resistant acid phosphatase 5b: A novel serum marker of bone resorption. *J. Bone Miner. Res.* **2000**, *15*, 1337–1345. [CrossRef] [PubMed]
31. Clements, J.D.; Herrich, M.V.; Singer, F.R.; Eyre, D. Evidence that serum NTx (collagen-type I N-telopeptides) can act as an immunochemical marker of bone resorption. *Clin. Chem.* **1997**, *43*, 2058–2063. [CrossRef]
32. Minguell, J.J.; Erices, A.; Conget, P. Mesenchymal stem cells. *Exp. Biol. Med.* **2001**, *226*, 507–520. [CrossRef]
33. Muruganandan, S.; Roman, A.A.; Sinal, C.J. Adipocyte differentiation of bone marrow-derived mesenchymal stem cells: Cross talk with the osteoblastogenic program. *Cell. Mol. Life Sci.* **2009**, *66*, 236–253. [CrossRef] [PubMed]
34. Gharibi, B.; Abraham, A.A.; Ham, J.; Evans, B.A.J. Adenosine receptor subtype expression and activation influence the differentiation of mesenchymal stem cells to osteoblasts and adipocytes. *J. Bone Miner. Res.* **2011**, *26*, 2112–2124. [CrossRef] [PubMed]
35. Yamaguchi, M.; Matsumoto, T. Marine algae *Sargassum horneri* bioactive factor suppresses adipogenesis in mouse bone marrow culture in vitro. *OA Biotechnol* **2014**, *3*, 7.
36. Lee, H.-G.; Kim, H.-S.; Je, J.-G.; Hwang, J.; Sanjeewa, K.K.A.; Lee, D.-S.; Song, K.-M.; Choi, Y.-S.; Kang, M.-C.; Jeon, Y.-J. Lipid inhibitory effect of (−)-loliolide isolated from *Sargassum horneri* in 3T3-L1 adipocytes: Inhibitory mechanism of adipose-specific proteins. *Mar. Drugs* **2021**, *19*, 96. [CrossRef] [PubMed]
37. Proietto, J. Obesity and Bone. *F1000Research* **2020**, *9*, 1111. [CrossRef] [PubMed]
38. Roy, B.; Curtis, M.E.; Fears, L.S.; Nahashon, S.N.; Fentress, H.M. Molecular mechanisms of obesity-induced osteoporosis and muscle atrophy. *Front. Physiol.* **2016**, *7*, 439. [CrossRef] [PubMed]
39. Kim, M.J.; Jo, H.G.; Ramakrishna, C.; Lee, S.-J.; Lee, D.-S.; Cheong, S.H. Anti-inflammatory and antioxidant activities of *Sargassum horneri* extract in RAW264.7 macrophages. *Phys. Act. Nutr.* **2021**, *25*, 45–53. [CrossRef]
40. Nagahawatta, D.P.; Kim, H.-S.; Jee, Y.-H.; Jayawardena, T.U.; Ahn, G.; Namgung, J.; Yeo, I.-K.; Sanjeewa, K.K.A.; Jeon, Y.-J. Sargachromenol isolated from *Sargassum horneri* inhibits particulate matter-induced inflammation in macrophages through toll-like receptor-mediated cell signaling pathways. *Mar. Drugs* **2021**, *20*, 28. [CrossRef] [PubMed]

41. Kim, E.-N.; Nabende, W.Y.; Jeong, H.; Hahn, D.; Jeong, G.-S. The marine-derived natural product epiloliolide isolated from *Sargassum horneri* regulates NLRP3 via PKA/CREB, promoting proliferation and anti-inflammatory effects of human periodontal ligament cells. *Mar. Drugs* **2021**, *19*, 388. [CrossRef]
42. Sanjeewa, K.A.; Jayawardena, T.U.; Kim, S.-Y.; Lee, H.G.; Je, J.-G.; Jee, Y.; Jeon, Y.-J. *Sargassum horneri* (Turner) inhibit urban particulate matter-induced inflammation in MH-S lung macrophages via blocking TLRs mediated NF-κB and MAPK activation. *J. Ethnopharmacol.* **2020**, *249*, 112363. [CrossRef]
43. Herath, K.H.I.N.M.; Cho, J.; Kim, H.J.; Dinh, D.T.T.; Kim, H.-S.; Ahn, G.; Jeon, Y.-J.; Jee, Y. Polyphenol containing *Sargassum horneri* attenuated Th2 differentiation in splenocytes of ovalbumin-sensitised mice: Involvement of the transcription factors GATA3/STAT5/NLRP3 in Th2 polarization. *Pharm. Biol.* **2021**, *59*, 1462–1470. [CrossRef] [PubMed]
44. Han, S.; Park, J.-S.; Umanzor, S.; Yarish, C.; Kim, J.K. Effects of extraction methods for a new source of biostimulant from *Sargassum horneri* on the growth of economically important red algae, Neopyropia yezoensis. *Sci. Rep.* **2022**, *12*, 11878. [CrossRef] [PubMed]
45. Kang, J.; Hyun, S.H.; Kim, H.; Park, S.; Lee, J.; Lee, I.; Bae, J. The effects of fucoidan-rich polysaccharides extracted from *Sargassum horneri* on enhancing collagen-related skin barrier function as a potential cosmetic product. *J. Cosmet. Dermatol.* **2024**, *23*, 1365–1373. [CrossRef] [PubMed]
46. Dias, M.K.H.M.; Madusanka, D.M.D.; Han, E.J.; Kim, H.-S.; Jeon, Y.-J.; Jee, Y.; Kim, K.-N.; Lee, K.; Fernando, I.P.S.; Ahn, G. *Sargassum horneri* (Turner) C. Agardh ethanol extract attenuates fine dust-induced inflammatory responses and impaired skin barrier functions in HaCaT keratinocytes. *J. Ethnopharmacol.* **2021**, *273*, 114003. [CrossRef]
47. Han, E.J.; Kim, H.-S.; Sanjeewa, K.K.A.; Jung, K.; Jee, Y.; Jeon, Y.-J.; Fernando, I.P.S.; Ahn, G. *Sargassum horneri* as a functional food ameliorated ige/bsa-induced mast cell activation and passive cutaneous anaphylaxis in mice. *Mar. Drugs* **2020**, *18*, 594. [CrossRef] [PubMed]
48. Han, E.J.; Fernando, I.P.S.; Kim, H.-S.; Jeon, Y.-J.; Madusanka, D.M.D.; Dias, M.K.H.M.; Jee, Y.; Ahn, G. Oral administration of *Sargassum horneri* improves the HDM/DNCB-Induced atopic dermatitis in NC/Nga mice. *Nutrients* **2020**, *12*, 2482. [CrossRef]
49. Herath, K.H.I.N.M.; Kim, H.J.; Jang, J.-H.; Kim, H.-S.; Kim, H.J.; Jeon, Y.-J.; Jee, Y. Mojabanchromanol isolated from *Sargassum horneri* attenuates particulate matter induced inflammatory responses via suppressing TLR2/4/7-MAPK signaling in MLE-12 cells. *Mar. Drugs* **2020**, *18*, 355. [CrossRef]
50. Wen, Z.-S.; Liu, L.-J.; OuYang, X.-K.; Qu, Y.-L.; Chen, Y.; Ding, G.-F. Protective effect of polysaccharides from *Sargassum horneri* against oxidative stress in RAW264.7 cells. *Int. J. Biol. Macromol.* **2014**, *68*, 98–106. [CrossRef]
51. Herath, K.H.I.N.M.; Cho, J.; Kim, A.; Kim, H.-S.; Han, E.J.; Kim, H.J.; Kim, M.S.; Ahn, G.; Jeon, Y.-J.; Jee, Y. Differential modulation of immune response and cytokine profiles of *Sargassum horneri* ethanol extract in murine spleen with or without Concanavalin A stimulation. *Biomed. Pharmacother.* **2019**, *110*, 930–942. [CrossRef]
52. Roodman, G.D. Mechanisms of Bone Metastasis. *N. Engl. J. Med.* **2004**, *350*, 1655–1664. [CrossRef]
53. Pang, L.; Gan, C.; Xu, J.; Jia, Y.; Chai, J.; Huang, R.; Li, A.; Ge, H.; Yu, S.; Cheng, H. Bone metastasis of breast cancer: Molecular mechanisms and therapeutic strategies. *Cancers* **2022**, *14*, 5727. [CrossRef] [PubMed]
54. Bray, F.; Ferlay, J.; Soerjomataram, I.; Siegel, R.L.; Torre, L.A.; Jemal, A. Global cancer statistics 2018: GLOBOCAN estimates of incidence and mortality worldwide for 36 cancers in 185 countries. *CA Cancer J. Clin.* **2018**, *68*, 394–424. [CrossRef]
55. Chang JC: Cancer stem cells: Role in tumor growth, recurrence, metastasis, and treatment resistance. *Medicine* **2016**, *95*, S20–S25. [CrossRef] [PubMed]
56. Boyce, B.F.; Yoneda, T.; Guise, T. Factors regulating the growth of metastasis cancer in bone. *Endocr. Relat. Cancer* **1999**, *6*, 333–347. [CrossRef] [PubMed]
57. Mundy, G. Metastasis to bone: Causes, consequences and therapeutic opportunities. *Nat. Rev. Cancer* **2002**, *2*, 584–593. [CrossRef] [PubMed]
58. Yeeravalli, R.; Das, A. Molecular mediators of breast cancer metastasis. *Hematol./Oncol. Stem Cell Ther.* **2021**, *14*, 275–289.
59. Yoneda, T.; Williams, P.J.; Hiraga, T.; Niewolna, M.; Nishimura, R. A bone-seeking clone exhibits different biological properties from the MDA-MB-231 parental human breast cancer cells and a brain-seeking clone in vivo and in vitro. *J. Bone Miner Res.* **2001**, *16*, 1486–1495. [CrossRef] [PubMed]
60. Yamaguchi, M.; Matsumoto, T. Marine algae *Sargassum horneri* bioactive factor suppresses proliferation and stimulates apoptotic cell death in human breast cancer MDA-MB-231 cells in vitro. *Integr. Mol. Med.* **2015**, *2*, 135–138. [CrossRef]
61. Shao, P.; Liu, J.; Chen, X.; Fang, Z.; Sun, P. Structural features and antitumor activity of a purified polysaccharide extracted from *Sargassum horneri*. *Int. J. Biol. Macromol.* **2015**, *73*, 1240130. [CrossRef] [PubMed]
62. Shao, P.; Chen, X.; Suri, P. Chemical characterization, antioxidant and antitumor activity o sulfated polysaccharide from *Sargassum horneri*. *Carbohydr. Polym.* **2014**, *105*, 260–269. [CrossRef]

MDPI AG
Grosspeteranlage 5
4052 Basel
Switzerland
Tel.: +41 61 683 77 34

Nutraceuticals Editorial Office
E-mail: nutraceuticals@mdpi.com
www.mdpi.com/journal/nutraceuticals

www.ingramcontent.com/pod-product-compliance
Lightning Source LLC
Chambersburg PA
CBHW041407210926
51850CB00013B/3206
* 9 7 8 3 7 2 5 8 3 6 4 3 7 *